Network-Based Control of Unmanned Marine Vehicles

Yu-Long Wang · Qing-Long Han · Chen Peng ·
Lang Ma

Network-Based Control of Unmanned Marine Vehicles

Springer

Yu-Long Wang
School of Mechatronic Engineering and Automation
Shanghai University
Shanghai, China

Qing-Long Han
School of Science, Computing and Engineering Technologies
Swinburne University of Technology
Melbourne, VIC, Australia

Chen Peng
School of Mechatronic Engineering and Automation
Shanghai University
Shanghai, China

Lang Ma
School of Mechatronic Engineering and Automation
Shanghai University
Shanghai, China

ISBN 978-3-031-28604-9 ISBN 978-3-031-28605-6 (eBook)
https://doi.org/10.1007/978-3-031-28605-6

This Springer imprint is published by the registered company Springer Nature Switzerland AG
The registered company address is: Gewerbestrasse 11, 6330 Cham, Switzerland

Preface

Networked Control Systems (NCSs) are spatially distributed systems in which the communication between sensors, controllers, and actuators occurs through shared band-limited communication networks. The flexibility in communication architectures, low cost in installation and maintenance, and high reliability make NCSs the future of industrial control systems. Network-based control systems have found a wide range of applications in areas including aircrafts, autonomous vehicles, transportation systems, Unmanned Marine Vehicles (UMVs), and power systems.

The development of UMVs is particularly significant in providing cost-effective solutions to coastal and offshore problems. UMVs are widely used in monitoring, oil and pollution clean-up, scientific characterization, exploration, and military operations such as mine sweeping and border surveillance. It should be mentioned that control for a UMV is usually based on a remote land-based or mother ship-based control station in network environments. The UMV's states such as yaw velocity, roll angle, and heading angle are sampled and transmitted to the control station through the sampler-to-control station communication network channel, while control instructions are constructed and transmitted to the steering machine through the control station-to-actuator communication network channel.

Note that the insertion of communication networks into control systems will inevitably induce network delays and packet dropouts. On the other hand, the occurrence of sensor faults and/or actuator faults in NCSs and UMVs is usually unavoidable. Considering the wide range of applications of NCSs, it is of paramount importance to investigate the stability analysis, stabilization, and fault detection for NCSs. Moreover, how to propose appropriate motion control and fault detection schemes for UMVs in network environments is attractive and practically valuable.

This monograph first presents systematic results for stability analysis, stabilization, and fault detection of NCSs. Then, within the framework of networked control, the problems of heading control, Fault Detection Filter (FDF) and controller coordinated design, dynamic positioning, and cooperative target tracking of UMVs are investigated in detail. Some fundamental concepts of stability analysis, stabilization, motion control, and fault detection are presented with insight and understanding.

Some benchmark examples are provided to show the merits and effectiveness of the network-based UMVs control schemes.

Structure and readership. This monograph is concerned with networked control and its applications in motion control and fault detection of UMVs. In Chap. 1, the importance of studying NCSs and network-based UMVs is first analyzed. Then the corresponding research developments and motivations are provided. Moreover, issues in stability analysis and stabilization of NCSs, and motion control and fault detection of UMVs are presented.

Stabilization and fault detection of NCSs: In Chap. 2, stability analysis and stabilization for an NCS under simultaneous consideration of non-uniformly distributed packet dropouts and interval time-varying sampling periods are investigated. Chapter 3 addresses observer-based FDF design for a continuous-time NCS by taking packet dropouts and network-induced delays into account. In Chap. 4, the output feedback controller design problem for NCSs under an independent and identically distributed (IID) scheduling protocol is discussed. Based on a stochastic impulsive delayed model, sufficient conditions for guaranteeing the stability of the studied system in the mean-square sense are achieved.

Motion control and fault detection of UMVs: In Chaps. 5–9, the problems of heading control, fault detection, dynamic positioning, dynamic output feedback (DOF) control, and cooperative target tracking for UMVs are investigated. In Chap. 5, a novel network-based model for a UMV is established by constructing a heading control error system and purposely dropping some control input packets. Then network-based heading control and rudder oscillation reduction are addressed. Chapter 6 deals with the network-based modeling, and observer-based FDF and controller coordinated design for a UMV. Chapter 7 investigates Takagi-Sugeno (T-S) fuzzy dynamic positioning controller design for a UMV in network environments. In Chap. 8, network-based models for a UMV subject to network-induced characteristics are established. Based on these models, dynamic output feedback controllers (DOFCs) are designed to attenuate the oscillation amplitudes of the yaw velocity error and the yaw angle. Chapter 9 deals with the cooperative target tracking problem of multiple UMVs under switching interaction topologies.

Shanghai, China — Yu-Long Wang
Melbourne, Australia — Qing-Long Han
Shanghai, China — Chen Peng
Shanghai, China — Lang Ma

Acknowledgments We would like to thank the support from the National Science Foundation of China (Grant Nos. 61873335 and 61833011); the Project of Science and Technology Commission of Shanghai Municipality, China (Grant Nos. 20ZR1420200, 21SQBS01600, 22JC1401400, 19510750300, 21190780300, and 21XD1401000). We are very grateful to Mr. Anthony Doyle, Executive Editor, Engineering, Springer, 236 Gray's Inn Road, Floor 6, London WC1X 8HL, UK, for his encouragement to write this monograph.

Contents

Acronyms

$\mathbb{R}^n$	Set of n-dimensional real vectors
$\mathbb{R}^{n \times m}$	Set of all $n \times m$ real matrices
A^{-1}	Inverse of an invertible matrix A
A^T	Transpose of a matrix A
$A \geq 0$	Matrix A is symmetric positive semi-definite
$A > 0$	Matrix A is symmetric positive definite
$A \leq 0$	Matrix A is symmetric negative semi-definite
$A < 0$	Matrix A is symmetric negative definite
$\lambda_{\max}(A)$	The maximum eigenvalue of a real symmetric matrix A
$\lambda_{\min}(A)$	The minimum eigenvalue of a real symmetric matrix A
$\|x\|$	2-norm of a real vector x
$\mathrm{diag}\{\cdots\}$	Block-diagonal matrix
$\mathrm{tr}(A)$	Trace of a matrix A
$\mathrm{Ones}(m, n)$	m-by-n matrix with all entries equal to 1
$\mathrm{Prob}\{\cdot\}$	Probability distribution
$E\{\cdot\}$	The expectation operation
I	Identity matrix with appropriate dimensions
0	Zero matrix with appropriate dimensions
I_m	$m \times m$ identity matrix
$*$	Entries of a matrix implied by symmetry
Sup	Supremum

Chapter 1
Introduction

In this chapter, we first introduce some underlying concepts of network-based control and motion control of unmanned marine vehicles (UMVs). Then we briefly review research motivations on scheduling protocol design and fault detection filter (FDF) design for networked control systems (NCSs). Moreover, we review how network-based heading control, fault detection, dynamic positioning, dynamic output feedback control, and cooperative target tracking for UMVs are motivated and promoted in the control community. Afterwards, we summarize future research topics. Finally, we give a brief outline of the monograph.

1.1 Stability and Stabilization of NCSs

NCSs are spatially distributed systems in which the communication between sensors, controllers and actuators occurs through shared band-limited communication networks [1–3]. Figure 1.1 depicts the typical structure for NCSs. The flexibility in communication architectures, low cost in installation and maintenance, and high reliability make NCSs the future of industrial control systems.

However, the insertion of communication networks into control systems will inevitably induce network delays, packet dropouts, and other network-induced characteristics [4–6], which possibly lead to system performance deterioration and even instability. Therefore, investigating the effects of network-induced delays and packet dropouts on NCSs has received considerable attention in the last decade [7–9]. There have been some nice results available in the existing literature. For example, the bounds on the maximum allowable transmission interval (MATI) and the maximum allowable delay (MAD) that guarantee the stability of NCSs in the presence of communication constraints were obtained in [10]. The design of proportion integration differentiation (PID) controllers for NCSs with polyhedral uncertainties was studied in [11]. The stability and state feedback controller design for continuous-time

Y.-L. Wang et al., *Network-Based Control of Unmanned Marine Vehicles*,
https://doi.org/10.1007/978-3-031-28605-6_1

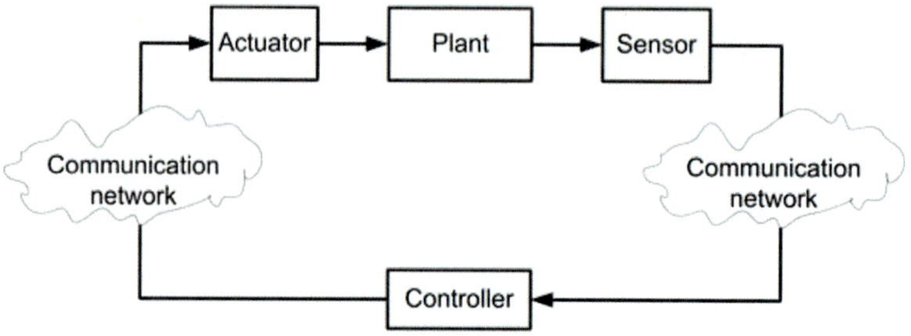

Fig. 1.1 The typical structure for an NCS

NCSs were considered in [12]. Based on information about the lower bound of network-induced delays, the design of robust H_∞ controllers for uncertain NCSs was addressed in [13]. The network-based H_∞ control problem was considered in [14] by applying a stochastic delay system approach. The stability analysis and controller design for NCSs with successive delay components in the state were investigated in [15]. A Lyapunov-Krasovskii functional was constructed in [16] to obtain some delay-dependent stability criteria. A controller design method for NCSs with nonlinearity was proposed in [17] by introducing an extended Jensen inequality. Integrated design of controller and communication sequences for NCSs under consideration of medium access limitations and measurement quantization was addressed in [18]. In [19], a general framework for analyzing the stability of general nonlinear NCSs with disturbances in the setting of L_p stability was provided. A tradeoff between the MATI, the MAD and the quantization parameters for an NCS was provided in [20]. The problem of prediction-based NCSs design has also been paid much attention in [21, 22].

It should be mentioned that the results in [14–16, 23] lumped network-induced delays and packet dropouts into one item $i_{k+1}h - i_k h + \tau_{k+1}$, and the upper bound of $i_{k+1}h - i_k h + \tau_{k+1}$ was employed in stability analysis and controller design. Thus, it is difficult to distinguish the effects of packet dropouts from network-induced delays on the stability and stabilization of NCSs. Establishing the quantitative relationship between packet dropout probability and stabilization of NCSs will help to reduce the negative effects of packet dropouts, which deserves further study.

By using a Markov chain-based method, stability/stabilization of NCSs with packet dropouts were studied in [24, 25]. The Markov chain-based packet dropouts were concerned in [24] with the jumping probabilities of packet dropouts from mode i to mode j being known. On the other hand, for the purpose of avoiding the occurrence of network congestion, one should allocate enough network bandwidth to NCSs. However, the actually needed network bandwidth may be smaller than the mean bandwidth at most time, and larger than the mean bandwidth at least time. Since the utilization of network bandwidth demonstrates such a non-uniform distribution characteristic, network-induced delays and packet dropouts, which are induced by network congestion and data packet collision, may be non-uniformly distributed. The non-uniform distribution characteristic of network-induced delays has been considered in [26–28]. To improve the admissible upper bound of consecutive packet dropouts, one should take the non-uniform distribution characteristic of packet dropouts into account.

Notice that the systems considered in [11, 18, 29, 30] are sampled with a constant sampling period. For control systems, the sensor supposedly samples over a fixed nominal period. However, computer loads, networks, and sporadic faults may lead to sampling period jitter inevitably. Recently, there have been considerable research efforts on time-varying sampling periods [31, 32]. For example, the problem of robust H_∞ control for sampled-data systems with probabilistic sampling periods was investigated in [33]. The output tracking control for NCSs with switched sampling periods was studied in [34]. However, if the sampling period is time-varying and varies on an interval, the switched sampling period-based design methods in [33, 34] are not applicable. On the other hand, the results in [31, 32, 34] are based on the discretization of continuous-time systems. If the actuator receives more than one control input during a sampling period, such a discretization process leads to much modeling and design complexity. For continuous-time NCSs under simultaneous consideration of network-induced delays, packet dropouts, and interval time-varying sampling periods, how to guarantee robustness to small variations of the sampling period and avoid the design complexity induced by discretization are of paramount importance.

1.2 FDF Design for NCSs

For NCSs, the occurrence of sensor faults or actuator faults is usually unavoidable. Then, it is of paramount importance to study how to detect the occurrence of faults in time. The fault detection for traditional control systems has been paid much attention and some interesting results have been reported, see [35–37] and the references therein. On the other hand, some nice results dealing with fault detection for NCSs have been obtained. For example, the network-based robust fault detection problem for discrete-time Takagi-Sugeno (T-S) fuzzy systems was addressed in [38]. The problems of fault detection, isolation, and estimation for discrete time-varying networked sensing systems were studied in [39]. Robust fault estimation for NCSs with network-induced delays was investigated in [40]. The fault detection of linear systems over networks with bounded packet dropouts was addressed in [41]. The problem of fault detection in finite frequency domain for NCSs was investigated in [42, 43]. A method for the fault detection of a nonlinear NCS was proposed in [44]. Fault detection of NCSs subject to uncertain time-varying network-induced delays was considered in [45]. A so-called packet-based periodic communication strategy for designing optimal observer-based residual generators was proposed in [46]. It should be mentioned that discrete-time systems are considered in [38, 39, 42, 44], while discretized NCSs are considered in [41, 43, 45]. Moreover, packet dropouts and/or network-induced delays are neglected in some of the above-mentioned works. For continuous-time NCSs considering packet dropouts and network-induced delays simultaneously, how to propose an appropriate observer-based FDF design scheme is interesting and practically valuable.

For NCSs with faults, the feasible fault signals are included in the measurement outputs $y(t_{k-1})$, $y(t_k)$, ..., where k denotes a positive integer, and $y(t_{k-1})$, $y(t_k)$, ... are adopted by the FDF to detect the occurrence of faults. For a specific time interval $[t_k + \tau_k,\ t_{k+1} + \tau_{k+1})$, where t_k and τ_k denote the sampling instant and the length of the network-induced delays, respectively, if we combine the recently received measurement output $y(t_k)$ and the formerly received measurement output $y(t_{k-1})$ together, which is named as data reconstruction in this monograph, and transmit the data reconstruction-based measurement outputs to the FDF, enlarged fault signals will be received by the FDF and the fault detection time will be reduced correspondingly. However, such a data reconstruction scheme has not been taken into full consideration.

For continuous-time NCSs with $t \in [t_k + \tau_k,\ t_{k+1} + \tau_{k+1})$, the artificial delay $\tau(t) = t - t_k$, which is named as an interval time-varying delay, is usually introduced to study stability/stabilization of considered systems. Suppose that $\tau(t) \in [\tau_m,\ \eta)$, where τ_m and η denote known constants. Divide $[\tau_m,\ \eta)$ into ρ equidistant time intervals with ρ denoting a given positive integer. One can see that for a specific instant t, $\tau(t)$ can not be included simultaneously in two different time intervals derived above. In what follows, we refer to such a phenomenon as mutually exclusive distribution. When dealing with integral inequalities for products of vectors, how to make full use of the mutually exclusive distribution characteristic of interval time-varying delays to derive less conservative results is of paramount importance.

1.3 Scheduling Protocol Design for NCSs

Note that most of the results mentioned above are concerned with stability and stabilization problems by assuming that system states are available. That is, the obtained results are only applicable for systems whose states are measurable. However, in the practical power systems and rougher flotation process, it is complicated to measure system states directly [47, 48]. As an alternative method, stability and stabilization problems of NCSs with output measurements have motivated a lot of interesting research, see [49, 50].

In the literature, single packet transmission is a general assumption [51]. This implies that all the measurements and control inputs are packaged into one data packet. However, it is generally unrealistic in many applications where the nodes are distributed over a large physical area. Therefore, it is interesting to study the problem of multiple packets transmission. In the scenario of multiple packets transmission, if the network bandwidth is limited, the node collision problem is unavoidable. To tackle the node collision problem, some scheduling protocols, such as try-once-discard (TOD) protocol [52], round-robin (RR) protocol [53, 54], and stochastic protocol [55, 56], are proposed to determine which nodes can access the communication network. In general, the TOD protocol and stochastic protocol can provide better system performance than the RR protocol. In the TOD protocol, the node with the largest weighted error gets the permission to access the communication network, whereas the stochastic protocol usually determines the transmission node via Bernoulli or Markov chain process (called Markovian protocol). With the utilization of a scheduling protocol, some fundamental challenges for the stability analysis will arise. Thus, it is worthwhile to study how the scheduling protocol affects the stability of NCSs.

Recently, a few preliminary results have been reported on the stability issue with a scheduling protocol. For example, stability of NCSs under RR and TOD protocols was considered in [10] by using the hybrid system approach. It should be mentioned that, in the presence of a scheduling protocol, the hybrid system approach does not allow large communication delays which are larger than the length of a sampling period. To incorporate such large communication delays, the time-delay system approach for NCSs under scheduling protocols was developed in [52, 57–59]. Note that the aforementioned results only provide sufficient conditions for exponential stability with a supposition that the controller feedback gain should be known. If the feedback gain of a controller is not known a priori, the aforementioned methods are no longer applicable. Moreover, the effects of packet dropouts on the studied systems are ignored in [58, 59]. Based on the discussions above, for NCSs under a stochastic scheduling protocol and the nonideal network quality of service (QoS) (network-induced delays and packet dropouts), how to analyze the effects of packet dropouts and how to design output feedback controllers are still a challenging issue.

1.4 Network-Based Heading Control of UMVs

Heading control of ships is one of main concerns for marine applications [60, 61]. Accurate heading control, which can be achieved through the regulation of the rudder angle, is important for the routine operation and safe navigation of ships. However, ships are exposed to wave-induced disturbances, which inevitably leads to heading angle deviation and frequent regulation of the rudder angle. Thus, it is of paramount importance to study how to achieve accurate heading control and rudder oscillation reduction for ships exposed to wave-induced disturbances. In the existing literature, there are some nice results available dealing with heading control and controller design for manned ships. For example, Kalman filtering-based positioning and heading control for ships and offshore rigs were addressed in [62]. A family of passivity-based controllers for dynamic positioning of ships were presented in [63]. The implementation of genetic programming in designing a controller structure for a surface ship was studied in [64]. The control of underactuated ships was studied in [65–67]. The thrust loss suppression algorithm and marine thruster control were considered in [68] and [69], respectively.

Network-based control systems have attracted considerable attention in the last decade [70–72]. For recent developments, we refer readers to two recent survey papers [4, 73]. Due to the introduction of modern high-speed communication networks between UMVs and remote land-based/mother ship-based control stations, network-induced delays are usually very small, and sometimes they are negligible. As observed from [4, 73], network-induced packet dropouts, which are referred to as passive packet dropouts in this monograph, are usually regarded as the source of system instability and performance degradation. In this monograph, we take a different and novel view. We investigate whether or not intentional packet dropouts, which are referred to as active packet dropouts, can stabilize the UMV in network environments and provide satisfactory heading control. For some damping-like systems such as offshore structures [72, 74, 75], by purposely introducing a proper time delay, the internal oscillation of the systems can be reduced. Note that packet dropouts in network-based control systems can be described as a type of time delay. Therefore, actively dropping some control input packets purposely may reduce the heading angle deviation and the oscillation of the rudder angle of the UMV in network environments, which deserves an in-depth investigation.

1.5 FDF and Controller Coordinated Design for UMVs

The motion control of marine vehicles emerges as a topic of significant interest due to an increasing demand for higher accuracy, higher performance, and reliability in practical applications [60]. Some interesting research results are reported in the literature on heading control [76], roll stabilization [77], and control of underactuated ships [78]. Note that manned marine vehicles are studied in [60, 76–78]. The development of unmanned marine vehicles is particularly significant in providing cost-effective solutions to coastal and offshore problems. UMVs are widely used in monitoring, oil and pollution clean-up, scientific characterization, exploration, military operations such as mine sweeping and border surveillance [79, 80]. There are some nice results available dealing with the motion control of UMVs. For example, an approach based on Theta* algorithm was suggested in [81] to create paths for UMVs in real-time. The problem of accurate identification and learning control of ocean surface ships in uncertain dynamical environments was addressed in [82].

It should be pointed out that filtering and control for UMVs is usually based on a remote land-based or mother ship-based control station in network environments. Despite of advantages such as lower cost, more flexibility and higher reliability, introducing communication networks into control systems inevitably induces network delays and packet dropouts [83–86]. Thus, if a communication network is introduced between the UMV and the remote control station, such network-induced characteristics should be taken into full consideration.

Moreover, for UMVs in network environments, the occurrence of faults, such as the saturation, the stuck steering machine-type faults, and the noise-type faults, is usually unavoidable. Then it is of paramount importance to study how to detect the occurrence of faults in time. On one hand, fault-tolerant control and fault detection for traditional control systems have been paid much attention and some interesting results have been reported, see [87, 88] and the references therein. Note that an integrated fault detection and robust control scheme was proposed in [87] to attenuate the effects of disturbances and detect the actuator stuck faults. Based on the scheme in [87], the simultaneous dynamic observer-based robust control and fault detection problem for linear systems was investigated in [36]. The simultaneous fault detection and robust control scheme gets further development in [89] to deal with continuous-time switched systems. On the other hand, for NCSs, there are some nice results dealing with fault detection in the literature. For example, the problem of fault detection, isolation, and estimation for networked sensing systems was considered in [39]. Fault detection filtering for a discrete-time system considering network-induced nonlinear characteristics was investigated in [90]. For a UMV subject to faults, if the sampler-to-control station communication network channel-induced and the control station-to-actuator communication network channel-induced characteristics such as network-induced delays and packet dropouts are considered simultaneously, the measured output available to the FDF during a specific time interval is variable. This induces some difficulty for describing the measured output in a uniform form and constructing the corresponding networked system. Thus, for a UMV subject to faults, and the sampler-to-control station communication network channel-induced and the control station-to-actuator communication network channel-induced characteristics, how to establish a network-based model is practically valuable and still unresolved, which motivates the current study. Moreover, network-based FDF and controller coordinated design for the UMV is of paramount importance and still unresolved, which deserves deep investigation.

One can see from Sect. 1.2 that taking into account the mutually exclusive distribution characteristic of the interval time-varying delay $\tau(t)$ is helpful. When investigating network-based FDF and controller coordinated design for the UMV in network environments, how to make full use of the mutually exclusive distribution characteristic of $\tau(t)$ to derive more relaxed design criteria is important.

1.6 Dynamic Positioning of UMVs

Marine vehicles have found applications in broad areas including transportation, military operations, hydrographic, fishing, oil and gas exploration and construction, oceanographic data collection, and scientific characterization [91, 92]. There are some interesting results available concerning the motion control of marine vehicles, which cover research topics in roll stabilization [77], tracking control [78], heading control [76, 93], containment maneuvering [94, 95], mooring control [96], fault detection [97], model predictive control [98], and path planning [99]. Besides the topics mentioned above, dynamic positioning, which aims at regulating the horizontal position and heading of marine vehicles, has also attracted much attention in the literature [100]. For example, Kalman filtering-based positioning and heading control of ships were investigated in [101]. The semiglobally practically asymptotically stabilizing controller design for dynamically positioned ships was proposed [102]. Adaptive robust output feedback control for a marine dynamic positioning system (DPS) was investigated [103]. Output feedback control for a marine DPS was addressed [104]. By using multiple unidirectional tugboats, robust dynamic positioning of an unactuated surface vessel was studied [105]. In [106], robust controller design for dynamic positioning of ships and offshore rigs using H_∞ and mixed-μ techniques was considered. A novel continuous robust controller for dynamically positioned surface vessels with added mass terms was constructed [107]. A robust nonlinear control law for the DPS of ships subject to unknown time-varying disturbances and input saturation was proposed [108]. Quadratic finite-horizon optimal controller design for T-S fuzzy-model-based dynamic ship positioning systems was addressed [109]. As one can see, the T-S fuzzy-control-based approach, which is different from those in [100, 105–108], was adopted in [109] to describe the dynamic positioning of a marine vehicle. There is a growing interest in applying the T-S fuzzy control approach to deal with nonlinear control systems. The main characteristic of T-S fuzzy control lies in utilizing a linear system model to describe the local dynamics of each fuzzy rule [109, 110]. Then the abundant linear control methodologies can be adopted to investigate each linear model. Since the DPS of a marine vehicle is a complex nonlinear system, how to propose an appropriate modeling and control scheme to improve the dynamic positioning performance is practically valuable and attractive.

Dynamic positioning for a UMV is usually based on a remote land-based/mother ship-based control station. The UMV, the remote land-based/mother ship-based control station, and communication networks constitute a network-based control system. Despite of advantages of network-based control, introducing communication networks into control systems may induce packet dropouts, delays, and packet disordering [111–114]. For the networked DPS of a UMV, how to take sampler-to-controller and controller-to-actuator packet dropouts, network-induced delays, and packet disordering into account, and to establish network-based T-S fuzzy models are of paramount importance and far from being resolved.

For a T-S fuzzy system, if a communication network is introduced between the controlled plant and the controller, the membership functions of the controlled plant and the controller are not synchronous, and such a characteristic is neglected in [23]. The asynchronous difference between the normalized membership function of the controlled plant and that of the controller was taken into consideration in [115, 116]. Moreover, for the T-S fuzzy-control-based DPS of a UMV, the T-S fuzzy model is closely related to the variation scope of the yaw angle. Then how to take into account the variation scope of the yaw angle and the asynchronous difference between the normalized membership function of the UMV and that of the controller, and to derive a novel stability criterion are practically valuable.

In practical situations, the controlled plant states are not always measurable. Thus, it is of paramount importance to study the observer-based control scheme for T-S fuzzy systems, and some nice results are available in the literature [117–120]. Note that only the sensor-to-observer network-induced delays are considered in [117], while packet dropouts and the controller-to-actuator network-induced delays are not considered. The work in [118] assumed that updating instants of the control input and the measured output are the same. In fact, if time-varying network-induced delays are considered, such an assumption may not be applicable. Moreover, packet dropouts are not considered in [118]. The Bernoulli stochastic process was adopted in [119] to describe packet dropouts in the sensor-to-observer channel and the controller-to-actuator channel with network-induced delays being neglected. The problem of observer-based output feedback control for T-S fuzzy systems under decentralized event-triggering communication was discussed in [120] with packet dropouts and network-induced delays being neglected. For the observer-based T-S fuzzy DPS of a UMV, taking into account sampler-to-controller and controller-to-actuator network-induced characteristics will lead to much modeling and controller design complexity. How to solve these issues is practically valuable.

1.7 Networked DOF Control of UMVs

Manned/unmanned marine platforms can provide cost-effective solutions to coastal and offshore problems. Compared with manned marine vehicles, unmanned ones can provide more flexibility in practical applications [76]. Some interesting results dealing with UMVs were reported in [121–124]. When carrying out tasks such as scientific characterization and exploration, a UMV may be stopped and anchored. However, external disturbances, such as waves, wind, and current, may induce the oscillation of the yaw velocity error and the yaw angle of the UMV, where the yaw velocity error denotes the difference between the actual yaw velocity and the constant yaw velocity reference. Obviously, the oscillation of the yaw velocity error and the yaw angle is not desired in practical applications. For a UMV, how to propose an appropriate control scheme to attenuate the oscillation amplitudes of the yaw velocity error and the yaw angle is practically valuable and attractive.

Based on a remote control station, one can control the motion of a UMV in network environments. Note that state feedback control of NCSs was studied in [125, 126]. In some practical situations, controlled plants' states may not be always measurable. Thus, it is significant to study observer-based control of systems under consideration [71, 97, 127] and dynamic output feedback control of systems under consideration [50, 116, 128]. For a UMV, if the surge velocity, sway velocity, and the yaw velocity are not measurable, how to propose an appropriate dynamic output feedback controller (DOFC) design scheme to attenuate the oscillation amplitudes of the yaw velocity error and the yaw angle is significant and has received little attention in the literature.

The non-uniform distribution characteristic of interval time-varying delays was considered in [129, 130] to study the stabilization of systems. For the UMV in network environments, packet dropouts and network-induced delays may be non-uniformly distributed. Note that the non-uniform distribution characteristic of packet dropouts and network-induced delays can be implied by the non-uniform distribution characteristic of an interval time-varying delay. If such a characteristic is considered, how to establish network-based models for the UMV, and how to construct appropriate integral inequalities for products of vectors which are introduced in DOFC design are challenging problems.

1.8 Cooperative Target Tracking of Multiple UMVs

Some typical motion control issues for UMVs such as the heading control [131, 132] and dynamic positioning [1, 108, 133, 134] have been reported in the literature. Recently, the target tracking [135–137], trajectory tracking [138, 139], and path following [140–142] have garnered widespread attention due to their extensive marine applications in military reconnaissance, environmental monitoring, ocean exploration, offshore inspections, and so on. In particular, maritime target tracking, whose purpose is to drive corresponding marine vehicles to track a moving target, has received much attention [143–146]. For an autonomous underwater vehicle, a prescribed performance bound method was presented to achieve target tracking and guarantee the transient performance [144]. For an autonomous robotic vehicle, a switched logic-based control strategy was proposed to address the target tracking problem by utilizing range-only measurements [145]. Note that the above-referenced results mainly involve one-to-one tracking, i.e., one marine vehicle tracks one target. However, some maritime target tracking missions require the cooperation of a fleet of UMVs, which can improve flexibility, robustness, and efficiency of mission completion. For a group of UMVs, the cooperative target tracking problem is far from being resolved. Therefore, how to achieve cooperative target tracking for multiple UMVs deserves in-depth investigation.

In most of cases, it is assumed that both position and velocity information of maritime targets can be either measured or received by follower UMVs [135]. For one UMV, however, it is costly and difficult to obtain the accurate velocity of the target [147, 148]. Thus, how to track a target by using only its position information measurement is of paramount importance. On the other hand, in a practical target tracking system composed of multiple UMVs, some of the UMVs can not obtain the target information due to restrictions of distances and/or external environments. To address this problem, some existing results adopt leader-follower tracking control schemes (see in [135, 136, 147, 149]), which can be sketched as "UMV 1 tracks the target, UMV 2 follows UMV 1, UMV 3 follows UMV 2$\cdots$". This means that one UMV only has one neighbor. Specifically, a UMV can only track the target or follow another UMV by measuring/receiving information from it. In this case, if a communication link or a measuring device fails, the cooperative target tracking mission may fail. In the field of multi-agent system control, the distributed control scheme provides a solution to this dilemma [150–153]. However, how to apply the distributed scheme to conduct the cooperative target tracking is still challenging, not to mention in the case of lacking the target's velocity information.

The interaction between the target and the UMVs can be described by a communication network. In some existing results concerning cooperative control of multiple UMVs, interaction network topologies are assumed to be fixed [154–156].

In practical maritime tracking applications, the connectivity of the corresponding interaction topology may switch due to link failures, sensor incapabilities, channel attenuation, and alteration of missions. This heightens the need for investigating the cooperative control problem of multiple UMVs under switching topologies. Actually, the cooperative control of multi-agent systems under switching topologies has been studied in the existing literature [157–160]. However, these results can not be directly extended to deal with the cooperative control of multiple UMVs. For a tracking system composed of multiple UMVs, the cooperative target tracking problem under switching topologies has not been adequately addressed hitherto.

1.9 Contributions of the Monograph

Based on systematic results for stability analysis, stabilization, and fault detection of NCSs, this monograph focuses on the heading control, fault detection, dynamic positioning, dynamic output feedback control, and cooperative target tracking of UMVs. The packet dropout separation method, the data reconstruction scheme, the active packet dropouts approach, and the cooperative target tracking scheme are proposed to deal with the stabilization of NCSs and the motion control of UMVs. More specifically, the main contributions of this monograph are highlighted as follows.

- A packet dropout separation method is proposed to separate packet dropouts from the lump sum of network-induced delays and packet dropouts. A new NCS model is established correspondingly by taking interval time-varying sampling periods and the non-uniform distribution characteristic of packet dropouts into full consideration. The quantitative relationship between packet dropout probability and stability/stabilization of the NCS is established by constructing a new packet dropout decomposition-based Lyapunov functional. The stability and stabilization criteria can guarantee robustness of the NCS to small variations of sampling periods.
- New closed-loop models for a continuous-time NCS with actuator faults are established by proposing a data reconstruction scheme. The mutually exclusive distribution characteristic (which introduces less conservatism) of interval time-varying delays is made full use to deal with integral inequalities for products of vectors. The observer-based FDF design criteria are achieved. The designed FDFs can guarantee the sensitivity of the residual signal to faults.
- A stochastic impulsive delayed system is constructed for an NCS under consideration of an IID scheduling protocol, network-induced delays, and packet dropouts. The effects of network-induced delays and packet dropouts are taken into full consideration. An optimization algorithm is presented to determine the optimal multi-sensor protocol scheduling parameters and the output feedback controller gains.
- A novel network-based model is established for a UMV by constructing a heading control error system and purposely dropping some control input packets received by the steering machine. Whether or not active packet dropouts can stabilize the

UMV and provide satisfactory heading control performance are investigated in depth. Algorithms for selecting the number of intentionally dropped control input packets are proposed.

- Network-based models for a UMV subject to actuator faults and wave-induced disturbances are established by considering network-induced delays and packet dropouts in the sampler-to-control station communication network channel and the control station-to-actuator communication network channel. The network-based FDF and controller coordinated design criteria are derived. These design criteria can guarantee the sensitivity of the residual signal to faults and the robustness of the UMV to external disturbances.
- Novel network-based T-S fuzzy DPS models for a UMV are established by taking into consideration the variation scope of the yaw angle, and the sampler-to-controller and controller-to-actuator network-induced characteristics. Appropriate observer-based controller design schemes are proposed to provide good dynamic positioning performance with the negative influences of wave-induced disturbances being reduced.
- A network-based model for a UMV with unavailable states is established by introducing a state tracking system, and taking sampler-to-control station packet dropouts, network-induced delays, and packet disordering into account. Then the model is extended to the UMV with control station-to-actuator, and both sampler-to-control station and control station-to-actuator packet dropouts, network-induced delays, and packet disordering. DOFCs are designed to attenuate the oscillation amplitudes of the yaw velocity error and the yaw angle.
- A distributed extended state observer (DESO) taking into account switching topologies of UMVs is designed to integrally estimate unknown target dynamics and neighboring UMVs' dynamics. The DESO is more suitable for estimating unknown dynamics including not only the target's but also neighboring UMVs' information. Then a novel kinematic controller design method is proposed. It can take full advantages of known information and avoid the approximation of some virtual control vectors. A distributed dynamic controller is designed to drive the involved UMVs to cooperatively track the target.

1.10 Future Research Topics

This monograph presents recent advances in network-based control and its applications in UMVs. There are still several challenging issues that deserve further investigation.

- *Event-triggered control for UMVs in network environments.* Note that the majority of existing results about motion control of UMVs are based on time-triggering transmission schemes [93, 97, 133]. This means that all the sampled data are transmitted to the control station and/or the actuator, which unavoidably leads to a waste of communication resources since some superfluous sampled data have virtually no influence over the control performance. For UMVs in network environments, event-triggering transmission strategy can reduce the data transmission frequency while preserving desired control performance [2, 3, 5, 47, 133]. Thus, how to well address the event-triggered control problem for UMVs deserves an in-depth investigation.

- *Network-based modeling and control for UMVs under cyber attacks.* For NCSs, cyber security is a critical issue due to the vulnerability of communication networks [161–163]. The motion control of UMVs is usually based on a remote land-based or mother ship-based control station in network environments. This means that cyber attacks on the sampler-to-control station and control station-to-actuator communication network channels may degrade UMVs' control performance significantly. Thus, how to achieve satisfying control performance of UMVs under cyber attacks is a promising research topic.
- *Fixed-time control for NCSs and networked UMVs.* Fixed-time control is currently a hot research topic in practical systems since it can provide a guaranteed settling time, which does not depend on initial conditions [164]. Over the past decades, by combining the nonlinear feedback method and the sliding-mode methodology, the fixed-time control problem has been studied for linear control systems [165], second-order nonlinear systems [166], and complex networks [167]. It should be mentioned that most of the existing fixed-time control results are obtained under an assumption that communication links between sensors, controllers, and actuators are perfect. This means that little attention has been paid to fixed-time control of systems subject to network-induced characteristics. Moreover, under complex marine environments, fixed-time control of networked UMVs is still an open problem.
- *Cooperative control of multiple UMVs in network environments.* There are increasing needs for deploying a fleet of UMVs instead of a single UMV to conduct complex marine tasks since employing multiple UMVs can offer many advantages. Several cooperative control methods for multiple UMVs have been provided in the literature, including target-guided ones [136], trajectory-guided ones [168], and path-guided ones [169]. However, for UMVs in network environments, the cooperative control problems are still challenging. Considering the influences of network-induced delays, packet dropouts, and packet disordering, how to systematically develop network-based cooperative control schemes for UMVs is an interesting research area.

1.11 Book Outline

The monograph is organized as follows. In this chapter briefly reviews recent advances in network-based control and its applications in UMVs. **Chapter** 2 conducts quantitative analysis and synthesis for an NCS under simultaneous consideration of non-uniformly distributed packet dropouts and interval time-varying sampling periods. **Chapter** 3 studies the observer-based FDF design problem for a continuous-time

NCS. **Chapter** 4 addresses the output feedback control problem for NCSs under a stochastic scheduling protocol. **Chapter** 5 is concerned with network-based heading control and rudder oscillation reduction for a UMV equipped with single rudder in network environments. **Chapter** 6 deals with network-based modeling, and observer-based FDF and controller coordinated design for a UMV in network environments. In **Chapter** 7, based on established network-based T-S fuzzy DPS models for a UMV, the T-S fuzzy dynamic positioning controller design scheme is proposed. In **Chapter** 8, network-based modeling and DOFC design methods are proposed for a UMV subject to sampler-to-control station and control station-to-actuator packet dropouts, network-induced delays, and packet disordering. In **Chapter** 9, a cooperative target tracking controller design method is proposed for multiple UMVs under switching network topologies.

References

1. Y.-L. Wang, Q.-L. Han, M.-R. Fei, C. Peng, Network-based T-S fuzzy dynamic positioning controller design for unmanned marine vehicles. IEEE Trans. Cybern. **48**(9), 2750–2763 (2018)
2. X. Ge, Q.-L. Han, Z. Wang, A dynamic event-triggered transmission scheme for distributed set-membership estimation over wireless sensor networks. IEEE Trans. Cybern. **49**(1), 171–183 (2019)
3. X.-M. Zhang, Q.-L. Han, B.-L. Zhang, An overview and deep investigation on sampled-data-based event-triggered control and filtering for networked systems. IEEE Trans. Ind. Inf. **13**(1), 4–16 (2017)
4. X. Ge, F. Yang, Q.-L. Han, Distributed networked control systems: a brief overview. Inf. Sci. **380**, 117–131 (2017)
5. S. Hu, D. Yue, X. Xie, Z. Du, Event-triggered H_∞ stabilization for networked stochastic systems with multiplicative noise and network-induced delays. Inf. Sci. **299**, 178–197 (2015)
6. J. Zhang, C. Peng, Guaranteed cost control of uncertain networked control systems with a hybrid communication scheme. IEEE Trans. Syst. Man Cybern. Syst. **50**(9), 3126–3135 (2020)
7. F. Yang, Z. Wang, D.W.C. Ho, M. Gani, Robust H_∞ control with missing measurements and time delays. IEEE Trans. Autom. Control **52**(9), 1666–1672 (2007)
8. A. Selivanov, E. Fridman, Predictor-based networked control under uncertain transmission delays. Automatica **70**, 101–108 (2016)
9. J. Chen, S. Meng, J. Sun, Stability analysis of networked control systems with aperiodic sampling and time-varying delay. IEEE Trans. Cybern. **47**(8), 2312–2320 (2017)
10. W.P.M.H. Heemels, A.R. Teel, N. van de Wouw, D. Nešić, Networked control systems with communication constraints: Tradeoffs between transmission intervals, delays and performance. IEEE Trans. Autom. Control **55**(8), 1781–1796 (2010)
11. H. Zhang, Y. Shi, A.S. Mehr, Robust H_∞ PID control for multivariable networked control systems with disturbance/noise attenuation. Int. J. Robust Nonlinear Control **22**(2), 183–204 (2011)
12. D. Yue, Q.-L. Han, C. Peng, State feedback controller design of networked control systems. IEEE Trans. Circuits Syst. II Express Briefs **51**(11), 640–644 (2004)
13. D. Yue, Q.-L. Han, J. Lam, Network-based robust H_∞ control of systems with uncertainty. Automatica **41**(6), 999–1007 (2005)
14. X. Meng, J. Lam, H. Gao, Network-based H_∞ control for stochastic systems. Int. J. Robust Nonlinear Control **19**(3), 295–312 (2009)

15. H. Gao, T. Chen, J. Lam, A new delay system approach to network-based control. Automatica **44**(1), 39–52 (2008)
16. X. Jiang, Q.-L. Han, New stability criteria for linear systems with interval time-varying delay. Automatica **44**(10), 2680–2685 (2008)
17. C. Peng, Y.-C. Tian, M.O. Tadé, State feedback controller design of networked control systems with interval time-varying delay and nonlinearity. Int. J. Robust Nonlinear Control **18**(12), 1285–1301 (2008)
18. G. Guo, H. Jin, A switching system approach to actuator assignment with limited channels. Int. J. Robust Nonlinear Control **20**(12), 1407–1426 (2010)
19. M. Tabbara, D. Nešić, A.R. Teel, Stability of wireless and wireline networked control systems. IEEE Trans. Autom. Control **52**(9), 1615–1630 (2007)
20. W.P.M.H. Heemels, D. Nešić, A.R. Teel, N. van de Wouw, Networked and quantized control systems with communication delays, in *Proceedings of the Joint 48th IEEE Conference on Decision and Control and the 28th Chinese Control Conference*, Shanghai, China, Dec. 2009, pp. 7929–7935
21. D.E. Quevedo, D. Nešić, Input-to-state stability of packetized predictive control over unreliable networks affected by packet-dropouts. IEEE Trans. Autom. Control **56**(2), 370–375 (2011)
22. A. Onat, T. Naskali, E. Parlakay, O. Mutluer, Control over imperfect networks: model-based predictive networked control systems. IEEE Trans. Ind. Electron. **58**(3), 905–913 (2011)
23. X. Jia, D. Zhang, X. Hao, N. Zheng, Fuzzy H_∞ tracking control for nonlinear networked control systems in T-S fuzzy model. IEEE Trans. Syst. Man Cybern. Part B Cybern. **39**(4), 1073–1079 (2009)
24. J. Wu, T. Chen, Design of networked control systems with packet dropouts. IEEE Trans. Autom. Control **52**(7), 1314–1319 (2007)
25. J. Yu, L. Wang, G. Zhang, M. Yu, Output feedback stabilisation of networked control systems via switched system approach. Int. J. Control **82**(9), 1665–1677 (2009)
26. C. Peng, D. Yue, E. Tian, Z. Gu, A delay distribution based stability analysis and synthesis approach for networked control systems. J. Frankl. Inst. **346**(4), 349–365 (2009)
27. Y. Tipsuwan, M.-Y. Chow, Gain scheduler middleware: a methodology to enable existing controllers for networked control and teleoperation-part I: networked control. IEEE Trans. Ind. Electron. **51**(6), 1218–1227 (2004)
28. D. Yue, E. Tian, Z. Wang, J. Lam, Stabilization of systems with probabilistic interval input delays and its applications to networked control systems. IEEE Trans. Syst. Man Cybern. Part A Syst. Hum. **39**(4), 939–945 (2009)
29. W.-A. Zhang, L. Yu, Modelling and control of networked control systems with both network-induced delay and packet-dropout. Automatica **44**(12), 3206–3210 (2008)
30. J. Xiong, J. Lam, Stabilization of networked control systems with a logic ZOH. IEEE Trans. Autom. Control **52**(2), 358–363 (2009)
31. R. Lozano, P. Castillo, P. Garcia, A. Dzul, Robust prediction-based control for unstable delay systems: application to the yaw control of a mini-helicopter. Automatica **40**(4), 603–612 (2004)
32. A. Sala, Computer control under time-varying sampling period: an LMI gridding approach. Automatica **41**(12), 2077–2082 (2005)
33. H. Gao, J. Wu, P. Shi, Robust sampled-data H_∞ control with stochastic sampling. Automatica **45**(7), 1729–1736 (2009)
34. Y.-L. Wang, G.-H. Yang, Output tracking control for networked control systems with time delay and packet dropout. Int. J. Control **81**(11), 1709–1719 (2008)
35. A. Johansson, M. Bask, T. Norlander, Dynamic threshold generators for robust fault detection in linear systems with parameter uncertainty. Automatica **42**(7), 1095–1106 (2006)
36. X.-J. Li, G.-H. Yang, Dynamic observer-based robust control and fault detection for linear systems. IET Control Theory Appl. **6**(17), 2657–2666 (2012)
37. Y. Zhao, J. Lam, H. Gao, Fault detection for fuzzy systems with intermittent measurements. IEEE Trans. Fuzzy Syst. **17**(2), 398–410 (2009)

38. H. Dong, Z. Wang, J. Lam, H. Gao, Fuzzy-model-based robust fault detection with stochastic mixed time delays and successive packet dropouts. IEEE Trans. Syst. Man Cybern. Part B Cybern. **42**(2), 365–376 (2012)
39. X. He, Z. Wang, Y. Liu, D.H. Zhou, Least-squares fault detection and diagnosis for networked sensing systems using a direct state estimation approach. IEEE Trans. Ind. Inf. **9**(3), 1670–1679 (2013)
40. D. Huang, S.K. Nguang, Robust fault estimator design for uncertain networked control systems with random time delays: an ILMI approach. Inf. Sci. **180**(3), 465–480 (2010)
41. B. Liu, Y. Xia, Y. Yang, M. Fu, Robust fault detection of linear systems over networks with bounded packet loss. J. Frankl. Inst. **349**(7), 2480–2499 (2012)
42. Y. Long, G.-H. Yang, Fault detection for a class of networked control systems with finite-frequency servo inputs and random packet dropouts. IET Control Theory Appl. **6**(15), 2397–2408 (2012)
43. Y. Long, G.-H. Yang, Fault detection in finite frequency domain for networked control systems with missing measurements. J. Frankl. Inst. **350**(9), 2605–2626 (2013)
44. Z. Mao, B. Jiang, P. Shi, Protocol and fault detection design for nonlinear networked control systems. IEEE Trans. Circuits Syst. II Express Briefs **56**(3), 255–259 (2009)
45. Y. Wang, S.X. Ding, H. Ye, G. Wang, A new fault detection scheme for networked control systems subject to uncertain time-varying delay. IEEE Trans. Signal Process. **56**(10), 5258–5268 (2008)
46. Y.Q. Wang, S.X. Ding, H. Ye, L. Wei, P. Zhang, G.Z. Wang, Fault detection of networked control systems with packet based periodic communication. Int. J. Adapt. Control Signal Process. **23**(8), 682–698 (2009)
47. C. Peng, J. Zhang, H. Yan, Adaptive event-triggering H_∞ load frequency control for network-based power systems. IEEE Trans. Ind. Electron. **65**(2), 1685–1694 (2018)
48. F. Liu, H. Gao, J. Qiu, S. Yin, J. Fan, T. Chai, Networked multirate output feedback control for setpoints compensation and its application to rougher flotation process. IEEE Trans. Ind. Electron. **61**(1), 460–468 (2014)
49. L. El Ghaoui, F. Oustry, M. AitRami, A cone complementarity linearization algorithm for static output-feedback and related problems. IEEE Trans. Autom. Control **42**(8), 1171–1176 (1997)
50. H. Gao, X. Meng, T. Chen, J. Lam, Stabilization of networked control systems via dynamic output-feedback controllers. SIAM J. Control Optim. **48**(5), 3643–3658 (2010)
51. L. Zhang, H. Gao, O. Kaynak, Network-induced constraints in networked control systems–a survey. IEEE Trans. Ind. Inf. **9**(1), 403–416 (2013)
52. D. Freirich, E. Fridman, Decentralized networked control of systems with local networks: a time-delay approach. Automatica **69**, 201–209 (2016)
53. K. Liu, E. Fridman, L. Hetel, Stability and L_2-gain analysis of networked control systems under round-robin scheduling: a time-delay approach. Syst. Control Lett. **61**(5), 666–675 (2012)
54. D. Ding, Z. Wang, Q.-L. Han, G. Wei, Neural-network-based output-feedback control under round-robin scheduling protocols. IEEE Trans. Cybern. **49**(6), 2372–2384 (2019)
55. V. Gupta, T.H. Chung, B. Hassibi, R.M. Murray, On a stochastic sensor selection algorithm with applications in sensor scheduling and sensor coverage. Automatica **42**(2), 251–260 (2006)
56. J. Zhang, C. Peng, M.-R. Fei, Y.-C. Tian, Output feedback control of networked systems with a stochastic communication protocol. J. Frankl. Inst. **354**(9), 3838–3853 (2017)
57. E. Fridman, *Introduction to Time-Delay Systems: Analysis and Control* (Birkhäuser, Basel, 2014)
58. K. Liu, E. Fridman, K.H. Johansson, Networked control with stochastic scheduling. IEEE Trans. Autom. Control **60**(11), 3071–3076 (2015)
59. G. Wen, Y. Wan, J. Cao, T. Huang, W. Yu, Master-slave synchronization of heterogeneous systems under scheduling communication. IEEE Trans. Syst. Man Cybern. Syst. **48**(3), 473–484 (2018)

60. T.I. Fossen, *Guidance and Control of Ocean Vehicles* (Wiley, Chichester, U.K., 1994)
61. Z. Li, J. Sun, Disturbance compensating model predictive control with application to ship heading control. IEEE Trans. Control Syst. Technol. **20**(1), 257–265 (2012)
62. T.I. Fossen, T. Perez, Kalman filtering for positioning and heading control of ships and offshore rigs. IEEE Control Syst. Mag. **29**(6), 32–46 (2009)
63. S. Muhammad, A. Dòria-Cerezo, Passivity-based control applied to the dynamic positioning of ships. IET Control Theory Appl. **6**(5), 680–688 (2012)
64. E. Alfaro-Cid, E.W. McGookin, D.J. Murray-Smith, T.I. Fossen, Genetic programming for the automatic design of controllers for a surface ship. IEEE Trans. Intell. Transp. Syst. **9**(2), 311–321 (2008)
65. J.-H. Li, P.-M. Lee, B.-H. Jun, Y.-K. Lim, Point-to-point navigation of underactuated ships. Automatica **44**(12), 3201–3205 (2008)
66. R. Yu, Q. Zhu, G. Xia, Z. Liu, Sliding mode tracking control of an underactuated surface vessel. IET Control Theory Appl. **6**(3), 461–466 (2012)
67. D. Chwa, Global tracking control of underactuated ships with input and velocity constraints using dynamic surface control method. IEEE Trans. Control Syst. Technol. **19**(6), 1357–1370 (2011)
68. S.-Y. Kim, Y.-D. Yoon, S.-K. Sul, Suppression of thrust loss for the maximum thrust operation in the electric propulsion ship. IEEE Trans. Ind. Appl. **45**(2), 756–762 (2009)
69. J. Bakkeheim, T.A. Johansen, O.N. Smogeli, A.J. Sorensen, Lyapunov-based integrator resetting with application to marine thruster control. IEEE Trans. Control Syst. Technol. **16**(5), 908–917 (2008)
70. Y.-L. Wang, Q.-L. Han, Quantitative analysis and synthesis for networked control systems with non-uniformly distributed packet dropouts and interval time-varying sampling periods. Int. J. Robust Nonlinear Control **25**(2), 282–300 (2015)
71. Y.-L. Wang, T.-B. Wang, Q.-L. Han, Fault detection filter design for data reconstruction-based continuous-time networked control systems. Inf. Sci. **328**, 577–594 (2016)
72. B.-L. Zhang, Q.-L. Han, Network-based modelling and active control for offshore steel jacket platform with TMD mechanisms. J. Sound Vib. **333**(25), 6796–6814 (2014)
73. X.-M. Zhang, Q.-L. Han, X. Yu, Survey on recent advances in networked control systems. IEEE Trans. Ind. Inf. **12**(5), 1740–1752 (2016)
74. B.-L. Zhang, Q.-L. Han, X.-M. Zhang, X. Yu, Sliding mode control with mixed current and delayed states for offshore steel jacket platforms. IEEE Trans. Control Syst. Technol. **22**(5), 1769–1783 (2014)
75. X.-M. Zhang, Q.-L. Han, D. Han, Effects of small time-delays on dynamic output feedback control of offshore steel jacket structures. J. Sound Vib. **330**(16), 3883–3900 (2011)
76. N.E. Kahveci, P.A. Ioannou, Adaptive steering control for uncertain ship dynamics and stability analysis. Automatica **49**(3), 685–697 (2013)
77. R.-Y. Ren, Z.-J. Zou, X.-G. Wang, A two-time scale control law based on singular perturbations used in rudder roll stabilization of ships. Ocean Eng. **88**, 488–498 (2014)
78. H. Katayama, H. Aoki, Straight-line trajectory tracking control for sampled-data underactuated ships. IEEE Trans. Control Syst. Technol. **22**(4), 1638–1645 (2014)
79. P. Mahacek, C.A. Kitts, I. Mas, Dynamic guarding of marine assets through cluster control of automated surface vessel fleets. IEEE/ASME Trans. Mechatron. **17**(1), 65–75 (2012)
80. S.-I. Sohn, J.-H. Oh, Y.-S. Lee, D.-H. Park, I.-K. Oh, Design of a fuel-cell-powered catamaran-type unmanned surface vehicle. IEEE J. Ocean. Eng. **40**(2), 388–396 (2015)
81. H. Kim, D. Kim, J.-U. Shin, H. Kim, H. Myung, Angular rate-constrained path planning algorithm for unmanned surface vehicles. Ocean Eng. **84**, 37–44 (2014)
82. S.-L. Dai, C. Wang, F. Luo, Identification and learning control of ocean surface ship using neural networks. IEEE Trans. Ind. Inf. **8**(4), 801–810 (2012)
83. B. Tavassoli, Stability of nonlinear networked control systems over multiple communication links with asynchronous sampling. IEEE Trans. Autom. Control **59**(2), 511–515 (2014)
84. C. Peng, D. Yue, M.-R. Fei, A higher energy efficient sampling scheme for networked control systems over IEEE 802.15.4 wireless networks. IEEE Trans. Ind. Inf. **12**(5), 1766–1774 (2016)

85. C.-L. Lai, P.-L. Hsu, The butterfly-shaped feedback loop in networked control systems for the unknown delay compensation. IEEE Trans. Ind. Inf. **10**(3), 1746–1754 (2014)
86. F. Li, P. Shi, L. Wu, M.V. Basin, C.-C. Lim, Quantized control design for cognitive radio networks modeled as nonlinear semi-Markovian jump systems. IEEE Trans. Ind. Electron. **62**(4), 2330–2340 (2015)
87. G.-H. Yang, H. Wang, L. Xie, Fault detection for output feedback control systems with actuator stuck faults: a steady-state-based approach. Int. J. Robust Nonlinear Control **20**(15), 1739–1757 (2010)
88. J. Zhao, B. Jiang, P. Shi, Z. He, Fault tolerant control for damaged aircraft based on sliding mode control scheme. Int. J. Innov. Comput. Inf. Control **10**(1), 293–302 (2014)
89. G.-X. Zhong, G.-H. Yang, Robust control and fault detection for continuous-time switched systems subject to a dwell time constraint. Int. J. Robust Nonlinear Control **25**(18), 3799–3817 (2015)
90. J. You, S. Yin, H. Gao, Fault detection for discrete systems with network-induced nonlinearities. IEEE Trans. Ind. Inf. **10**(4), 2216–2223 (2014)
91. T.I. Fossen, *Handbook of Marine Craft Hydrodynamics and Motion Control* (Wiley, Chichester, U.K., 2011)
92. Y.-L. Wang, Q.-L. Han, Network-based modelling and dynamic output feedback control for unmanned marine vehicles in network environments. Automatica **91**, 43–53 (2018)
93. Y.-L. Wang, Q.-L. Han, Network-based heading control and rudder oscillation reduction for unmanned surface vehicles. IEEE Trans. Control Syst. Technol. **25**(5), 1609–1620 (2017)
94. Z. Peng, J. Wang, D. Wang, Distributed containment maneuvering of multiple marine vessels via neurodynamics-based output feedback. IEEE Trans. Ind. Electron. **64**(5), 3831–3839 (2017)
95. Z. Peng, J. Wang, D. Wang, Containment maneuvering of marine surface vehicles with multiple parameterized paths via spatial-temporal decoupling. IEEE/ASME Trans. Mechatron. **22**(2), 1026–1036 (2017)
96. M. Chen, S.S. Ge, B.V.E. How, Y.S. Choo, Robust adaptive position mooring control for marine vessels. IEEE Trans. Control Syst. Technol. **21**(2), 395–409 (2013)
97. Y.-L. Wang, Q.-L. Han, Network-based fault detection filter and controller coordinated design for unmanned surface vehicles in network environments. IEEE Trans. Ind. Inf. **12**(5), 1753–1765 (2016)
98. Z. Yan, J. Wang, Model predictive control for tracking of underactuated vessels based on recurrent neural networks. IEEE J. Ocean. Eng. **37**(4), 717–726 (2012)
99. Y. Wang, S. Wang, M. Tan, Path generation of autonomous approach to a moving ship for unmanned vehicles. IEEE Trans. Ind. Electron. **62**(9), 5619–5629 (2015)
100. T.A. Johansen, T.I. Bø, E. Mathiesen, A. Veksler, A.J. Sørensen, Dynamic positioning system as dynamic energy storage on diesel-electric ships. IEEE Trans. Power Syst. **29**(6), 3086–3091 (2014)
101. Z. Peng, D. Wang, J. Wang, Cooperative dynamic positioning of multiple marine offshore vessels: a modular design. IEEE/ASME Trans. Mechatron. **21**(3), 1210–1221 (2016)
102. H. Katayama, Nonlinear sampled-data stabilization of dynamically positioned ships. IEEE Trans. Control Syst. Technol. **18**(2), 463–468 (2010)
103. J. Du, X. Hu, H. Liu, C.L.P. Chen, Adaptive robust output feedback control for a marine dynamic positioning system based on a high-gain observer. IEEE Trans. Neural Netw. Learn. Syst. **26**(11), 2775–2786 (2015)
104. Z. Peng, D. Wang, J. Wang, Predictor-based neural dynamic surface control for uncertain nonlinear systems in strict-feedback form. IEEE Trans. Neural Netw. Learn. Syst. **28**(9), 2156–2167 (2017)
105. B. Bidikli, E. Tatlicioglu, E. Zergeroglu, Robust dynamic positioning of surface vessels via multiple unidirectional tugboats. Ocean Eng. **113**, 237–245 (2016)
106. V. Hassani, A.J. Sørensen, A.M. Pascoal, M. Athans, Robust dynamic positioning of offshore vessels using mixed-μ synthesis modeling, design, and practice. Ocean Eng. **129**, 389–400 (2017)

107. B. Bidikli, E. Tatlicioglu, E. Zergeroglu, Compensating of added mass terms in dynamically positioned surface vehicles: a continuous robust control approach. Ocean Eng. **139**, 198–204 (2017)
108. J. Du, X. Hu, M. Krstić, Y. Sun, Robust dynamic positioning of ships with disturbances under input saturation. Automatica **73**, 207–214 (2016)
109. W.-H. Ho, S.-H. Chen, J.-H. Chou, Optimal control of Takagi-Sugeno fuzzy-model-based systems representing dynamic ship positioning systems. Appl. Soft Comput. **13**(7), 3197–3210 (2013)
110. W.-B. Xie, Y.-L. Wang, J. Zhang, M.-Y. Fu, Novel separation principle based H_∞ observer-controller design for a class of T-S fuzzy systems. IEEE Trans. Fuzzy Syst. **26**(6), 3206–3221 (2018)
111. N.A.M. Subha, G.-P. Liu, Design and practical implementation of external consensus protocol for networked multiagent systems with communication delays. IEEE Trans. Control Syst. Technol. **23**(2), 619–631 (2015)
112. C. Peng, E. Tian, J. Zhang, D. Du, Decentralized event-triggering communication scheme for large-scale systems under network environments. Inf. Sci. **380**, 132–144 (2017)
113. Z.-H. Pang, G.-P. Liu, D. Zhou, Design and performance analysis of incremental networked predictive control systems. IEEE Trans. Cybern. **46**(6), 1400–1410 (2016)
114. G.-P. Liu, Consensus and stability analysis of networked multiagent predictive control systems. IEEE Trans. Cybern. **47**(4), 1114–1119 (2017)
115. D. Zhang, Q.-L. Han, X. Jia, Network-based output tracking control for a class of T-S fuzzy systems that can not be stabilized by nondelayed output feedback controllers. IEEE Trans. Cybern. **45**(8), 1511–1524 (2015)
116. D. Zhang, Q.-L. Han, X. Jia, Network-based output tracking control for T-S fuzzy systems using an event-triggered communication scheme. Fuzzy Sets Syst. **273**, 26–48 (2015)
117. C. Peng, S. Ma, X. Xie, Observer-based non-PDC control for networked T-S fuzzy systems with an event-triggered communication. IEEE Trans. Cybern. **47**(8), 2279–2287 (2017)
118. Z. Zhong, Y. Zhu, Observer-based output-feedback control of large-scale networked fuzzy systems with two-channel event-triggering. J. Frankl. Inst. **354**(13), 5398–5420 (2017)
119. H. Li, C. Wu, S. Yin, H.-K. Lam, Observer-based fuzzy control for nonlinear networked systems under unmeasurable premise variables. IEEE Trans. Fuzzy Syst. **24**(5), 1233–1245 (2016)
120. C. Zhang, J. Hu, J. Qiu, Q. Chen, Reliable output feedback control for T-S fuzzy systems with decentralized event triggering communication and actuator failures. IEEE Trans. Cybern. **47**(9), 2592–2602 (2017)
121. T. Adamek, C.A. Kitts, I. Mas, Gradient-based cluster space navigation for autonomous surface vessels. IEEE/ASME Trans. Mechatron. **20**(2), 506–518 (2015)
122. N. Fischer, D. Hughes, P. Walters, E.M. Schwartz, W.E. Dixon, Nonlinear RISE-based control of an autonomous underwater vehicle. IEEE Trans. Robot. **30**(4), 845–852 (2014)
123. Y. Liu, R. Bucknall, Path planning algorithm for unmanned surface vehicle formations in a practical maritime environment. Ocean Eng. **97**, 126–144 (2015)
124. N. Wang, Y. Gao, H. Zhao, C.K. Ahn, Reinforcement learning-based optimal tracking control of an unknown unmanned surface vehicle. IEEE Trans. Neural Netw. Learn. Syst. **32**(7), 3034–3045 (2021)
125. C. Peng, Q.-L. Han, A novel event-triggered transmission scheme and $\mathcal{L}_2$ control co-design for sampled-data control systems. IEEE Trans. Autom. Control **58**(10), 2620–2626 (2013)
126. C. Peng, Q.-L. Han, On designing a novel self-triggered sampling scheme for networked control systems with data losses and communication delays. IEEE Trans. Ind. Electron. **63**(2), 1239–1248 (2016)
127. S.-L. Du, X.-M. Sun, W. Wang, Guaranteed cost control for uncertain networked control systems with predictive scheme. IEEE Trans. Autom. Sci. Eng. **11**(3), 740–748 (2014)
128. M. Jungers, E.B. Castelan, V.M. Moraes, U.F. Moreno, A dynamic output feedback controller for NCS based on delay estimates. Automatica **49**(3), 788–792 (2013)

129. D. Yue, E. Tian, Q.-L. Han, A delay system method for designing event-triggered controllers of networked control systems. IEEE Trans. Autom. Control **58**(2), 475–481 (2013)
130. Y.-L. Wang, Q.-L. Han, Modelling and observer-based H_∞ controller design for networked control systems. IET Control Theory Appl. **8**(15), 1478–1486 (2014)
131. C. Lv, H. Yu, J. Chi, T. Xu, H. Zang, H. Jiang, Z. Zhang, A hybrid coordination controller for speed and heading control of underactuated unmanned surface vehicles system. Ocean Eng. **176**, 222–230 (2019)
132. Z. Gao, G. Guo, Command filtered finite/fixed-time heading tracking control of surface vehicles. IEEE/CAA J. Automatica Sinica **8**(10), 1667–1676 (2021)
133. L. Ma, Y.-L. Wang, Q.-L. Han, Event-triggered dynamic positioning for mass-switched unmanned marine vehicles in network environments. IEEE Trans. Cybern. **52**(5), 3159–3171 (2022)
134. J. Ye, S. Roy, M. Godjevac, V. Reppa, S. Baldi, Robustifying dynamic positioning of crane vessels for heavy lifting operation. IEEE/CAA J. Automatica Sinica **8**(4), 753–765 (2021)
135. M. Breivik, V.E. Hovstein, T.I. Fossen, Straight-line target tracking for unmanned surface vehicles. Model. Identif. Control **29**(4), 131–149 (2008)
136. L. Liu, D. Wang, Z. Peng, C.L.P. Chen, T. Li, Bounded neural network control for target tracking of underactuated autonomous surface vehicles in the presence of uncertain target dynamics. IEEE Trans. Neural Netw. Learn. Syst. **30**(4), 1241–1249 (2019)
137. S. Gao, Z. Peng, L. Liu, H. Wang, D. Wang, Coordinated target tracking by multiple unmanned surface vehicles with communication delays based on a distributed event-triggered extended state observer. Ocean Eng. **227**, Article no. 108283 (2021)
138. Y. Yang, J. Du, H. Liu, C. Guo, A. Abraham, A trajectory tracking robust controller of surface vessels with disturbance uncertainties. IEEE Trans. Control Syst. Technol. **22**(4), 1511–1518 (2014)
139. S. Dai, S. He, M. Wang, C. Yuan, Adaptive neural control of underactuated surface vessels with prescribed performance guarantees. IEEE Trans. Neural Netw. Learn. Syst. **30**(12), 3686–3698 (2019)
140. Z. Zheng, L. Sun, L. Xie, Error-constrained LOS path following of a surface vessel with actuator saturation and faults. IEEE Trans. Syst. Man Cybern. Syst. **48**(10), 1794–1805 (2018)
141. L. Liu, D. Wang, Z. Peng, ESO-based line-of-sight guidance law for path following of underactuated marine surface vehicles with exact sideslip compensation. IEEE J. Ocean. Eng. **42**(2), 477–487 (2017)
142. C. Liu, D. Wang, Y. Zhang, X. Meng, Model predictive control for path following and roll stabilization of marine vessels based on neurodynamic optimization. Ocean Eng. **217**, Article no. 107524 (2020)
143. Z. Peng, J. Wang, D. Wang, Q.-L. Han, An overview of recent advances in coordinated control of multiple autonomous surface vehicles. IEEE Trans. Ind. Inf. **17**(2), 732–745 (2021)
144. O. Elhaki, K. Shojaei, Neural network-based target tracking control of underactuated autonomous underwater vehicles with a prescribed performance. Ocean Eng. **167**, 239–256 (2018)
145. O. Namaki-Shoushtari, A.P. Aguiar, A. Khaki-Sedigh, Target tracking of autonomous robotic vehicles using range-only measurements: a switched logic-based control strategy. Int. J. Robust Nonlinear Control **22**(17), 1983–1998 (2011)
146. K. Shojaei, Three-dimensional neural network tracking control of a moving target by underactuated autonomous underwater vehicles. Neural Comput. Appl. **31**, 509–521 (2019)
147. K. Choi, S.J. Yoo, J.B. Park, Y.H. Choi, Adaptive formation control in absence of leader's velocity information. IET Control Theory Appl. **4**(4), 521–528 (2010)
148. Z. Peng, D. Wang, Robust adaptive formation control of autonomous surface vehicles with uncertain dynamics. IET Control Theory Appl. **5**(12), 1378–1387 (2011)
149. R. Cui, S.S. Ge, B.V.E. How, Y.S. Choo, Leader-follower formation control of underactuated autonomous underwater vehicles. Ocean Eng. **37**(17–18), 1491–1502 (2010)
150. L. Ma, Y.-L. Wang, Q.-L. Han, H_∞ cluster formation control of networked multiagent systems with stochastic sampling. IEEE Trans. Cybern. **51**(12), 5761–5772 (2021)

151. X. Ge, Q.-L. Han, L. Ding, Y.-L. Wang, X.-M. Zhang, Dynamic event-triggered distributed coordination control and its applications: a survey of trends and techniques. IEEE Trans. Syst. Man Cybern. Syst. **50**(9), 3112–3125 (2020)
152. Y. Wang, Y. Liu, Z. Wang, Theory and experiments on enclosing control of multi-agent systems. IEEE/CAA J. Automatica Sinica **8**(10), 1677–1685 (2021)
153. X. Ge, Q.-L. Han, J. Wang, X.-M. Zhang, A scalable adaptive approach to multi-vehicle formation control with obstacle avoidance. IEEE/CAA J. Automatica Sinica **9**(6), 990–1004 (2022)
154. B. Xiao, X. Yang, X. Huo, A novel disturbance estimation scheme for formation control of ocean surface vessels. IEEE Trans. Ind. Electron. **64**(6), 4994–5003 (2017)
155. T. Li, R. Zhao, C.L.P. Chen, L. Fang, C. Liu, Finite-time formation control of under-actuated ships using nonlinear sliding mode control. IEEE Trans. Cybern. **48**(11), 3243–3253 (2018)
156. Z. Peng, J. Wang, D. Wang, Distributed maneuvering of autonomous surface vehicles based on neurodynamic optimization and fuzzy approximation. IEEE Trans. Control Syst. Technol. **26**(3), 1083–1090 (2018)
157. X. Dong, Y. Zhou, Z. Ren, Y. Zhong, Time-varying formation tracking for second-order multi-agent systems subjected to switching topologies with application to quadrotor formation flying. IEEE Trans. Ind. Electron. **64**(6), 5014–5024 (2017)
158. J. Dai, G. Guo, Event-triggered leader-following consensus for multi-agent systems with semi-Markov switching topologies. Inf. Sci. **459**, 290–301 (2018)
159. Y. Cheng, Y. Zhang, L. Shi, J. Shao, Y. Xiao, Consensus seeking in heterogeneous second-order multi-agent systems with switching topologies and random link failures. Neurocomputing **319**, 188–195 (2018)
160. M. Meng, G. Xiao, C. Zhai, G. Li, Z. Wang, Distributed consensus of heterogeneous multi-agent systems subject to switching topologies and delays. J. Frankl. Inst. **357**(11), 6899–6917 (2020)
161. H. Sun, C. Peng, W. Zhang, T. Yang, Z. Wang, Security-based resilient event-triggered control of networked control systems under denial of service attacks. J. Frankl. Inst. **356**(7), 10277–10295 (2019)
162. L. Ma, Z. Wang, Q.-L. Han, H.-K. Lam, Variance-constrained distributed filtering for time-varying systems with multiplicative noises and deception attacks over sensor networks. IEEE Sens. J. **17**(7), 2279–2288 (2017)
163. D. Ding, Z. Wang, Q.-L. Han, G. Wei, Security control for discrete-time stochastic nonlinear systems subject to deception attacks. IEEE Trans. Syst. Man Cybern. Syst. **48**(5), 779–789 (2018)
164. Z. Zuo, Q.-L. Han, B. Ning, X. Ge, X.-M. Zhang, An overview of recent advances in fixed-time cooperative control of multiagent systems. IEEE Trans. Ind. Inf. **14**(6), 2322–2334 (2018)
165. A. Polyakov, Nonlinear feedback design for fixed-time stabilization of linear control systems. IEEE Trans. Autom. Control **57**(8), 2106–2110 (2012)
166. Z. Zuo, Non-singular fixed-time terminal sliding mode control of non-linear systems. IET Control Theory Appl. **9**(4), 545–552 (2015)
167. X. Yang, J. Lam, D.W.C. Ho, Z. Feng, Fixed-time synchronization of complex networks with impulsive effects via nonchattering control. IEEE Trans. Autom. Control **62**(11), 5511–5521 (2017)
168. B.S. Park, S.J. Yoo, Robust trajectory tracking with adjustable performance of underactuated surface vessels via quantized state feedback. Ocean Eng. **246**, Article no. 110475 (2022)
169. N. Wang, C.K. Ahn, Hyperbolic-tangent LOS guidance-based finite-time path following of underactuated marine vehicles. IEEE Trans. Ind. Electron. **67**(10), 8566–8575 (2020)

Chapter 2
Quantitative Analysis and Synthesis of NCSs

This chapter is concerned with quantitative analysis and synthesis for an NCS under simultaneous consideration of non-uniformly distributed packet dropouts and interval time-varying sampling periods. A packet dropout separation method is proposed to separate packet dropouts from the lump sum of network-induced delays and packet dropouts. An interval time-varying sampling period approach, which is more general than a switched sampling period approach, is presented to model the variation of the sampling period. Then a packet dropout decomposition based Lyapunov functional is constructed to derive some stability and controller design criteria. Numerical examples are given to illustrate the effectiveness of the obtained results.

2.1 Packet Dropout Separation-Based Modeling

Consider a linear time-invariant system described by

$$\begin{cases} \dot{x}(t) = Ax(t) + B_1 u(t) + B_2 \omega(t), \\ z(t) = Cx(t) + Du(t), \\ x(t_0) = x_0, \end{cases} \tag{2.1}$$

where $x(t) \in \mathbb{R}^n$, $u(t) \in \mathbb{R}^m$, $z(t) \in \mathbb{R}^r$, and $\omega(t) \in \mathbb{R}^q$ are the state vector, the control input vector, the controlled output, and the disturbance, respectively; $\omega(t)$ is assumed to belong to $L_2[t_0, \infty)$; $x_0 \in \mathbb{R}^n$ denotes the initial condition; A, B_1, B_2, C and D are known constant matrices of appropriate dimensions.

Throughout this chapter, we assume that the sensor is clock-driven while the controller and the actuator are event-driven; the system (2.1) is controlled through a shared band-limited network; network-induced delays and packet dropouts occur

Y.-L. Wang et al., *Network-Based Control of Unmanned Marine Vehicles*,
https://doi.org/10.1007/978-3-031-28605-6_2

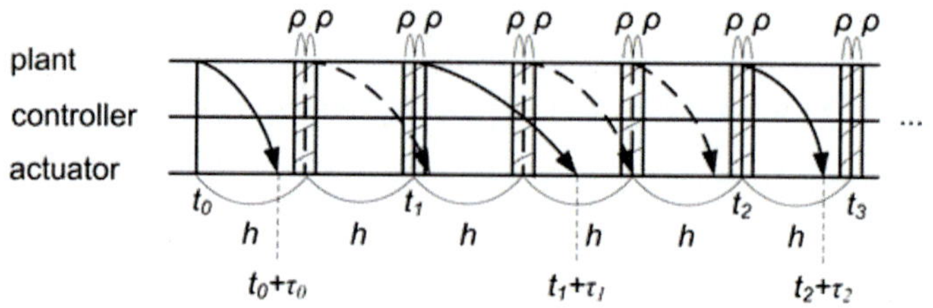

Fig. 2.1 An interval time-varying sampling mode

in both sensor-to-controller and controller-to-actuator channels; and the sampling period of the sensor varies on an interval.

Figure 2.1 depicts an interval time-varying sampling mode for an NCS under simultaneous consideration of network-induced delays and packet dropouts, where the dashed lines denote that the corresponding control inputs are dropped. Suppose that h is the ideal length of a sampling period and h_j is the actual sampling period, where $j = 1, 2, \ldots$. Notice that the sampling instant t_0 and the sampling instants t_j $(j = 1, 2, \ldots)$ are invariable and variable, respectively. Then, the first sampling period is $h_1 = h + \Delta h_1$, where $\Delta h_1 \in [-\rho, \rho]$ with ρ denoting the fluctuation of the sampling period and $\rho > 0$. The sequential sampling periods are $h_\sigma = h + \Delta h_\sigma$ $(\sigma = 2, 3, \ldots)$, where $\Delta h_\sigma \in [-2\rho, 2\rho]$. In practical situations, the fluctuation ρ of the sampling period h_j is usually much smaller than the ideal sampling period h. Without loss of generality, in this chapter, we assume that $\rho < \frac{h}{4}$. Denote $\mathscr{N}$ as the upper bound of consecutive packet dropouts. As shown in Fig. 2.1, the control inputs, which are based on the plant states at the sampling instants $t_k, t_{k+1}, \ldots$ $(k = 0, 1, 2, \ldots)$ are transmitted to the actuator successfully, while the ones which are based on the plant states between the instants t_k and t_{k+1} are dropped. Then the number of consecutive packet dropouts (denoted as $N_{0,1}$) during the time interval $[t_0, t_1)$ can be described as

$$N_{0,1} = \begin{cases} 0, & t_1 - t_0 \in [h - \rho, h + \rho], \\ 1, & t_1 - t_0 \in [2h - \rho, 2h + \rho], \\ \vdots & \vdots \\ \mathscr{N}, & t_1 - t_0 \in [(\mathscr{N} + 1)h - \rho, (\mathscr{N} + 1)h + \rho]. \end{cases} \tag{2.2}$$

The number of consecutive packet dropouts (denoted as $N_{j,j+1}$) during the time interval $[t_j, t_{j+1})$ can be described as

$$N_{j,j+1} = \begin{cases} 0, & t_{j+1} - t_j \in [h - 2\rho, h + 2\rho], \\ 1, & t_{j+1} - t_j \in [2h - 2\rho, 2h + 2\rho], \\ \vdots & \vdots \\ \mathscr{N}, & t_{j+1} - t_j \in [(\mathscr{N} + 1)h - 2\rho, (\mathscr{N} + 1)h + 2\rho], \end{cases} \tag{2.3}$$

where $j = 1, 2, \ldots$.

One can see that (2.2) and (2.3) hold for the time intervals $[t_0,\ t_1)$ and $[t_j,\ t_{j+1})$, respectively. Notice that $[h-\rho,\ h+\rho] \subset [h-2\rho,\ h+2\rho]$, ..., $[(\mathcal{N}+1)h-\rho,\ (\mathcal{N}+1)h+\rho] \subset [(\mathcal{N}+1)h-2\rho,\ (\mathcal{N}+1)h+2\rho]$. We rewrite (2.2) and (2.3) in the following uniform form

$$N_{k,\ k+1} = \begin{cases} 0, & t_{k+1}-t_k \in [h-2\rho,\ h+2\rho], \\ 1, & t_{k+1}-t_k \in [2h-2\rho,\ 2h+2\rho], \\ \vdots & \vdots \\ \mathcal{N}, & t_{k+1}-t_k \in [(\mathcal{N}+1)h-2\rho,\ (\mathcal{N}+1)h+2\rho], \end{cases} \tag{2.4}$$

where $N_{k,\ k+1}$ denotes the number of consecutive packet dropouts during the time interval $[t_k,\ t_{k+1})$, and $k = 0,\ 1,\ 2,\ \ldots$.

Let τ_k (named as network-induced delays in the literature) be the time from the instant t_k when the sensor samples data from the plant to the instant when the actuator transmits data to the plant. Then

$$u(t) = Kx(t_k), \tag{2.5}$$

where $t \in [t_k + \tau_k,\ t_{k+1} + \tau_{k+1})$, $k = 0,\ 1,\ 2,\ \ldots$, and K is a state feedback controller gain which will be designed later in this chapter.

For $t \in [t_k + \tau_k,\ t_{k+1} + \tau_{k+1})$, define $\tau(t) = t - t_k$. Then $t_k = t - \tau(t)$, $u(t) = Kx(t - \tau(t))$, and $\tau(t) \in [\tau_k,\ t_{k+1} - t_k + \tau_{k+1})$. Define $\eta_k = t_{k+1} - t_k + \tau_{k+1}$. Suppose that $\tau_m \le \tau_k < \tau_M$ and $\eta_k \le \eta$. Then $\tau_m \le \tau_k \le \tau(t) < \eta_k \le \eta$. Based on τ_m and η, one can get some stability/stabilization criteria for the NCS under consideration of network-induced delays and packet dropouts. However, network-induced delays and packet dropouts are lumped into one item η_k, and the upper bound of η_k, that is η, is employed to deal with stability analysis and stabilizing controller design. Then it is difficult to obtain the quantitative relationship between packet dropout probability and stability/stabilization of the NCS.

In the following, we will propose a packet dropout separation method to separate packet dropouts from the lump sum of network-induced delays and packet dropouts.

Suppose that $\tau_M = h - 4\rho + \tau_m$. Then $\tau_m \le \tau_{k+1} < h - 4\rho + \tau_m$. Combining (2.4) and $\tau_m \le \tau_{k+1} < h - 4\rho + \tau_m$ together, one has

$$N_{k,\ k+1} = \begin{cases} 0, & \eta_k \in \Psi_0, \\ 1, & \eta_k \in \Psi_1, \\ \vdots & \vdots \\ \mathcal{N}, & \eta_k \in \Psi_{\mathcal{N}}, \end{cases} \tag{2.6}$$

where

$$\begin{cases} \Psi_0 = [\tau_m + h - 2\rho,\ \tau_m + 2h - 2\rho), \\ \Psi_1 = [\tau_m + 2h - 2\rho,\ \tau_m + 3h - 2\rho), \\ \quad \vdots \\ \Psi_{\mathscr{N}} = [\tau_m + (\mathscr{N} + 1)h - 2\rho,\ \tau_m + (\mathscr{N} + 2)h - 2\rho). \end{cases} \tag{2.7}$$

Note that $Kx(t - \eta_k)$ is different from $Kx(t_k)$ in (2.5). Thus, $Kx(t - \eta_k)$ can not be used as the actual control input. From the definition of $\tau(t)$, one can see that $Kx(t - \tau(t))$ can be used as the actual control input. Considering that $\tau_m \leq \tau(t)$, $\tau_M = h - 4\rho + \tau_m$, and (2.6), one can obtain the relationship between $\tau(t)$ and the number of accumulated packet dropouts

$$N_{k,\ k+1} = \begin{cases} 0, & \tau(t) \in \Gamma_0, \\ 1, & \tau(t) \in \Gamma_1, \\ \vdots & \quad \vdots \\ \mathscr{N}, & \tau(t) \in \Gamma_{\mathscr{N}}, \end{cases} \tag{2.8}$$

where

$$\begin{cases} \Gamma_0 = [\tau_m,\ \tau_M + h + 2\rho), \\ \Gamma_1 = [\tau_M + h + 2\rho,\ \tau_M + 2h + 2\rho), \\ \quad \vdots \\ \Gamma_{\mathscr{N}} = [\tau_M + \mathscr{N} h + 2\rho,\ \tau_M + (\mathscr{N} + 1)h + 2\rho). \end{cases} \tag{2.9}$$

Obviously, $\bigcup_{l=0}^{\mathscr{N}} \Gamma_l = [\tau_m,\ \tau_M + (\mathscr{N} + 1)h + 2\rho)$. From $\tau(t) \in [\tau_k,\ t_{k+1} - t_k + \tau_{k+1})$, one can see that $\tau(t)$ increases from τ_k to $t_{k+1} - t_k + \tau_{k+1}$. If $\tau(t) \in \Gamma_l$ $(l = 1,\ \ldots,\ \mathscr{N})$, $\tau(t)$ varies from $\Gamma_0, \ldots, \Gamma_{l-1}$ to Γ_l.

Remark 2.1 The quantitative relationship between the number of packet dropouts and $\tau(t)$ is established in (2.8), and packet dropouts are separated from the lump sum of network-induced delays and packet dropouts. The quantitative relationship established in (2.8) plays an important role in establishing the quantitative relationship between packet dropout probability and stability/stabilization of the NCS.

Remark 2.2 One can see clearly that (2.8) is applicable for every feasible value of the interval time-varying sampling period h_j. However, in [1, 2], the sampling period h_j is assumed to be switched in a finite set. Then the time-varying sampling periods considered in (2.8) are more general than the ones considered in [1, 2]. Moreover, in this chapter, the discretization process in [2–4] is avoided to reduce modeling and design complexity.

For convenience of analysis and design, define $\mathscr{A} = \lfloor \frac{\mathscr{N}}{2} \rfloor$, $\beta_1 = \tau_M + (\mathscr{A} + 1)h + 2\rho$, $\beta_2 = \tau_M + (\mathscr{N} + 1)h + 2\rho$, where $\mathscr{A}$ is the largest integer smaller than or equal to $\frac{\mathscr{N}}{2}$. Without loss of generality, we divide $[\tau_m,\ \beta_2)$ into two subintervals Φ_1 and Φ_2, where $\Phi_1 = [\tau_m,\ \beta_1)$, $\Phi_2 = [\beta_1,\ \beta_2)$. Then,

$$N_{k,\,k+1} = \begin{cases} 0, \ \ldots, \ \mathscr{A}, & \tau(t) \in \Phi_1, \\ \mathscr{A}+1, \ \ldots, \ \mathscr{N}, & \tau(t) \in \Phi_2. \end{cases} \tag{2.10}$$

Throughout this chapter, suppose that packet dropouts are non-uniformly distributed, and the probability of 0 to $\mathscr{A}$ packets dropped is $\bar{\lambda}$, where $\bar{\lambda} \in [0,\ 1]$. Then the probability of $\mathscr{A}+1$ to $\mathscr{N}$ packets dropped is $1-\bar{\lambda}$. Such a statistic characteristic can be described by

$$\begin{cases} \text{Prob}\{\tau(t) \in \Phi_1\} = \bar{\lambda}, \\ \text{Prob}\{\tau(t) \in \Phi_2\} = 1-\bar{\lambda}. \end{cases} \tag{2.11}$$

Define the following stochastic variable $\lambda(t)$

$$\lambda(t) = \begin{cases} 1, & \tau(t) \in \Phi_1, \\ 0, & \tau(t) \in \Phi_2. \end{cases} \tag{2.12}$$

By using the Bernoulli distributed white sequence to describe the stochastic variable $\lambda(t)$, one has

$$\begin{cases} \text{Prob}\{\lambda(t)=1\} = E\{\lambda(t)\} = \bar{\lambda}, \\ \text{Prob}\{\lambda(t)=0\} = 1-E\{\lambda(t)\} = 1-\bar{\lambda}. \end{cases} \tag{2.13}$$

Taking into account the non-uniform distribution characteristic of packet dropouts, one has the control law described as

$$u(t) = \lambda(t)Kx(t-\tau_1(t)) + (1-\lambda(t))Kx(t-\tau_2(t)), \tag{2.14}$$

where

$$\tau_1(t) = \begin{cases} \tau(t), & \tau(t) \in \Phi_1, \\ \bar{\tau}_1, & \tau(t) \in \Phi_2, \end{cases}$$

$$\tau_2(t) = \begin{cases} \tau(t), & \tau(t) \in \Phi_2, \\ \bar{\tau}_2, & \tau(t) \in \Phi_1, \end{cases}$$

with $\bar{\tau}_1$ and $\bar{\tau}_2$ being constants, and $\bar{\tau}_1 \in \Phi_1$, $\bar{\tau}_2 \in \Phi_2$. Then, for $t \in [t_k+\tau_k,\ t_{k+1}+\tau_{k+1})$, one has

$$\begin{cases} \dot{x}(t) = \varphi_1(t) + (\lambda(t)-\bar{\lambda})\varphi_2(t) + B_2\omega(t), \\ z(t) = \varphi_3(t) + (\lambda(t)-\bar{\lambda})\varphi_4(t), \end{cases} \tag{2.15}$$

where

$$\begin{aligned}\varphi_1(t) &= Ax(t) + \bar{\lambda}B_1Kx(t-\tau_1(t)) + (1-\bar{\lambda})B_1Kx(t-\tau_2(t)),\\ \varphi_2(t) &= B_1K[x(t-\tau_1(t)) - x(t-\tau_2(t))],\\ \varphi_3(t) &= Cx(t) + \bar{\lambda}DKx(t-\tau_1(t)) + (1-\bar{\lambda})DKx(t-\tau_2(t)),\\ \varphi_4(t) &= DK[x(t-\tau_1(t)) - x(t-\tau_2(t))].\end{aligned}$$

The initial condition of the state $x(t)$ on $[t_0 - \beta_2,\ t_0]$ is supplemented as

$$x(\xi) = \psi(\xi),\ \xi \in [t_0 - \beta_2,\ t_0],$$

where $\psi \in \mathbb{W}$ and $\mathbb{W}$ denotes the Banach space of absolutely continuous functions $[-\beta_2,\ 0] \to \mathbb{R}^n$ with square-integrable derivative and with the norm

$$\|\psi\|_W^2 = \|\psi(0)\|^2 + \int_{-\beta_2}^{0} \|\psi(\xi)\|^2 d\xi + \int_{-\beta_2}^{0} \|\dot{\psi}(\xi)\|^2 d\xi \tag{2.16}$$

where the vector norm $\|\cdot\|$ represents the Euclidean norm.

Remark 2.3 As shown in the system (2.15), the non-uniform distribution characteristic of packet dropouts is taken into full consideration. In fact, if we choose $\mathscr{A} = \mathscr{N}$ and $\lambda(t) = 1$, the system (2.15) reduces to the one without considering the non-uniform distribution characteristic of packet dropouts. However, since more information about packet dropouts is utilized, considering the non-uniform distribution characteristic of packet dropouts will lead to better results.

Remark 2.4 For the system (2.15), if packet dropouts are not separated from the lump sum of network-induced delays and packet dropouts, the sampling period is constant, and the non-uniform distribution characteristic of packet dropouts is not considered, then (2.15) reduces to the corresponding models in [5–7]. The analysis presented above demonstrates that the system (2.15) is more general than some existing ones.

2.2 Stability Analysis and Controller Design

This section is concerned with performance analysis and controller design for the system (2.15). For this purpose, construct the following Lyapunov functional

$$V(t, x_t) = \sum_{i=1}^{5} V_i(t, x_t), \tag{2.17}$$

where

$$
\begin{aligned}
V_1(t,x_t) &= x^T(t)Px(t),\\
V_2(t,x_t) &= \int_{t-\beta_2}^{t-\beta_1} x^T(s)Q_1x(s)ds + \int_{t-\beta_1}^{t} x^T(s)Q_2x(s)ds + \int_{t-\tau_m}^{t} x^T(s)Q_3x(s)ds,\\
V_3(t,x_t) &= \frac{\beta_2}{2}\int_{-\frac{\beta_2}{2}}^{0}\int_{t+s}^{t} \dot{x}^T(\theta)R_1\dot{x}(\theta)d\theta ds + \frac{\beta_1}{2}\int_{-\frac{\beta_1}{2}}^{0}\int_{t+s}^{t} \dot{x}^T(\theta)R_2\dot{x}(\theta)d\theta ds\\
&\quad + \frac{\tau_m}{2}\int_{-\frac{\tau_m}{2}}^{0}\int_{t+s}^{t} \dot{x}^T(\theta)R_3\dot{x}(\theta)d\theta ds,\\
V_4(t,x_t) &= (\beta_2-\beta_1)\int_{-\beta_2}^{-\beta_1}\int_{t+s}^{t} \dot{x}^T(\theta)M_1\dot{x}(\theta)d\theta ds\\
&\quad + (\beta_1-\tau_m)\int_{-\beta_1}^{-\tau_m}\int_{t+s}^{t} \dot{x}^T(\theta)M_2\dot{x}(\theta)d\theta ds,\\
V_5(t,x_t) &= \int_{t-\frac{\tau_m}{2}}^{t} \begin{bmatrix} x(\theta) \\ x(\theta-\frac{\tau_m}{2}) \end{bmatrix}^T \begin{bmatrix} W_1 & W_2 \\ * & W_3 \end{bmatrix} \begin{bmatrix} x(\theta) \\ x(\theta-\frac{\tau_m}{2}) \end{bmatrix} d\theta\\
&\quad + \int_{t-\frac{\beta_1}{2}}^{t} \begin{bmatrix} x(\theta) \\ x(\theta-\frac{\beta_1}{2}) \end{bmatrix}^T \begin{bmatrix} W_4 & W_5 \\ * & W_6 \end{bmatrix} \begin{bmatrix} x(\theta) \\ x(\theta-\frac{\beta_1}{2}) \end{bmatrix} d\theta\\
&\quad + \int_{t-\frac{\beta_2}{2}}^{t} \begin{bmatrix} x(\theta) \\ x(\theta-\frac{\beta_2}{2}) \end{bmatrix}^T \begin{bmatrix} W_7 & W_8 \\ * & W_9 \end{bmatrix} \begin{bmatrix} x(\theta) \\ x(\theta-\frac{\beta_2}{2}) \end{bmatrix} d\theta,
\end{aligned}
$$

where $x_t = x(t+\xi)$, $\xi \in [t_0-\beta_2,\ t_0]$, P, Q_1, Q_2, Q_3, R_1, R_2, R_3, M_1, and M_2 are symmetric positive definite matrices with appropriate dimensions, and W_κ $(\kappa = 1, \ldots, 9)$ are matrices satisfying

$$
\begin{bmatrix} W_1 & W_2 \\ * & W_3 \end{bmatrix} > 0,\ \begin{bmatrix} W_4 & W_5 \\ * & W_6 \end{bmatrix} > 0,\ \begin{bmatrix} W_7 & W_8 \\ * & W_9 \end{bmatrix} > 0. \tag{2.18}
$$

Then we state and establish the following result.

Theorem 2.1 *For given scalars $\bar{\lambda}$ $(0 \le \bar{\lambda} \le 1)$, $\tau_m > 0$, $\tau_M > 0$, $h > 0$, $\mathcal{N} > 0$, $\rho > 0$, $\gamma > 0$, and the controller gain K, the system (2.15) is asymptotically stable in the sense of mean-square with an H_∞ norm bound γ if there exist symmetric positive definite matrices P, Q_1, Q_2, Q_3, R_1, R_2, R_3, M_1, M_2, W_1, W_3, W_4, W_6, W_7, W_9, and matrices W_2, W_5, W_8 such that (2.18) and*

$$
\begin{bmatrix} \Pi_{11} & \Pi_{12} & \Pi_{13} \\ * & \Pi_{22} & 0 \\ * & * & \Pi_{33} \end{bmatrix} < 0, \tag{2.19}
$$

where $\Pi_{11} = \begin{bmatrix} \Lambda_{11} & \Lambda_{12} \\ * & \Lambda_{22} \end{bmatrix}$ *with*

$$\Lambda_{11} = \begin{bmatrix} \Omega_{11} & \Omega_{12} & 0 & \Omega_{14} & \Omega_{15} \\ * & \Omega_{22} & -W_2 & 0 & 0 \\ * & * & \Omega_{33} & M_2 & 0 \\ * & * & * & -2M_2 & 0 \\ * & * & * & * & \Omega_{55} \end{bmatrix}, \ \Lambda_{12} = \begin{bmatrix} 0 & \Omega_{17} & \Omega_{18} & 0 & PB_2 \\ 0 & 0 & 0 & 0 & 0 \\ 0 & 0 & 0 & 0 & 0 \\ M_2 & 0 & 0 & 0 & 0 \\ -W_5 & 0 & 0 & 0 & 0 \end{bmatrix},$$

$$\Lambda_{22} = \begin{bmatrix} \Omega_{66} & M_1 & 0 & 0 & 0 \\ * & -2M_1 & 0 & M_1 & 0 \\ * & * & \Omega_{88} & -W_8 & 0 \\ * & * & * & \Omega_{99} & 0 \\ * & * & * & * & -\gamma I \end{bmatrix},$$

and

$$\Pi_{12} = \begin{bmatrix} \Theta A \ 0 \ 0 \ \Upsilon_1 \ 0 \ 0 \ \ \Upsilon_2 \ \ 0 \ 0 \ \Theta B_2 \\ 0 \ \ 0 \ 0 \ \Upsilon_3 \ 0 \ 0 \ -\Upsilon_3 \ 0 \ 0 \ \ 0 \end{bmatrix}^T,$$

$$\Pi_{13} = \begin{bmatrix} C \ 0 \ 0 \ \bar{\lambda} DK \ 0 \ 0 \ \ \Upsilon_4 \ \ 0 \ 0 \ 0 \\ 0 \ 0 \ 0 \ \ \Upsilon_5 \ \ 0 \ 0 \ -\Upsilon_5 \ 0 \ 0 \ 0 \end{bmatrix}^T,$$

$$\Pi_{22} = \mathrm{diag}\{-\Theta, \ -\bar{\lambda}(1-\bar{\lambda})\Theta\}, \ \Pi_{33} = \mathrm{diag}\{-\gamma I, \ -\bar{\lambda}(1-\bar{\lambda})\gamma I\},$$

$$\Omega_{11} = PA + A^T P + Q_2 + Q_3 - R_1 - R_2 - R_3 + W_1 + W_4 + W_7, \ \Omega_{12} = R_3 + W_2,$$

$$\Omega_{14} = \bar{\lambda} PB_1 K, \ \Omega_{15} = R_2 + W_5, \ \Omega_{17} = (1-\bar{\lambda}) PB_1 K, \ \Omega_{18} = R_1 + W_8,$$

$$\Omega_{22} = -R_3 + W_3 - W_1, \ \Omega_{33} = -Q_3 - M_2 - W_3, \ \Omega_{55} = -R_2 + W_6 - W_4,$$

$$\Omega_{66} = Q_1 - Q_2 - M_1 - M_2 - W_6, \ \Omega_{88} = -R_1 + W_9 - W_7, \ \Omega_{99} = -Q_1 - M_1 - W_9,$$

$$\Upsilon_1 = \bar{\lambda}\Theta B_1 K, \ \Upsilon_2 = (1-\bar{\lambda})\Theta B_1 K, \ \Upsilon_3 = \bar{\lambda}(1-\bar{\lambda})\Theta B_1 K, \ \Upsilon_4 = (1-\bar{\lambda}) DK,$$

$$\Upsilon_5 = \bar{\lambda}(1-\bar{\lambda}) DK, \ \Theta = \frac{\beta_2^2}{4} R_1 + \frac{\beta_1^2}{4} R_2 + \frac{\tau_m^2}{4} R_3 + (\beta_2 - \beta_1)^2 M_1 + (\beta_1 - \tau_m)^2 M_2.$$

Proof Taking the time derivative of the Lyapunov functional $V(t, x_t)$ given in (2.17) along the trajectory of the system (2.15), and considering that

$$E\{\dot{V}(t, x_t)\} = \sum_{i=1}^{5} E\{\dot{V}_i(t, x_t)\}, \ E\{\lambda(t) - \bar{\lambda}\} = 0, \ E\{(\lambda(t) - \bar{\lambda})^2\} = \bar{\lambda}(1-\bar{\lambda}),$$

one obtains

$$E\{\dot{V}_1(t, x_t)\} = 2E\{x^T(t) P \dot{x}(t)\} = 2x^T(t) P \varphi_1(t) + 2x^T(t) PB_2 \omega(t), \tag{2.20}$$

$$\begin{aligned} E\{\dot{V}_2(t, x_t)\} = & x^T(t-\beta_1) Q_1 x(t-\beta_1) + x^T(t) Q_2 x(t) + x^T(t) Q_3 x(t) \\ & - x^T(t-\beta_2) Q_1 x(t-\beta_2) - x^T(t-\beta_1) Q_2 x(t-\beta_1) \\ & - x^T(t-\tau_m) Q_3 x(t-\tau_m), \end{aligned} \tag{2.21}$$

$$
\begin{aligned}
E\{\dot{V}_3(t,x_t)\} = E\{\dot{x}^T(t)\Theta_1\dot{x}(t) &- \frac{\beta_2}{2}\int_{t-\frac{\beta_2}{2}}^{t}\dot{x}^T(\theta)R_1\dot{x}(\theta)d\theta \\
&- \frac{\beta_1}{2}\int_{t-\frac{\beta_1}{2}}^{t}\dot{x}^T(\theta)R_2\dot{x}(\theta)d\theta - \frac{\tau_m}{2}\int_{t-\frac{\tau_m}{2}}^{t}\dot{x}^T(\theta)R_3\dot{x}(\theta)d\theta\},
\end{aligned}
\tag{2.22}
$$

where $\Theta_1 = \frac{\beta_2^2}{4}R_1 + \frac{\beta_1^2}{4}R_2 + \frac{\tau_m^2}{4}R_3$,

$$
\begin{aligned}
E\{\dot{V}_4(t,x_t)\} = E\{\dot{x}^T(t)\Theta_2\dot{x}(t) &- (\beta_2-\beta_1)\int_{t-\beta_2}^{t-\beta_1}\dot{x}^T(\theta)M_1\dot{x}(\theta)d\theta \\
&- (\beta_1-\tau_m)\int_{t-\beta_1}^{t-\tau_m}\dot{x}^T(\theta)M_2\dot{x}(\theta)d\theta\},
\end{aligned}
\tag{2.23}
$$

where $\Theta_2 = (\beta_2-\beta_1)^2M_1 + (\beta_1-\tau_m)^2M_2$,

$$
\begin{aligned}
&E\{\dot{V}_5(t,x_t)\} \\
&= \begin{bmatrix} x(t) \\ x(t-\frac{\tau_m}{2}) \end{bmatrix}^T \begin{bmatrix} W_1 & W_2 \\ * & W_3 \end{bmatrix} \begin{bmatrix} x(t) \\ x(t-\frac{\tau_m}{2}) \end{bmatrix} - \begin{bmatrix} x(t-\frac{\tau_m}{2}) \\ x(t-\tau_m) \end{bmatrix}^T \begin{bmatrix} W_1 & W_2 \\ * & W_3 \end{bmatrix} \begin{bmatrix} x(t-\frac{\tau_m}{2}) \\ x(t-\tau_m) \end{bmatrix} \\
&+ \begin{bmatrix} x(t) \\ x(t-\frac{\beta_1}{2}) \end{bmatrix}^T \begin{bmatrix} W_4 & W_5 \\ * & W_6 \end{bmatrix} \begin{bmatrix} x(t) \\ x(t-\frac{\beta_1}{2}) \end{bmatrix} - \begin{bmatrix} x(t-\frac{\beta_1}{2}) \\ x(t-\beta_1) \end{bmatrix}^T \begin{bmatrix} W_4 & W_5 \\ * & W_6 \end{bmatrix} \begin{bmatrix} x(t-\frac{\beta_1}{2}) \\ x(t-\beta_1) \end{bmatrix} \\
&+ \begin{bmatrix} x(t) \\ x(t-\frac{\beta_2}{2}) \end{bmatrix}^T \begin{bmatrix} W_7 & W_8 \\ * & W_9 \end{bmatrix} \begin{bmatrix} x(t) \\ x(t-\frac{\beta_2}{2}) \end{bmatrix} - \begin{bmatrix} x(t-\frac{\beta_2}{2}) \\ x(t-\beta_2) \end{bmatrix}^T \begin{bmatrix} W_7 & W_8 \\ * & W_9 \end{bmatrix} \begin{bmatrix} x(t-\frac{\beta_2}{2}) \\ x(t-\beta_2) \end{bmatrix}.
\end{aligned}
\tag{2.24}
$$

Then by using the Jensen integral inequality in [8], the Schur complement and the definition of H_∞ norm bound, one can conclude that if (2.18)–(2.19) are satisfied, the system (2.15) is asymptotically stable in the sense of mean square with an H_∞ norm bound γ. This completes the proof. □

It should be pointed out that the system (2.15) can be rewritten as the following simplified form

$$
\begin{cases}
\dot{x}(t) = Ax(t) + \lambda(t)B_1Kx(t-\tau_1(t)) + (1-\lambda(t))B_1Kx(t-\tau_2(t)) + B_2\omega(t), \\
z(t) = Cx(t) + \lambda(t)DKx(t-\tau_1(t)) + (1-\lambda(t))DKx(t-\tau_2(t)).
\end{cases}
$$

Notice that $E\{\dot{x}^T(t)\Theta_1\dot{x}(t)\}$ and $E\{\dot{x}^T(t)\Theta_2\dot{x}(t)\}$ are introduced in $E\{\dot{V}_3(t,x_t)\}$ and $E\{\dot{V}_4(t,x_t)\}$, respectively, in the proof of Theorem 2.1. For convenience of dealing with the items $E\{\dot{x}^T(t)\Theta_1\dot{x}(t)\}$ and $E\{\dot{x}^T(t)\Theta_2\dot{x}(t)\}$, the closed-loop system (2.15) instead of the simplified system is considered.

Remark 2.5 In [9, 10], a discrete delay decomposition approach was proposed to study the stability of linear retarded and neutral systems with constant time delays. The delay decomposition approach is adopted in Theorem 2.1 to deal with time-varying network-induced delays and packet dropouts. From the Lyapunov functional (2.17), one can see that the time interval $[\tau_m,\ \beta_2]$ is decomposed as $[\tau_m,\ \beta_1]$ and $[\beta_1,\ \beta_2]$; $[-\beta_1,\ 0]$ is decomposed as $[-\beta_1,\ -\frac{\beta_1}{2}]$ and $[-\frac{\beta_1}{2},\ 0]$; $[-\beta_2,\ 0]$ is decomposed as $[-\beta_2,\ -\frac{\beta_2}{2}]$ and $[-\frac{\beta_2}{2},\ 0]$. If $\tau_m = 0$ and $M_1 = M_2$, $\int_{-\beta_2}^{-\beta_1}\int_{t+s}^{t}\dot{x}^T(\theta)M_1\dot{x}(\theta)d\theta ds + \int_{-\beta_1}^{-\tau_m}\int_{t+s}^{t}\dot{x}^T(\theta)M_2\dot{x}(\theta)d\theta ds$ in $V_4(t, x_t)$ in (2.17) reduces to $\int_{t-\eta}^{t}\int_{s}^{t}\dot{x}^T(v)\ T\dot{x}(v)dvds$ in the Lyapunov functional (7) in [5]. If $Q_1 = Q_2$, then $\int_{t-\beta_2}^{t-\beta_1}x^T(s)Q_1x(s)ds + \int_{t-\beta_1}^{t}x^T(s)Q_2x(s)ds$ reduces to $V_3(t)$ in the Lyapunov functional (11) in [11]. Similarly, $V_2(t, x_t)$ and $V_4(t, x_t)$ in (2.17) can be reduced to the corresponding items of the Lyapunov functional (2) in [7]. The analysis presented above illustrates the merits of the Lyapunov functional (2.17).

Remark 2.6 Notice that Theorem 2.1 is formulated in terms of linear matrix inequalities (LMIs), which can be tested by an efficient interior-point algorithm. The algorithm is of polynomial-time complexity. The total number of scalar decision variables of Theorem 2.1 is $\mathscr{M}_1 = 10.5n^2 + 7.5n$ and the total row size of the LMIs is $\mathscr{L}_1 = 17n + q + 2r$. Then the computational complexity of Theorem 2.1 is proportional to $\mathscr{L}_1\mathscr{M}_1^3$.

Based on the packet dropout separation method, Theorem 2.1 presents the quantitative relationship between packet dropout probability and H_∞ performance of the system (2.15). We turn to consider the case that the packet dropout separation method is not adopted, and the sampling period is constant. By replacing the upper bound of $\tau_1(t)$, the lower bound of $\tau_2(t)$, and the upper bound of $\tau_2(t)$ with $\bar{\beta}_1$, $\bar{\beta}_1$, and $\bar{\beta}_2$, respectively, where $\bar{\beta}_1$ and $\bar{\beta}_2$ are given scalars, the system (2.15) is converted into a new system (2.15)'. If setting $\omega(t) = 0$ in (2.15)', then the following stability criterion can be derived immediately.

Corollary 2.1 *For given scalars $\bar{\lambda}$ $(0 \le \bar{\lambda} \le 1)$, $\tau_m > 0$, $\bar{\beta}_1 > 0$, $\bar{\beta}_2 > 0$, and the controller gain K, the system (2.15)' with $\omega(t) = 0$ is asymptotically stable in the sense of mean square if there exist symmetric positive definite matrices P, Q_1, Q_2, Q_3, R_1, R_2, R_3, M_1, M_2, W_1, W_3, W_4, W_6, W_7, W_9, and matrices W_2, W_5, W_8 such that (2.18) and*

$$\begin{bmatrix} \Pi_{11}^0 & \Pi_{12}^0 \\ * & \Pi_{22}^0 \end{bmatrix} < 0, \tag{2.25}$$

where Π_{11}^0 is derived by deleting the $10th$ row and the $10th$ column of Π_{11} in (2.19); Π_{22}^0 is derived by replacing β_1 and β_2 in Π_{22} with $\bar{\beta}_1$ and $\bar{\beta}_2$, respectively, in (2.19);

$$\Pi_{12}^0 = \begin{bmatrix} \bar{\Theta}A & 0 & 0 & \bar{\Upsilon}_1 & 0 & 0 & \bar{\Upsilon}_2 & 0 & 0 \\ 0 & 0 & 0 & \bar{\Upsilon}_3 & 0 & 0 & -\bar{\Upsilon}_3 & 0 & 0 \end{bmatrix}^T,$$

$\bar{\Upsilon}_1$, $\bar{\Upsilon}_2$, $\bar{\Upsilon}_3$, and $\bar{\Theta}$ are derived by replacing β_1 and β_2 in Υ_1, Υ_2, Υ_3, and Θ with $\bar{\beta}_1$ and $\bar{\beta}_2$, respectively, in (2.19).

Compared with some existing results, more scalar decision variables are involved in the LMI (2.25). As a consequence, the computational complexity of Corollary 2.1 is "higher". In order to reduce the computational complexity, we choose a simplified Lyapunov functional by replacing $V_3(t, x_t) + V_5(t, x_t)$ in (2.17) with $\tau_m \int_{-\tau_m}^{0} \int_{t+s}^{t} \dot{x}^T(\theta) M_3 \dot{x}(\theta) d\theta ds$. Then, based on Corollary 2.1, we have the following stability criterion with "lower" computational complexity.

Corollary 2.2 *For given scalars $\bar{\lambda}$ ($0 \leq \bar{\lambda} \leq 1$), $\tau_m > 0$, $\bar{\beta}_1 > 0$, $\bar{\beta}_2 > 0$, and the controller gain K, the system (2.15)' with $\omega(t) = 0$ is asymptotically stable in the sense of mean square if there exist symmetric positive definite matrices P, Q_1, Q_2, Q_3, M_1, M_2, M_3, such that*

$$\begin{bmatrix} \hat{\Omega}_{11} & M_3 & \hat{\Omega}_{13} & 0 & \hat{\Omega}_{15} & 0 \\ * & \hat{\Omega}_{22} & 3M_2 & 0 & 0 & 0 \\ * & * & \hat{\Omega}_{33} & 3M_2 & 0 & 0 \\ * & * & * & \hat{\Omega}_{44} & 3M_1 & 0 \\ * & * & * & * & \hat{\Omega}_{55} & 3M_1 \\ * & * & * & * & * & \hat{\Omega}_{66} \end{bmatrix} < 0, \quad (2.26)$$

where

$$\begin{aligned} \hat{\Omega}_{11} &= PA + A^T P + Q_2 + Q_3 + A^T \Lambda A - M_3, \ \hat{\Omega}_{13} = \bar{\lambda}(PB_1K + A^T \Lambda B_1 K), \\ \hat{\Omega}_{15} &= (1-\bar{\lambda})(PB_1K + A^T \Lambda B_1 K), \ \hat{\Omega}_{22} = -Q_3 - 3M_2 - M_3, \\ \hat{\Omega}_{33} &= \bar{\lambda} K^T B_1^T \Lambda B_1 K - 6M_2, \ \hat{\Omega}_{44} = Q_1 - Q_2 - 3M_1 - 3M_2, \\ \hat{\Omega}_{55} &= (1-\bar{\lambda}) K^T B_1^T \Lambda B_1 K - 6M_1, \ \hat{\Omega}_{66} = -Q_1 - 3M_1, \\ \Lambda &= (\bar{\beta}_2 - \bar{\beta}_1)^2 M_1 + (\bar{\beta}_1 - \tau_m)^2 M_2 + \tau_m^2 M_3. \end{aligned}$$

We turn to analyze the computational complexity of the stability criteria in Corollaries 2.1, 2.2 and the existing literature, which is listed in Table 2.1. From Table 2.1, one can see that the computational complexity of the stability criterion in Corollary 2.1 is higher than the one of the stability criteria in [6, 12] and Corollary 2.2. Moreover, as shown in Sect. 2.3 of this chapter, the stability criterion in Corollary 2.1 provides a larger MATI than the ones in [6, 12, 13] and Corollary 2.2. Then the results in this chapter provide the flexibility that allows us to trade off between the computational complexity and the conservatism of the stability criteria.

Table 2.1 The computational complexity of different methods

Methods	The computational complexity
[13]	$(9n)(17n^2 + 4n)^3$
[6]	$(6n)(6n^2 + 4n)^3$
[12]	$(16n)(4.5n^2 + 4.5n)^3$
Corollary 2.2	$(6n)(3.5n^2 + 3.5n)^3$
Corollary 2.1	$(17n)(10.5n^2 + 7.5n)^3$

Based on Theorem 2.1, we are now in a position to design the controller gain K such that the system (2.15) is asymptotically stable in the sense of mean square with an H_∞ norm bound γ.

Theorem 2.2 *For given scalars $\bar{\lambda}$ $(0 \le \bar{\lambda} \le 1)$, $\tau_m > 0$, $\tau_M > 0$, $h > 0$, $\mathcal{N} > 0$, $\rho > 0$, $\gamma > 0$, the system (2.15) is asymptotically stable in the sense of mean square with an H_∞ norm bound γ and the controller gain is given by $K = L^T S^{-1}$, if there exist symmetric positive definite matrices S, $\widetilde{Q}_1$, $\widetilde{Q}_2$, $\widetilde{Q}_3$, $\widetilde{R}_1$, $\widetilde{R}_2$, $\widetilde{R}_3$, $\widetilde{M}_1$, $\widetilde{M}_2$, $\widetilde{W}_1$, $\widetilde{W}_3$, $\widetilde{W}_4$, $\widetilde{W}_6$, $\widetilde{W}_7$, $\widetilde{W}_9$, and matrices $\widetilde{W}_2$, $\widetilde{W}_5$, $\widetilde{W}_8$, L such that*

$$\begin{bmatrix} \widetilde{W}_1 & \widetilde{W}_2 \\ * & \widetilde{W}_3 \end{bmatrix} > 0, \quad \begin{bmatrix} \widetilde{W}_4 & \widetilde{W}_5 \\ * & \widetilde{W}_6 \end{bmatrix} > 0, \quad \begin{bmatrix} \widetilde{W}_7 & \widetilde{W}_8 \\ * & \widetilde{W}_9 \end{bmatrix} > 0, \tag{2.27}$$

$$\begin{bmatrix} \widetilde{\Pi}_{11} & \widetilde{\Pi}_{12} & \widetilde{\Pi}_{13} & \widetilde{\Pi}_{14} \\ * & \widetilde{\Pi}_{22} & 0 & 0 \\ * & * & \widetilde{\Pi}_{33} & 0 \\ * & * & * & \widetilde{\Pi}_{44} \end{bmatrix} < 0, \tag{2.28}$$

where $\widetilde{\Pi}_{11} = \begin{bmatrix} \widetilde{\Lambda}_{11} & \widetilde{\Lambda}_{12} \\ * & \widetilde{\Lambda}_{22} \end{bmatrix}$ *with*

$$\widetilde{\Lambda}_{11} = \begin{bmatrix} \widetilde{\Omega}_{11} & \widetilde{\Omega}_{12} & 0 & \widetilde{\Omega}_{14} & \widetilde{\Omega}_{15} \\ * & \widetilde{\Omega}_{22} & -\widetilde{W}_2 & 0 & 0 \\ * & * & \widetilde{\Omega}_{33} & \widetilde{M}_2 & 0 \\ * & * & * & \widetilde{\Omega}_{44} & 0 \\ * & * & * & * & \widetilde{\Omega}_{55} \end{bmatrix}, \quad \widetilde{\Lambda}_{12} = \begin{bmatrix} 0 & \widetilde{\Omega}_{17} & \widetilde{\Omega}_{18} & 0 & B_2 \\ 0 & 0 & 0 & 0 & 0 \\ 0 & 0 & 0 & 0 & 0 \\ \widetilde{M}_2 & 0 & 0 & 0 & 0 \\ -\widetilde{W}_5 & 0 & 0 & 0 & 0 \end{bmatrix},$$

$$\widetilde{\Lambda}_{22} = \begin{bmatrix} \widetilde{\Omega}_{66} & \widetilde{M}_1 & 0 & 0 & 0 \\ * & \widetilde{\Omega}_{77} & 0 & \widetilde{M}_1 & 0 \\ * & * & \widetilde{\Omega}_{88} & -\widetilde{W}_8 & 0 \\ * & * & * & \widetilde{\Omega}_{99} & 0 \\ * & * & * & * & -\gamma I \end{bmatrix},$$

and

$$\begin{aligned}
\widetilde{\Pi}_{12} &= \mathrm{diag}\{SA^T,\ 0,\ 0,\ \bar{\lambda}LB_1^T,\ 0,\ 0,\ (1-\bar{\lambda})LB_1^T,\ 0,\ 0,\ B_2^T\}\mathrm{Ones}(10,\ 5),\\
\widetilde{\Pi}_{13} &= \mathrm{diag}\{0,\ 0,\ 0,\ \bar{\lambda}(1-\bar{\lambda})LB_1^T,\ 0,\ 0,\ -\bar{\lambda}(1-\bar{\lambda})LB_1^T,\ 0,\ 0,\ 0\}\mathrm{Ones}(10,\ 5),\\
\widetilde{\Pi}_{14} &= \begin{bmatrix} CS & 0 & 0 & \widetilde{\Upsilon}_1 & 0 & 0 & \widetilde{\Upsilon}_2 & 0 & 0 & 0 \\ 0 & 0 & 0 & \widetilde{\Upsilon}_3 & 0 & 0 & -\widetilde{\Upsilon}_3 & 0 & 0 & 0 \end{bmatrix}^T,\\
\widetilde{\Pi}_{22} &= \mathrm{diag}\{(\frac{\beta_2^2}{4})^{-1}(\widetilde{R}_1 - 2S),\ (\frac{\beta_1^2}{4})^{-1}(\widetilde{R}_2 - 2S),\ (\frac{\tau_m^2}{4})^{-1}(\widetilde{R}_3 - 2S),\\
&\qquad (\beta_2 - \beta_1)^{-2}(\widetilde{M}_1 - 2S),\ (\beta_1 - \tau_m)^{-2}(\widetilde{M}_2 - 2S)\},\\
\widetilde{\Pi}_{33} &= \bar{\lambda}(1-\bar{\lambda})\widetilde{\Pi}_{22},\ \widetilde{\Pi}_{44} = \mathrm{diag}\{-\gamma I,\ -\bar{\lambda}(1-\bar{\lambda})\gamma I\},\\
\widetilde{\Upsilon}_1 &= \bar{\lambda}DL^T,\ \widetilde{\Upsilon}_2 = (1-\bar{\lambda})DL^T,\ \widetilde{\Upsilon}_3 = \bar{\lambda}(1-\bar{\lambda})DL^T,
\end{aligned}$$

$$\widetilde{\Omega}_{11} = AS + SA^T + \widetilde{Q}_2 + \widetilde{Q}_3 - \widetilde{R}_1 - \widetilde{R}_2 - \widetilde{R}_3 + \widetilde{W}_1 + \widetilde{W}_4 + \widetilde{W}_7,\ \widetilde{\Omega}_{12} = \widetilde{R}_3 + \widetilde{W}_2,$$
$$\widetilde{\Omega}_{14} = \bar{\lambda} B_1 L^T,\ \widetilde{\Omega}_{15} = \widetilde{R}_2 + \widetilde{W}_5,\ \widetilde{\Omega}_{17} = (1-\bar{\lambda}) B_1 L^T,\ \widetilde{\Omega}_{18} = \widetilde{R}_1 + \widetilde{W}_8,$$
$$\widetilde{\Omega}_{22} = -\widetilde{R}_3 + \widetilde{W}_3 - \widetilde{W}_1,\ \widetilde{\Omega}_{33} = -\widetilde{Q}_3 - \widetilde{M}_2 - \widetilde{W}_3,\ \widetilde{\Omega}_{44} = -2\widetilde{M}_2,$$
$$\widetilde{\Omega}_{55} = -\widetilde{R}_2 + \widetilde{W}_6 - \widetilde{W}_4,\ \widetilde{\Omega}_{66} = \widetilde{Q}_1 - \widetilde{Q}_2 - \widetilde{M}_1 - \widetilde{M}_2 - \widetilde{W}_6,\ \widetilde{\Omega}_{77} = -2\widetilde{M}_1,$$
$$\widetilde{\Omega}_{88} = -\widetilde{R}_1 + \widetilde{W}_9 - \widetilde{W}_7,\ \widetilde{\Omega}_{99} = -\widetilde{Q}_1 - \widetilde{M}_1 - \widetilde{W}_9,$$

while Ones(10, 5) *is a 10-by-5 matrix with all entries equal to 1.*

Proof By considering the inequalities (2.18)–(2.19), and using Schur complement and the controller design method in [2], one can conclude that if (2.27)–(2.28) are satisfied, the system (2.15) is asymptotically stable in the sense of mean square with an H_∞ norm bound γ, and the controller gain is given by $K = L^T S^{-1}$. This completes the proof. □

Remark 2.7 The total number of scalar decision variables of Theorem 2.2 is $\mathcal{M}_2 = 10.5n^2 + 7.5n + nm$, and the total row size of the LMIs is $\mathcal{L}_2 = 25n + q + 2r$. Then the computational complexity of Theorem 2.2 is proportional to $\mathcal{L}_2\mathcal{M}_2^3$.

2.3 Performance Analysis and Discussion

In this section, we give two examples to show the effectiveness of the obtained results.

Example 2.1 Consider the following system

$$\dot{x}(t) = \begin{bmatrix} 0 & 1 \\ 0 & -0.1 \end{bmatrix} x(t) + \begin{bmatrix} 0 \\ 0.1 \end{bmatrix} u(t), \tag{2.29}$$

where the controller gain is $K = \begin{bmatrix} -3.75 & -11.5 \end{bmatrix}$. For Corollaries 2.1 and 2.2, suppose that $\bar{\beta}_1 = 0.5$, $\bar{\lambda} = 0.5$, $\tau_m = 0$. By using some existing stability criteria, Corollaries 2.1 and 2.2, one can obtain the MATIs ensuring the stability of the NCS (2.29), which are listed in Table 2.2. It should be mentioned that the computation of the MATIs for Corollaries 2.1 and 2.2 is based on a trial-and-error algorithm. From Table 2.2, one can see clearly that both Corollaries 2.1 and 2.2 provide larger MATIs than the stability criteria in [6, 7, 13].

For comparison with the result in [12], suppose that $\bar{\beta}_1 = 0.5$, $\bar{\lambda} = 0.8$. Then the MATIs corresponding to different τ_m are given in Table 2.3. If $\tau_m = 0.2$, $\bar{\beta}_1 = 1$,

Table 2.2 MATIs for different methods

Methods	[13]	[6, 7]	Corollary 2.2	Corollary 2.1
MATIs	0.8871	1.0081	1.3361	1.4298

Table 2.3 MATIs for different τ_m

MATIs/τ_m	0	0.05	0.1	0.2	0.3	0.4
Theorem 1 [12]	2.1859	2.1913	2.1977	2.2135	2.2294	2.2449
Corollary 2.1	2.2671	2.2684	2.2695	2.2730	2.2787	2.2871

Table 2.4 MATIs for different $\bar{\lambda}$

MATIs/$\bar{\lambda}$	0.5	0.7	0.9	0.99
Theorem 1 [12]	1.1002	1.1593	1.3775	2.5447
Corollary 2.1	1.1563	1.2010	1.3915	2.6309

Table 2.5 β_1, β_2 and H_∞ norm bound γ for different $\mathcal{N}$

$\beta_1, \beta_2, \gamma/\mathcal{N}$	5	7	11	19	35
β_1	1.5880	1.8880	2.4880	3.6880	6.0880
β_2	2.4880	3.0880	4.2880	6.6880	11.4880
γ	0.6333	0.8599	1.4718	3.3285	653.9443

we obtain the MATIs for different $\bar{\lambda}$, which are listed in Table 2.4. One can see that Corollary 2.1 provides larger MATIs than Theorem 1 in [12].

We now give the following example to illustrate the effectiveness and advantage of the proposed packet dropout separation based controller design.

Example 2.2 Consider the system

$$\begin{aligned} \dot{x}(t) &= \begin{bmatrix} 0 & 1 \\ 0 & -0.1 \end{bmatrix} x(t) + \begin{bmatrix} 0 \\ 0.1 \end{bmatrix} u(t) + \begin{bmatrix} 0.1 \\ -0.1 \end{bmatrix} \omega(t), \\ z(t) &= \begin{bmatrix} 0.1 & 1 \end{bmatrix} x(t) - 0.1u(t). \end{aligned} \tag{2.30}$$

Suppose that $\bar{\lambda} = 0.8$, $h = 0.3s$, $\rho = 0.02h$, $\tau_m = 0.4$. Then, $\tau_M = \tau_m + h - 4\rho = 0.6760$. Solving the LMIs in Theorem 2.2, one can obtain β_1, β_2 and γ corresponding to different $\mathcal{N}$ (the upper bound of consecutive packet dropouts), and the corresponding results are listed in Table 2.5. It should be pointed out that if $\mathcal{N} = 36$, the LMIs in Theorem 2.2 are infeasible, which implies that the admissible upper bound of consecutive packet dropouts is 35.

In the case of $\mathcal{N} = 5$, solving the LMIs in Theorem 2.2, one can obtain the stabilizing controller gain $K = [-0.3296 \quad -3.2959]$. Suppose that the initial state of the system is $x_0 = [0.2 \quad -0.2]^T$, and the interval time-varying delay $\tau(t)$ is given in Fig. 2.2. The disturbance $\omega(t)$ is defined by

$$\omega(t) = \begin{cases} 0.5sin(t), & 1s \le t \le 4s, \\ 0, & otherwise. \end{cases}$$

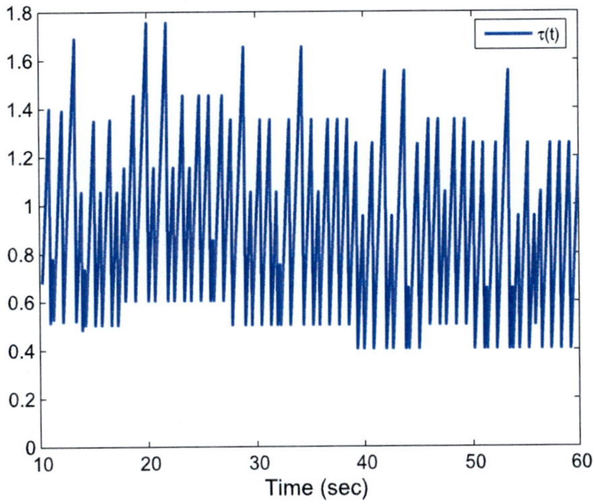

Fig. 2.2 Curve of $\tau(t)$

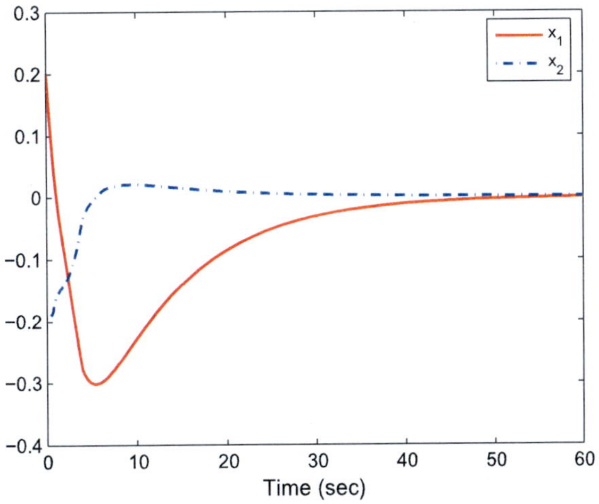

Fig. 2.3 Curves of plant state

Then curves of the plant state are pictured in Fig. 2.3, which illustrates the effectiveness of the proposed controller design. The advantage of the proposed packet dropout separation based controller design is analyzed as follows. By combining (2.8) and the curve of $\tau(t)$ in Fig. 2.2 together, one can obtain the number of accumulated packet dropouts, which is presented in Fig. 2.4. Then, by combining (2.8), Figs. 2.2, 2.3, and 2.4 together, one can obtain the quantitative relationship between packet dropout probability and stabilization of the NCS (2.30). That is, the proposed

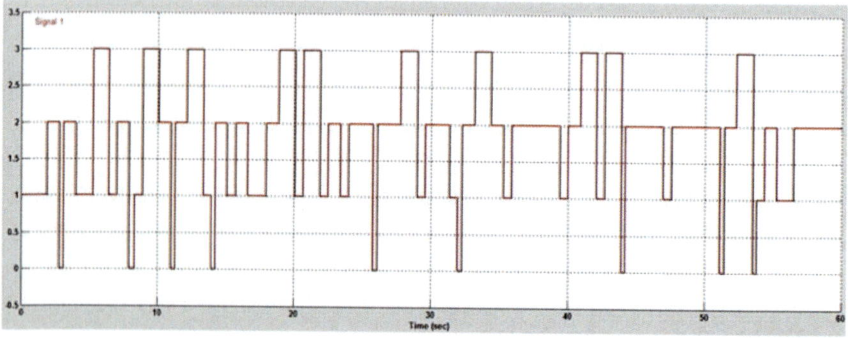

Fig. 2.4 The number of accumulated packet dropouts

packet dropout separation based controller design helps to establish the quantitative relationship between packet dropout probability and stabilization of the considered NCS.

2.4 Conclusions

The quantitative analysis and controller synthesis for an NCS under simultaneous consideration of packet dropouts and interval time-varying sampling periods have been studied. The packet dropout separation method and the packet dropout decomposition based Lyapunov functional have been proposed, and the non-uniform distribution characteristic of packet dropouts has been taken into full consideration to establish the quantitative relationship between packet dropout probability and stability/stabilization of the NCS. The interval time-varying sampling period approach, which is more general than the switched sampling period approach, has been proposed to guarantee robustness of the NCS to small variations of the sampling period. Even for an NCS considering a constant sampling period and unseparated packet dropouts, the obtained results are still less conservative than some existing ones.

2.5 Notes

Note that network-induced delays and packet dropouts are lumped into one item $i_{k+1}h - i_k h + \tau_{k+1}$ in [6, 11, 14, 15], which will introduce much difficulty for distinguishing the effects of packet dropouts from the effects of network-induced delays. Moreover, the non-uniform distribution characteristic of network-induced delays is made full use in [12, 16, 17]. In this chapter, the packet dropout separation method is proposed, and the non-uniform distribution characteristic of packet dropouts is taken into full consideration correspondingly. The interval time-varying sampling periods, which are more general than the switched sampling periods in [1, 2], are considered. The results in this chapter are based mainly on [18]. For more details about the corresponding analysis and design methods, one can also refer to [9, 13, 19, 20], etc.

References

1. H. Gao, J. Wu, P. Shi, Robust sampled-data H_∞ control with stochastic sampling. Automatica **45**(7), 1729–1736 (2009)
2. Y.-L. Wang, G.-H. Yang, Output tracking control for networked control systems with time delay and packet dropout. Int. J. Control **81**(11), 1709–1719 (2008)
3. R. Lozano, P. Castillo, P. Garcia, A. Dzul, Robust prediction-based control for unstable delay systems: application to the yaw control of a mini-helicopter. Automatica **40**(4), 603–612 (2004)
4. A. Sala, Computer control under time-varying sampling period: an LMI gridding approach. Automatica **41**(12), 2077–2082 (2005)
5. D. Yue, Q.-L. Han, C. Peng, State feedback controller design of networked control systems. IEEE Trans. Circuits Syst. II: Express Briefs **51**(11), 640–644 (2004)
6. X. Jiang, Q.-L. Han, New stability criteria for linear systems with interval time-varying delay. Automatica **44**(10), 2680–2685 (2008)
7. X. Jiang, Q.-L. Han, S. Liu, A. Xue, A new H_∞ stabilization criterion for networked control systems. IEEE Trans. Autom. Control **53**(4), 1025–1032 (2008)
8. Q.-L. Han, Absolute stability of time-delay systems with sector-bounded nonlinearity. Automatica **41**(12), 2171–2176 (2005)
9. Q.-L. Han, A discrete delay decomposition approach to stability of linear retarded and neutral systems. Automatica **45**(2), 517–524 (2009)
10. Q.-L. Han, Improved stability criteria and controller design for linear neutral systems. Automatica **45**(8), 1948–1952 (2009)
11. H. Gao, T. Chen, J. Lam, A new delay system approach to network-based control. Automatica **44**(1), 39–52 (2008)
12. C. Peng, D. Yue, E. Tian, Z. Gu, A delay distribution based stability analysis and synthesis approach for networked control systems. J. Frankl. Inst. **346**(4), 349–365 (2009)
13. D. Yue, Q.-L. Han, J. Lam, Network-based robust H_∞ control of systems with uncertainty. Automatica **41**(6), 999–1007 (2005)
14. X. Meng, J. Lam, H. Gao, Network-based H_∞ control for stochastic systems. Int. J. Robust Nonlinear Control **19**(3), 295–312 (2009)
15. X. Jia, D. Zhang, X. Hao, N. Zheng, Fuzzy H_∞ tracking control for nonlinear networked control systems in T-S fuzzy model. IEEE Trans. Syst. Man Cybern. Part B Cybern. **39**(4), 1073–1079 (2009)
16. Y. Tipsuwan, M.-Y. Chow, Gain scheduler middleware: a methodology to enable existing controllers for networked control and teleoperation-part I: networked control. IEEE Trans. Ind. Electron. **51**(6), 1218–1227 (2004)
17. D. Yue, E. Tian, Z. Wang, J. Lam, Stabilization of systems with probabilistic interval input delays and its applications to networked control systems. IEEE Trans. Syst. Man Cybern. Part A Syst. Hum. **39**(4), 939–945 (2009)
18. Y.-L. Wang, Q.-L. Han, Quantitative analysis and synthesis for networked control systems with non-uniformly distributed packet dropouts and interval time-varying sampling periods. Int. J. Robust Nonlinear Control **25**(2), 282–300 (2015)

19. X.-M. Zhang, Q.-L. Han, B.-L. Zhang, An overview and deep investigation on sampled-data-based event-triggered control and filtering for networked systems. IEEE Trans. Ind. Inf. **13**(1), 4–16 (2017)
20. X.-M. Zhang, Q.-L. Han, X. Yu, Survey on recent advances in networked control systems. IEEE Trans. Ind. Inf. **12**(5), 1740–1752 (2016)

Chapter 3
FDF Design for Data Reconstruction-Based NCSs

This chapter addresses the observer-based FDF design for a continuous-time NCS by taking packet dropouts and network-induced delays into account. An observer-based FDF and a reference residual model are introduced to construct a model for the continuous-time NCS. To reduce the time for fault detection, a new data reconstruction scheme is proposed and the corresponding closed-loop model is established. Based on the established models, FDF design criteria are derived to asymptotically stabilize the residual system. The mutually exclusive distribution characteristic of interval time-varying delays is made full use to deal with integral inequalities for products of vectors. The designed FDFs can guarantee the sensitivity of the residual signal to faults. A numerical example is presented to demonstrate the effectiveness of the proposed observer-based FDF design.

3.1 Data Reconstruction-Based Modeling for an NCS with Faults

Consider a continuous-time NCS, whose faults are to be detected, described by

$$\begin{cases} \dot{x}(t) = Ax(t) + Bu(t) + D_1\omega(t) + E_1 f(t), \\ y(t) = Cx(t), \\ x(t_0) = x_0, \end{cases} \tag{3.1}$$

where $x(t) \in \mathbb{R}^n$, $u(t) \in \mathbb{R}^m$, $y(t) \in \mathbb{R}^s$, $\omega(t) \in \mathbb{R}^p$, $f(t) \in \mathbb{R}^q$ are the state vector, control input vector, measurement output, disturbance input, and actuator fault signal, respectively; $\omega(t)$ is assumed to belong to $L_2[t_0, \infty)$; $x_0 \in \mathbb{R}^n$ denotes the initial condition; A, B, D_1, E_1, and C are known constant matrices of appropriate dimensions; C^T is assumed to be full column rank and (A, C) is detectable.

Y.-L. Wang et al., *Network-Based Control of Unmanned Marine Vehicles*,
https://doi.org/10.1007/978-3-031-28605-6_3

Throughout this chapter, we introduce a buffer at the FDF to store the outputs of the FDF and the recently received measurement outputs. Let t_k, $t_k + h$, $t_k + 2h$, $t_k + \cdots, t_{k+1}, t_{k+1} + h, t_{k+1} + 2h, t_{k+1} + \cdots$ $(k = 0,\ 1,\ 2,\ \ldots)$ denote the sampling instants, where h is the length of the sampling period. Suppose that the measurement outputs which are sampled at the instants t_k, t_{k+1}, ... are transmitted to the FDF successfully, while the measurement outputs sampled between the instants t_k and t_{k+1} $(k = 0,\ 1,\ 2,\ \ldots)$ are dropped, and δ denotes the upper bound of consecutive packet dropouts. Define τ_k as the time from the instant t_k when the sensor samples data from the plant to the instant when the FDF receives the data. Suppose that $\tau_m \leq \tau_k < \tau_M$, where τ_m and τ_M are given scalars satisfying $\tau_m > 0$, $\tau_M > 0$.

In this chapter, the following FDF is adopted to generate the residual signal

$$\begin{cases} \dot{\hat{x}}(t) = A\hat{x}(t) + Bu(t) + L(y(t) - \hat{y}(t)), \\ \hat{y}(t) = C\hat{x}(t), \\ r(t) = S(y(t) - \hat{y}(t)), \end{cases} \tag{3.2}$$

where $\hat{x}(t) \in \mathbb{R}^n$, $\hat{y}(t) \in \mathbb{R}^s$, and $r(t) \in \mathbb{R}^{q_r}$ are the state of the FDF, the output of the FDF and the residual signal, respectively; L and S are to be determined.

A reference residual model is usually needed to describe the desired behavior of the residual signal $r(t)$. In this chapter, we introduce the following reference residual model [1, 2]

$$\begin{cases} \dot{e}_r(t) = A_r e_r(t) + B_r \omega(t) + C_r f(t), \\ r_{ref}(t) = D_r e_r(t) + E_r \omega(t) + F_r f(t), \end{cases} \tag{3.3}$$

where $e_r(t) \in \mathbb{R}^{n_r}$ and $r_{ref}(t) \in \mathbb{R}^{q_r}$ are the state and the output of the reference residual model, respectively; A_r, B_r, C_r, D_r, E_r and F_r are given constant matrices of appropriate dimensions; A_r is a stable matrix.

Over the time interval $[t_k + \tau_k,\ t_{k+1} + \tau_{k+1})$, the available measurement output for the FDF is $y(t_k)$. At the instant t_k, the output of the FDF, i.e., $\hat{y}(t_k)$, is stored in the buffer. Define $e(t) = x(t) - \hat{x}(t)$, $\xi(t) = [x^T(t)\ e^T(t)\ e_r^T(t)]^T$, $\nu(t) = [u^T(t)\ \omega^T(t)\ f^T(t)]^T$, $r_e(t) = r_{ref}(t) - r(t)$, $\tau(t) = t - t_k$. For $t \in [t_k + \tau_k,\ t_{k+1} + \tau_{k+1})$, one has $\tau(t) \in [\tau_k,\ t_{k+1} - t_k + \tau_{k+1})$. Considering that $\tau_m \leq \tau_k < \tau_M$ and $t_{k+1} - t_k \leq (\delta + 1)h$, one has $\tau(t) \in [\tau_m,\ \eta)$, where $\eta = (\delta + 1)h + \tau_M$. Based on (3.1)–(3.3) and the above given parameters, one can derive the following closed-loop system

$$\begin{cases} \dot{\xi}(t) = \tilde{A}\xi(t) + \tilde{B}\xi(t - \tau(t)) + \tilde{D}\nu(t), \\ r_e(t) = \tilde{C}\xi(t) + \tilde{E}\xi(t - \tau(t)) + \tilde{F}\nu(t), \end{cases} \tag{3.4}$$

where

$$\tilde{A} = \begin{bmatrix} A & 0 & 0 \\ 0 & A & 0 \\ 0 & 0 & A_r \end{bmatrix},\ \tilde{B} = \begin{bmatrix} 0 & 0 & 0 \\ 0 & -LC & 0 \\ 0 & 0 & 0 \end{bmatrix},\ \tilde{D} = \begin{bmatrix} B & D_1 & E_1 \\ 0 & D_1 & E_1 \\ 0 & B_r & C_r \end{bmatrix}, \tag{3.5}$$
$$\tilde{C} = \begin{bmatrix} 0 & 0 & D_r \end{bmatrix},\ \tilde{E} = \begin{bmatrix} 0 & -SC & 0 \end{bmatrix},\ \tilde{F} = \begin{bmatrix} 0 & E_r & F_r \end{bmatrix}.$$

Remark 3.1 For a fault detection system, how to detect the occurrence of faults in time is quite important. Notice that the feasible fault signals are included in the measurement outputs $y(t_{k-1}),\ y(t_k),\ \ldots$. For $t \in [t_k + \tau_k,\ t_{k+1} + \tau_{k+1})$, if we combine the recently received measurement output $y(t_k)$ and the formerly received measurement output $y(t_{k-1})$ together, which method is named as data reconstruction in this chapter, and transmit the data reconstruction-based measurement outputs to the FDF, the fault signals generated at the instants t_k and t_{k-1} can affect simultaneously the residual signal $r(t)$ during the time interval $[t_k + \tau_k,\ t_{k+1} + \tau_{k+1})$. Compared with the fault detection scheme which adopts only $y(t_k)$ over the time interval $[t_k + \tau_k,\ t_{k+1} + \tau_{k+1})$, the data reconstruction-based fault detection scheme helps to reduce the needed time for fault detection.

In what follows, we turn to modeling for a fault detection system considering data reconstruction. Notice that a buffer is equipped at the FDF to store the recently received measurement outputs. The outputs $\hat{y}(t_{k-1}),\ \hat{y}(t_k),\ \ldots$ of the FDF are also stored in the buffer. It should be mentioned that the storage capacity of the buffer is limited, and only the recent data can be stored in the buffer. Then, for $t \in [t_k + \tau_k,\ t_{k+1} + \tau_{k+1})$, the measurement outputs, which are actually adopted by the FDF, can be described by

$$\tilde{y}(t) = \frac{1}{\tilde{a}}(a_1 y(t_k) + a_2 y(t_{k-1})), \tag{3.6}$$

where $\tilde{a} = a_1 + a_2$ with a_1 and a_2 being given parameters.

Remark 3.2 For the data reconstruction scheme presented in (3.6), if $a_1 = 1$ and $a_2 = 0$, the actually adopted measurement output at the FDF reduces to $y(t_k)$. Then, the data reconstruction scheme in (3.6) provides more flexibility for fault detection. On the other hand, to reduce the time for fault detection, fault information, which is included in $y(t_{k-1})$, should be assigned a higher weight than fault information included in $y(t_k)$. That is, one should set $a_1 < a_2$. Moreover, if the storage capacity of the buffer is enough, one can construct the measurement outputs in (3.6) as $\tilde{y}(t) = \frac{1}{\tilde{a}}(a_1 y(t_k) + a_2 y(t_{k-2}))$, or $\tilde{y}(t) = \frac{1}{\tilde{a}}(a_1 y(t_k) + a_2 y(t_{k-3}))$, etc. For given a_1 and a_2, if choosing $\tilde{y}(t) = \frac{1}{\tilde{a}}(a_1 y(t_k) + a_2 y(t_{k-\rho_1}))$ with $\rho_1 = 1,\ 2,\ \ldots$ can provide shorter fault detection time than $\tilde{y}(t) = \frac{1}{\tilde{a}}(a_1 y(t_k) + a_2 y(t_{k-\rho_2}))$ with $\rho_2 = 1,\ 2,\ \ldots$ and $\rho_1 \neq \rho_2$, one should choose $\tilde{y}(t) = \frac{1}{\tilde{a}}(a_1 y(t_k) + a_2 y(t_{k-\rho_1}))$.

Define $d(t) = t - t_{k-1}$. Considering that $t - t_{k-1} = (t - t_k) + (t_k - t_{k-1})$ and $t_k - t_{k-1} \in \{h,\ 2h,\ \ldots,\ (\delta + 1)h\}$, one has $d(t) \in [\tau_m + h,\ \tilde{\eta})$, where $\tilde{\eta} = 2(\delta + 1)h + \tau_M$. Based on the data reconstruction scheme in (3.6) and the closed-loop system (3.4), and for $t \in [t_k + \tau_k,\ t_{k+1} + \tau_{k+1})$, one can establish the following closed-loop system

$$\begin{cases} \dot{\xi}(t) = \tilde{A}\xi(t) + \tilde{B}_1\xi(t - \tau(t)) + \tilde{B}_2\xi(t - d(t)) + \tilde{D}v(t), \\ r_e(t) = \tilde{C}\xi(t) + \tilde{E}_1\xi(t - \tau(t)) + \tilde{E}_2\xi(t - d(t)) + \tilde{F}v(t), \end{cases} \tag{3.7}$$

where $\tilde{A}$, $\tilde{D}$, $\tilde{C}$, and $\tilde{F}$ are the same as the corresponding items in (3.4), and

$$\tilde{B}_1 = \begin{bmatrix} 0 & 0 & 0 \\ 0 & -\frac{a_1}{\tilde{a}}LC & 0 \\ 0 & 0 & 0 \end{bmatrix}, \; \tilde{B}_2 = \begin{bmatrix} 0 & 0 & 0 \\ 0 & -\frac{a_2}{\tilde{a}}LC & 0 \\ 0 & 0 & 0 \end{bmatrix},$$
$$\tilde{E}_1 = \begin{bmatrix} 0 & -\frac{a_1}{\tilde{a}}SC & 0 \end{bmatrix}, \; \tilde{E}_2 = \begin{bmatrix} 0 & -\frac{a_2}{\tilde{a}}SC & 0 \end{bmatrix}.$$

To detect the occurrence of faults, one should construct a residual evaluation function $J(t)$. When the value of the residual evaluation function $J(t)$ is larger than a given threshold J_{th}, an alarm of faults will be generated. Define the residual evaluation function as

$$J(t) \triangleq \left(\int_0^t r_e^T(s) r_e(s) ds\right)^{\frac{1}{2}}. \tag{3.8}$$

Choose a residual evaluation function threshold J_{th} as follows (see [3] for the selection criterion of J_{th})

$$J_{th} = \mathop{\mathrm{Sup}}_{v(t)\in L_2,\; f(t)=0} J(t). \tag{3.9}$$

The fault detection logic is

$$\begin{cases} J(t) > J_{th}, & with\ faults, \\ J(t) \le J_{th}, & without\ fault. \end{cases} \tag{3.10}$$

Based on the systems (3.4) and (3.7), and the fault detection logic (3.10), we will investigate the problem of FDF design for the continuous-time NCS considering packet dropouts and network-induced delays.

It should be pointed out that $\tau(t)$ and $d(t)$, which are named as interval time-varying delays in the literature, are different from the network-induced delays τ_k. In this chapter, we take the mutually exclusive distribution characteristic of interval time-varying delays $\tau(t)$ and $d(t)$ into full consideration to derive less conservative FDF design criteria. Take the interval time-varying delay $\tau(t)$ for example. Considering that $\tau(t) \in [\tau_m,\ \eta)$, we divide $[\tau_m,\ \eta)$ into ρ equidistant time intervals with ρ denoting a given positive integer, and define $\sigma_1 = (\eta - \tau_m)/\rho$, $\eta_1 = \tau_m + \sigma_1$, $\eta_2 = \tau_m + 2\sigma_1$, ..., $\eta_{\rho-1} = \tau_m + (\rho - 1)\sigma_1$. Then, one can conclude that at any instant t with $t \in [t_k + \tau_k,\ t_{k+1} + \tau_{k+1})$, $\tau(t) \in [\tau_m,\ \eta_1)$ or $\tau(t) \in [\eta_1,\ \eta_2)$, ..., or $\tau(t) \in [\eta_{\rho-1},\ \eta)$. On the other hand, for the specific instant t, $\tau(t) \in [\tau_m,\ \eta_1)$ or $\tau(t) \in [\eta_1,\ \eta_2)$, ..., or $\tau(t) \in [\eta_{\rho-1},\ \eta)$ can not occur simultaneously. In this chapter, we refer to such a phenomenon as mutually exclusive distribution. The mutually exclusive distribution characteristic of $d(t)$ can be achieved similarly. To make full use of the mutually exclusive distribution characteristic of $\tau(t)$ and $d(t)$, define $\sigma_3 = (\tilde{\eta} - \tau_m - h)/\rho$, $\tilde{\eta}_1 = \tau_m + h + \sigma_3$, $\tilde{\eta}_2 = \tau_m + h + 2\sigma_3$, ..., $\tilde{\eta}_{\rho-1} = \tau_m + h + (\rho - 1)\sigma_3$. Define scalars $\lambda_1, \lambda_2, \ldots, \lambda_\rho$ and $\tilde{\lambda}_1, \tilde{\lambda}_2, \ldots, \tilde{\lambda}_\rho$, where

$$\lambda_1 = \begin{cases} 1, & \tau(t) \in [\tau_m, \ \eta_1), \\ 0, & otherwise, \end{cases}$$
$$\lambda_2 = \begin{cases} 1, & \tau(t) \in [\eta_1, \ \eta_2), \\ 0, & otherwise, \end{cases} \tag{3.11}$$
$$\vdots$$
$$\lambda_\rho = \begin{cases} 1, & \tau(t) \in [\eta_{\rho-1}, \ \eta), \\ 0, & otherwise, \end{cases}$$

and

$$\tilde{\lambda}_1 = \begin{cases} 1, & d(t) \in [\tau_m + h, \ \tilde{\eta}_1), \\ 0, & otherwise, \end{cases}$$
$$\tilde{\lambda}_2 = \begin{cases} 1, & d(t) \in [\tilde{\eta}_1, \ \tilde{\eta}_2), \\ 0, & otherwise, \end{cases} \tag{3.12}$$
$$\vdots$$
$$\tilde{\lambda}_\rho = \begin{cases} 1, & d(t) \in [\tilde{\eta}_{\rho-1}, \ \tilde{\eta}), \\ 0, & otherwise. \end{cases}$$

Remark 3.3 Note that the mutually exclusive distribution characteristic of interval time-varying delays is used to deal with integral inequalities for products of vectors in this chapter, while the delay decomposition approach in [4] is used to construct the Lyapunov functional, they are different in essence.

3.2 Observer-Based FDF Design

In this section, we first present an observer-based FDF design scheme for the closed-loop system (3.4). Then, we analyze the merits for considering the mutually exclusive distribution characteristic of the interval time-varying delay $\tau(t)$.

3.2.1 FDF Design

This section is concerned with observer-based FDF design for the closed-loop system (3.4). Construct the following Lyapunov functional

$$V(t, \xi_t) = \sum_{i=1}^{4} V_i(t, \xi_t), \tag{3.13}$$

where

$$V_1(t,\xi_t) = \xi(t)^T P \xi(t),$$
$$V_2(t,\xi_t) = (\eta - \tau(t)) \int_{t-\tau(t)}^{t} \dot{\xi}^T(s) Q \dot{\xi}(s) ds,$$
$$V_3(t,\xi_t) = \int_{t-\tau_m}^{t} \xi^T(s) R_1 \xi(s) ds + \int_{t-\eta}^{t-\tau_m} \xi^T(s) R_2 \xi(s) ds,$$
$$V_4(t,\xi_t) = \int_{-\tau_m}^{0} \int_{t+s}^{t} \dot{\xi}^T(\theta) S_1 \dot{\xi}(\theta) d\theta ds + \int_{-\eta}^{-\tau_m} \int_{t+s}^{t} \dot{\xi}^T(\theta) S_2 \dot{\xi}(\theta) d\theta ds,$$

P, Q, R_1, R_2, S_1, and S_2 are symmetric positive definite matrices with appropriate dimensions. We state and establish the following result.

Theorem 3.1 *For given scalars τ_m, τ_M, δ, h, and ρ, the residual system (3.4) is asymptotically stable with an H_∞ norm bound γ and the FDF gains $L = V_2^T N^{-T}$ and $S = V_1^T N^{-T}$ if there exist symmetric positive definite matrices W_1, W_2, W_3, $\widetilde{Q}$, $\widetilde{R}_1$, $\widetilde{R}_2$, $\widetilde{S}_1$, $\widetilde{S}_2$, and matrices V_1, V_2, N, such that (3.14) and (3.15) hold for every feasible value of λ_i $(i = 1, 2, \ldots, \rho)$*

$$\begin{bmatrix} \widetilde{H}_{11} & \widetilde{H}_{12} \\ * & \widetilde{H}_{22} \end{bmatrix} < 0, \tag{3.14}$$

$$W_2 C^T = C^T N, \tag{3.15}$$

where

$$\widetilde{H}_{11} = \begin{bmatrix} \widetilde{\Pi}_{11} & \widetilde{\Pi}_{12} & \widetilde{\Pi}_{13} & 0 & \widetilde{D} \\ * & \widetilde{\Pi}_{22} & \widetilde{\Pi}_{23} & 0 & 0 \\ * & * & \widetilde{\Pi}_{33} & \widetilde{\Pi}_{34} & 0 \\ * & * & * & \widetilde{\Pi}_{44} & 0 \\ * & * & * & * & -\gamma I \end{bmatrix},$$

$$\widetilde{H}_{12} = \left[\widetilde{\Phi}_1 \ \widetilde{\Phi}_2 \ \widetilde{\Phi}_2 \ \widetilde{\Phi}_2\right], \widetilde{H}_{22} = diag\{-\gamma I, \mathscr{X}_1, \mathscr{X}_2, \mathscr{X}_3\},$$

$$\widetilde{\Phi}_{11} = \widetilde{A}W + W\widetilde{A}^T + \widetilde{R}_1 - \frac{\widetilde{S}_1}{\tau_m} - (\frac{\lambda_1}{\eta_1} + \frac{\lambda_2}{\eta_2} + \cdots + \frac{\lambda_\rho}{\eta})\widetilde{Q},$$

$$\widetilde{\Pi}_{12} = \frac{\widetilde{S}_1}{\tau_m}, \ \widetilde{\Pi}_{13} = \widetilde{V}_2^T + (\frac{\lambda_1}{\eta_1} + \frac{\lambda_2}{\eta_2} + \cdots + \frac{\lambda_\rho}{\eta})\widetilde{Q},$$

$$\widetilde{\Pi}_{22} = \widetilde{R}_2 - \widetilde{R}_1 - \frac{\widetilde{S}_1}{\tau_m} - (\frac{\lambda_1}{\sigma_1} + \frac{\lambda_2}{2\sigma_1} + \cdots + \frac{\lambda_\rho}{\rho\sigma_1})\widetilde{S}_2,$$

$$\widetilde{\Pi}_{33} = -(\frac{\lambda_1}{\eta_1} + \frac{\lambda_2}{\eta_2} + \cdots + \frac{\lambda_\rho}{\eta})\widetilde{Q} - (\frac{\lambda_1}{\sigma_1} + \frac{\lambda_2}{2\sigma_1} + \cdots + \frac{\lambda_\rho}{\rho\sigma_1})\widetilde{S}_2,$$
$$-(\frac{\lambda_1}{\rho\sigma_1} + \frac{\lambda_2}{(\rho-1)\sigma_1} + \cdots + \frac{\lambda_\rho}{\sigma_1})\widetilde{S}_2, \ \widetilde{\Pi}_{23} = (\frac{\lambda_1}{\sigma_1} + \frac{\lambda_2}{2\sigma_1} + \cdots + \frac{\lambda_\rho}{\rho\sigma_1})\widetilde{S}_2,$$

$$\widetilde{\Pi}_{34} = (\frac{\lambda_1}{\rho\sigma_1} + \frac{\lambda_2}{(\rho-1)\sigma_1} + \cdots + \frac{\lambda_\rho}{\sigma_1})\widetilde{S}_2,$$

$$\widetilde{\Pi}_{44} = -\widetilde{R}_2 - (\frac{\lambda_1}{\rho\sigma_1} + \frac{\lambda_2}{(\rho-1)\sigma_1} + \cdots + \frac{\lambda_\rho}{\sigma_1})\widetilde{S}_2,$$

$$\widetilde{\Phi}_1 = \left[\widetilde{C}W \ 0 \ \widetilde{\Pi}_{36}^T \ 0 \ \widetilde{F}\right]^T, \widetilde{\Phi}_2 = \left[\widetilde{A}W \ 0 \ \widetilde{\Pi}_{37}^T \ 0 \ \widetilde{D}\right]^T,$$

$$\mathscr{X}_1 = \psi^{-1}(\widetilde{Q} - 2W), \ \mathscr{X}_2 = \tau_m^{-1}(\widetilde{S}_1 - 2W), \ \mathscr{X}_3 = (\eta - \tau_m)^{-1}(\widetilde{S}_2 - 2W),$$

$$W = diag\{W_1, W_2, W_3\},\ \psi = \lambda_1\rho\sigma_1 + \lambda_2(\rho-1)\sigma_1 + \cdots + \lambda_\rho\sigma_1,$$

$$\widetilde{\Pi}_{36} = \widetilde{V}_1,\ \widetilde{\Pi}_{37} = \widetilde{V}_2,\ \widetilde{V}_1 = \begin{bmatrix} 0 \\ -C^T V_1 \\ 0 \end{bmatrix},\ \widetilde{V}_2 = \begin{bmatrix} 0 & 0 & 0 \\ 0 & -C^T V_2 & 0 \\ 0 & 0 & 0 \end{bmatrix}.$$

Proof Taking the time derivative of the Lyapunov functional $V(t, \xi_t)$ given in (3.13) along the trajectory of the system (3.4), one has

$$\dot{V}_1(t, \xi_t) = 2\xi^T(t)P\dot{\xi}(t), \tag{3.16}$$

$$\dot{V}_2(t, \xi_t) = (\eta - \tau(t))\dot{\xi}^T(t)Q\dot{\xi}(t) - \int_{t-\tau(t)}^{t} \dot{\xi}^T(s)Q\dot{\xi}(s)ds, \tag{3.17}$$

$$\dot{V}_3(t, \xi_t) = \xi^T(t)R_1\xi(t) + \xi^T(t-\tau_m)(R_2 - R_1)\xi(t-\tau_m) - \xi^T(t-\eta)R_2\xi(t-\eta), \tag{3.18}$$

$$\dot{V}_4(t, \xi_t) = \dot{\xi}^T(t)\Theta\dot{\xi}(t) - \int_{t-\tau_m}^{t} \dot{\xi}^T(\theta)S_1\dot{\xi}(\theta)d\theta - \int_{t-\eta}^{t-\tau_m} \dot{\xi}^T(\theta)S_2\dot{\xi}(\theta)d\theta, \tag{3.19}$$

where $\Theta = \tau_m S_1 + (\eta - \tau_m)S_2$.

Notice that $\tau(t) \in [\tau_m,\ \eta)$. To make full use of the mutually exclusive distribution characteristic of the interval time-varying delay $\tau(t)$, we divide $[\tau_m,\ \eta)$ into ρ equidistant time intervals. Define $\sigma_1 = (\eta - \tau_m)/\rho$, $\eta_1 = \tau_m + \sigma_1$, $\eta_2 = \tau_m + 2\sigma_1$, ..., $\eta_{\rho-1} = \tau_m + (\rho-1)\sigma_1$. From the definition of σ_1, one can see that η is equal to $\tau_m + \rho\sigma_1$.

Considering the mutually exclusive distribution characteristic of $\tau(t)$, and adopting the Jensen integral inequality in [5], one obtains

$$\begin{aligned}
&(\eta - \tau(t))\dot{\xi}^T(t)Q\dot{\xi}(t) \\
=&\lambda_1[(\eta - \tau(t))\dot{\xi}^T(t)Q\dot{\xi}(t)] + \lambda_2[(\eta - \tau(t))\dot{\xi}^T(t)Q\dot{\xi}(t)] \\
&+ \cdots + \lambda_\rho[(\eta - \tau(t))\dot{\xi}^T(t)Q\dot{\xi}(t)] \\
\leq&\psi\dot{\xi}^T(t)Q\dot{\xi}(t),
\end{aligned} \tag{3.20}$$

where $\psi = \lambda_1\rho\sigma_1 + \lambda_2(\rho-1)\sigma_1 + \cdots + \lambda_\rho\sigma_1$, and

$$\begin{aligned}
&-\int_{t-\tau(t)}^{t} \dot{\xi}^T(s)Q\dot{\xi}(s)ds \\
=&\lambda_1[\frac{1}{\tau(t)}(-\tau(t))\int_{t-\tau(t)}^{t} \dot{\xi}^T(s)Q\dot{\xi}(s)ds] + \lambda_2[\frac{1}{\tau(t)}(-\tau(t))\int_{t-\tau(t)}^{t} \dot{\xi}^T(s)Q\dot{\xi}(s)ds] \\
&+ \cdots + \lambda_\rho[\frac{1}{\tau(t)}(-\tau(t))\int_{t-\tau(t)}^{t} \dot{\xi}^T(s)Q\dot{\xi}(s)ds] \\
\leq& -\frac{\lambda_1}{\eta_1}\varphi_1^T Q\varphi_1 - \frac{\lambda_2}{\eta_2}\varphi_1^T Q\varphi_1 - \cdots - \frac{\lambda_\rho}{\eta}\varphi_1^T Q\varphi_1 \\
=& -(\frac{\lambda_1}{\eta_1} + \frac{\lambda_2}{\eta_2} + \cdots + \frac{\lambda_\rho}{\eta})\varphi_1^T Q\varphi_1,
\end{aligned} \tag{3.21}$$

where $\varphi_1 = [\xi(t) - \xi(t - \tau(t))]$, and

$$-\int_{t-\tau_m}^{t} \dot{\xi}^T(\theta)S_1\dot{\xi}(\theta)d\theta \leq -\frac{1}{\tau_m}\varphi_2^T S_1\varphi_2, \tag{3.22}$$

where $\varphi_2 = [\xi(t) - \xi(t-\tau_m)]$.

Similarly, one has

$$\begin{aligned}
&-\int_{t-\eta}^{t-\tau_m} \dot{\xi}^T(\theta)S_2\dot{\xi}(\theta)d\theta \\
&= \lambda_1[-\int_{t-\tau(t)}^{t-\tau_m} \dot{\xi}^T(\theta)S_2\dot{\xi}(\theta)d\theta - \int_{t-\eta}^{t-\tau(t)} \dot{\xi}^T(\theta)S_2\dot{\xi}(\theta)d\theta] \\
&\quad + \lambda_2[-\int_{t-\tau(t)}^{t-\tau_m} \dot{\xi}^T(\theta)S_2\dot{\xi}(\theta)d\theta - \int_{t-\eta}^{t-\tau(t)} \dot{\xi}^T(\theta)S_2\dot{\xi}(\theta)d\theta] \\
&\quad + \cdots + \lambda_\rho[-\int_{t-\tau(t)}^{t-\tau_m} \dot{\xi}^T(\theta)S_2\dot{\xi}(\theta)d\theta - \int_{t-\eta}^{t-\tau(t)} \dot{\xi}^T(\theta)S_2\dot{\xi}(\theta)d\theta] \\
&\leq -(\frac{\lambda_1}{\sigma_1} + \frac{\lambda_2}{2\sigma_1} + \cdots + \frac{\lambda_\rho}{\rho\sigma_1})\varphi_3^T S_2\varphi_3 - (\frac{\lambda_1}{\rho\sigma_1} + \frac{\lambda_2}{(\rho-1)\sigma_1} + \cdots + \frac{\lambda_\rho}{\sigma_1})\varphi_4^T S_2\varphi_4,
\end{aligned} \tag{3.23}$$

where $\varphi_3 = [\xi(t-\tau_m) - \xi(t-\tau(t))]$, $\varphi_4 = [\xi(t-\tau(t)) - \xi(t-\eta)]$.

Then, considering the $r_e(t)$ in (3.4), and combining (3.16)–(3.23) together, one has

$$\dot{V}(t, \xi_t) + \gamma^{-1}r_e^T(t)r_e(t) - \gamma v^T(t)v(t) \leq \tilde{\xi}^T(t)(\Pi + \Xi)\tilde{\xi}(t), \tag{3.24}$$

where $\tilde{\xi}(t) = [\xi^T(t)\ \ \xi^T(t-\tau_m)\ \ \xi^T(t-\tau(t))\ \ \xi^T(t-\eta)\ \ v^T(t)]^T$, and

$$\Pi = \begin{bmatrix} \Pi_{11} & \Pi_{12} & \Pi_{13} & 0 & P\widetilde{D} \\ * & \Pi_{22} & \Pi_{23} & 0 & 0 \\ * & * & \Pi_{33} & \Pi_{34} & 0 \\ * & * & * & \Pi_{44} & 0 \\ * & * & * & * & -\gamma I \end{bmatrix},$$

$$\Xi = \gamma^{-1}\Phi_1\Phi_1^T + \Phi_2(\psi Q + \tau_m S_1 + (\eta - \tau_m)S_2)\Phi_2^T,$$

$$\Pi_{11} = P\widetilde{A} + \widetilde{A}^T P + R_1 - \frac{S_1}{\tau_m} - (\frac{\lambda_1}{\eta_1} + \frac{\lambda_2}{\eta_2} + \cdots + \frac{\lambda_\rho}{\eta})Q,$$

$$\Pi_{12} = \frac{S_1}{\tau_m},\ \Pi_{13} = P\widetilde{B} + (\frac{\lambda_1}{\eta_1} + \frac{\lambda_2}{\eta_2} + \cdots + \frac{\lambda_\rho}{\eta})Q,$$

$$\Pi_{22} = R_2 - R_1 - \frac{S_1}{\tau_m} - (\frac{\lambda_1}{\sigma_1} + \frac{\lambda_2}{2\sigma_1} + \cdots + \frac{\lambda_\rho}{\rho\sigma_1})S_2,$$

$$\Pi_{23} = (\frac{\lambda_1}{\sigma_1} + \frac{\lambda_2}{2\sigma_1} + \cdots + \frac{\lambda_\rho}{\rho\sigma_1})S_2,$$

$$\begin{aligned}
\Pi_{33} =& -(\frac{\lambda_1}{\eta_1} + \frac{\lambda_2}{\eta_2} + \cdots + \frac{\lambda_\rho}{\eta})Q - (\frac{\lambda_1}{\sigma_1} + \frac{\lambda_2}{2\sigma_1} + \cdots + \frac{\lambda_\rho}{\rho\sigma_1})S_2 \\
& -(\frac{\lambda_1}{\rho\sigma_1} + \frac{\lambda_2}{(\rho-1)\sigma_1} + \cdots + \frac{\lambda_\rho}{\sigma_1})S_2, \\
\Pi_{34} =& (\frac{\lambda_1}{\rho\sigma_1} + \frac{\lambda_2}{(\rho-1)\sigma_1} + \cdots + \frac{\lambda_\rho}{\sigma_1})S_2, \\
\Pi_{44} =& -R_2 - (\frac{\lambda_1}{\rho\sigma_1} + \frac{\lambda_2}{(\rho-1)\sigma_1} + \cdots + \frac{\lambda_\rho}{\sigma_1})S_2, \\
\Phi_1 =& \left[\widetilde{C}\ 0\ \widetilde{E}\ 0\ \widetilde{F}\right]^T, \Phi_2 = \left[\widetilde{A}\ 0\ \widetilde{B}\ 0\ \widetilde{D}\right]^T.
\end{aligned}$$

From the inequality in (3.24), one can see that if $\Pi + \Xi < 0$, one has $\dot{V}(t, \xi_t) + \gamma^{-1} r_e^T(t) r_e(t) - \gamma \nu^T(t)\nu(t) < 0$. By using Schur complement, $\Pi + \Xi < 0$ is equivalent to

$$\begin{bmatrix} \Pi & H_{12} \\ * & H_{22} \end{bmatrix} < 0, \tag{3.25}$$

where Π is the same as the corresponding item presented above, and

$$\begin{aligned}
H_{12} =& \left[\Phi_1\ \Phi_2\ \Phi_2\ \Phi_2\right], \\
H_{22} =& \text{diag}\{-\gamma I,\ -\psi^{-1}Q^{-1},\ -\tau_m^{-1}S_1^{-1},\ -(\eta-\tau_m)^{-1}S_2^{-1}\}.
\end{aligned}$$

Pre- and post-multiplying both sides of (3.25) with $\text{diag}\{\underbrace{P^{-1}, \ldots, P^{-1}}_{4}, \underbrace{I, \ldots, I}_{5}\}$ and its transpose, supposing that there exists a matrix N such that $W_2 C^T = C^T N$, defining $P^{-1} = W$, $W = \text{diag}\{W_1, W_2, W_3\}$, $P^{-1}QP^{-1} = \widetilde{Q}$, $P^{-1}R_1P^{-1} = \widetilde{R}_1$, $P^{-1}R_2P^{-1} = \widetilde{R}_2$, $P^{-1}S_1P^{-1} = \widetilde{S}_1$, $P^{-1}S_2P^{-1} = \widetilde{S}_2$, $NS^T = V_1$, $NL^T = V_2$, and considering that $-Q^{-1} \le \widetilde{Q} - 2W$, $-S_1^{-1} \le \widetilde{S}_1 - 2W$, $-S_2^{-1} \le \widetilde{S}_2 - 2W$, one can see that if (3.14) and (3.15) are satisfied simultaneously, $\Pi + \Xi < 0$ is also satisfied, which implies that $\dot{V}(t, \xi_t) + \gamma^{-1} r_e^T(t) r_e(t) - \gamma \nu^T(t)\nu(t) < 0$.

By using the definition for H_∞ performance, one can prove that if $\nu(t) = 0$, (3.14) and (3.15) can ensure the asymptotic stability of the system (3.4); if $\nu(t) \neq 0$, we have $||r_e(t)||_2 < \gamma||\nu(t)||_2$. Therefore, if (3.14)–(3.15) are satisfied, the system (3.4) is asymptotically stable with an H_∞ norm bound γ. This completes the proof. □

Remark 3.4 As shown in (3.11) and (3.12), λ_i and $\tilde{\lambda}_i$ $(i = 1, 2, \ldots, \rho)$ are not fixed but variable. The variations of λ_i and $\tilde{\lambda}_i$ are determined by the variations of $\tau(t)$ and $d(t)$, respectively. To guarantee the asymptotic stability of the residual system (3.4), the FDF design criterion in Theorem 3.1 should hold for every feasible value of λ_i $(i = 1, 2, \ldots, \rho)$ with $\lambda_i = 1$ or $\lambda_i = 0$. Moreover, for any $\lambda_i = 1$, one has $\lambda_j = 0$, where $j = 1, 2, \ldots, \rho$, and $j \neq i$.

3.2.2 Merits of Considering Mutually Exclusive Distribution

The mutually exclusive distribution characteristic of the interval time-varying delay $\tau(t)$ is made full use in Theorem 3.1 to deal with integral inequalities for products of vectors. If such a mutually exclusive distribution characteristic is not considered, the proposed FDF design method in Theorem 3.1 is still applicable, and the following FDF design criterion is followed immediately.

Corollary 3.1 *For given scalars τ_m, τ_M, δ, and h, the residual system (3.4) is asymptotically stable with an H_∞ norm bound γ and the FDF gains $L = V_2^T N^{-T}$ and $S = V_1^T N^{-T}$ if there exist symmetric positive definite matrices W_1, W_2, W_3, $\widetilde{Q}$, $\widetilde{R}_1$, $\widetilde{R}_2$, $\widetilde{S}_1$, $\widetilde{S}_2$, and matrices V_1, V_2, N, such that (3.26) and (3.27) hold*

$$\begin{bmatrix} \hat{H}_{11} & \widetilde{H}_{12}, \\ * & \hat{H}_{22} \end{bmatrix} < 0, \tag{3.26}$$

$$W_2 C^T = C^T N \tag{3.27}$$

where

$$\hat{H}_{11} = \begin{bmatrix} \hat{\Pi}_{11} & \widetilde{\Pi}_{12} & \hat{\Pi}_{13} & 0 & \widetilde{D} \\ * & \hat{\Pi}_{22} & \hat{\Pi}_{23} & 0 & 0 \\ * & * & \hat{\Pi}_{33} & \hat{\Pi}_{34} & 0 \\ * & * & * & \hat{\Pi}_{44} & 0 \\ * & * & * & * & -\gamma I \end{bmatrix},$$

$$\hat{H}_{22} = diag\{-\gamma I, \hat{\mathscr{X}}_1, \mathscr{X}_2, \mathscr{X}_3\},\ \hat{\Pi}_{11} = \widetilde{A}W + W\widetilde{A}^T + \widetilde{R}_1 - \frac{\widetilde{S}_1}{\tau_m} - \frac{\widetilde{Q}}{\eta},$$

$$\hat{\Pi}_{13} = \widetilde{V}_2^T + \frac{\widetilde{Q}}{\eta},\ \hat{\Pi}_{22} = \widetilde{R}_2 - \widetilde{R}_1 - \frac{\widetilde{S}_1}{\tau_m} - \frac{\widetilde{S}_2}{\eta - \tau_m},\ \hat{\Pi}_{23} = \frac{\widetilde{S}_2}{\eta - \tau_m},$$

$$\hat{\Pi}_{33} = -\frac{\widetilde{Q}}{\eta} - \frac{2}{\eta - \tau_m}\widetilde{S}_2,\ \hat{\Pi}_{34} = \frac{\widetilde{S}_2}{\eta - \tau_m},\ \hat{\Pi}_{44} = -\widetilde{R}_2 - \frac{\widetilde{S}_2}{\eta - \tau_m},$$

$$\hat{\mathscr{X}}_1 = (\eta - \tau_m)^{-1}(\widetilde{Q} - 2W),$$

while $\widetilde{H_{12}}$, $\widetilde{\Pi}_{12}$, $\mathscr{X}_2$, $\mathscr{X}_3$, $\widetilde{V}_2$ and W are the same as the corresponding items in Theorem 3.1.

Proof If the mutually exclusive distribution characteristic of the interval time-varying delay $\tau(t)$ is not considered, the bounding inequalities in (3.20), (3.21) and (3.23) are converted into (3.28), (3.29) and (3.30), respectively,

$$(\eta - \tau(t))\dot{\xi}^T(t)Q\dot{\xi}(t) \le (\eta - \tau_m)\dot{\xi}^T(t)Q\dot{\xi}(t), \tag{3.28}$$

$$-\int_{t-\tau(t)}^{t} \dot{\xi}^T(s)Q\dot{\xi}(s)ds \le -\frac{1}{\eta}\varphi_1^T Q\varphi_1, \tag{3.29}$$

$$-\int_{t-\eta}^{t-\tau_m} \dot{\xi}^T(\theta)S_2\dot{\xi}(\theta)d\theta \le -\frac{1}{\eta-\tau_m}(\varphi_3^T S_2\varphi_3 + \varphi_4^T S_2\varphi_4), \tag{3.30}$$

where φ_1 in (3.29) is the same as the corresponding item in (3.21), φ_3 and φ_4 in (3.30) are the same as the corresponding items in (3.23).

The rest of the proof is similar to the proof of Theorem 3.1, here it is omitted. This completes the proof. □

The following theorem establishes the relationship between the FDF design criteria in Theorem 3.1 and Corollary 3.1.

Theorem 3.2 *For the residual system (3.4), if the FDF design criterion in Corollary 3.1 is satisfied, then the FDF design criterion in Theorem 3.1 is also satisfied.*

Proof Notice that $(\eta - \tau_m)\dot{\xi}^T(t)Q\dot{\xi}(t)$ in (3.28) can be rewritten as

$$(\eta - \tau_m)\dot{\xi}^T(t)Q\dot{\xi}(t) = \Gamma_1 + \Delta_1,$$

where Γ_1 denotes the right side of the inequality (3.20), $\Delta_1 = [(\eta - \tau_m) - \psi]\dot{\xi}^T(t) Q\dot{\xi}(t)$, and ψ is the same as the corresponding item in (3.20).

Moreover, $-\frac{1}{\eta}\varphi_1^T Q\varphi_1$ in (3.29) can be rewritten as $-\frac{1}{\eta}\varphi_1^T Q\varphi_1 = \Gamma_2 + \Delta_2$, where Γ_2 denotes the right side of the inequality (3.21), $\Delta_2 = [-\frac{1}{\eta} + (\frac{\lambda_1}{\eta_1} + \frac{\lambda_2}{\eta_2} + \cdots + \frac{\lambda_\rho}{\eta})]\varphi_1^T Q\varphi_1$.

Furthermore, $-\frac{1}{\eta-\tau_m}(\varphi_3^T S_2\varphi_3 + \varphi_4^T S_2\varphi_4)$ in (3.30) can be rewritten as $-\frac{1}{\eta-\tau_m}\times (\varphi_3^T S_2\varphi_3 + \varphi_4^T S_2\varphi_4) = \Gamma_3 + \Delta_3 + \Delta_4$, where Γ_3 denotes the right side of the inequality (3.23), $\Delta_3 = [-\frac{1}{\eta-\tau_m} + (\frac{\lambda_1}{\sigma_1} + \frac{\lambda_2}{2\sigma_1} + \cdots + \frac{\lambda_\rho}{\rho\sigma_1})]\varphi_3^T S_2\varphi_3$, $\Delta_4 = [-\frac{1}{\eta-\tau_m} + (\frac{\lambda_1}{\rho\sigma_1} + \frac{\lambda_2}{(\rho-1)\sigma_1} + \cdots + \frac{\lambda_\rho}{\sigma_1})]\varphi_4^T S_2\varphi_4$, while φ_1, φ_2, φ_3, and φ_4 are the same as the corresponding items in the proof of Theorem 3.1.

Considering that $\Delta_1 \ge 0$, $\Delta_2 \ge 0$, and Δ_1 and Δ_2 are not equal to zero simultaneously, one has $\Delta_1 + \Delta_2 > 0$. Similarly, one can see that $\Delta_3 + \Delta_4 > 0$. Then the integral inequalities in (3.20), (3.21) and (3.23) are more easier to be satisfied than the ones in (3.28), (3.29) and (3.30), respectively. From the proof of Theorem 3.1 and Corollary 3.1, one can conclude that if the FDF design criterion in Corollary 3.1 is satisfied, the FDF design criterion in Theorem 3.1 is also satisfied. This completes the proof. □

Remark 3.5 It has been proved that if the FDF design criterion in Corollary 3.1 is satisfied, then the FDF design criterion in Theorem 3.1 is also satisfied, which implies that considering the mutually exclusive distribution characteristic of the interval time-varying delay $\tau(t)$ can introduce less conservatism. If the mutually exclusive distribution characteristic of interval time-varying delays is made full use to deal with the problems in [6–8], less conservative results are expected to be obtained.

Notice that the reciprocally convex approach is proposed in [9, 10] to deal with integral inequalities for products of vectors. The theoretical comparison between the reciprocally convex approach and the mutually exclusive distribution characteristic-based approach is a subject for further research.

In the proof of Theorem 3.1, the mutually exclusive distribution characteristic of $\tau(t)$ has been made full use by dividing the time interval $[\tau_m,\ \eta)$ into ρ equidistant time intervals, and it has been proved in Theorem 3.2 that considering the mutually exclusive distribution characteristic of the interval time-varying delay $\tau(t)$ can introduce less conservatism. Then, a natural question is whether dividing $[\tau_m,\ \eta)$ into $\rho+1$ equidistant time intervals will introduce less conservatism than dividing $[\tau_m,\ \eta)$ into ρ equidistant time intervals or not.

If the time interval $[\tau_m,\ \eta)$ is divided into $\rho+1$ equidistant time intervals, define $\sigma_2=(\eta-\tau_m)/(\rho+1)$. Then, the bounding inequality in (3.20) is converted into

$$\begin{aligned}
&(\eta-\tau(t))\dot{\xi}^T(t)Q\dot{\xi}(t)\\
&=\hat{\lambda}_1[(\eta-\tau(t))\dot{\xi}^T(t)Q\dot{\xi}(t)]+\hat{\lambda}_2[(\eta-\tau(t))\dot{\xi}^T(t)Q\dot{\xi}(t)]\\
&\quad+\cdots+\hat{\lambda}_\rho[(\eta-\tau(t))\dot{\xi}^T(t)Q\dot{\xi}(t)]+\hat{\lambda}_{\rho+1}[(\eta-\tau(t))\dot{\xi}^T(t)Q\dot{\xi}(t)]\\
&\le\hat{\psi}\dot{\xi}^T(t)Q\dot{\xi}(t),
\end{aligned}\tag{3.31}$$

where $\hat{\psi}=\hat{\lambda}_1(\rho+1)\sigma_2+\hat{\lambda}_2\rho\sigma_2+\cdots+2\hat{\lambda}_\rho\sigma_2+\hat{\lambda}_{\rho+1}\sigma_2$, and the definition for $\hat{\lambda}_1,\hat{\lambda}_2,\cdots$, and $\hat{\lambda}_{\rho+1}$ is similar to the definition for $\lambda_1,\lambda_2,\ldots$, and λ_ρ in (3.11).

Since σ_1 is different from σ_2, it is difficult to prove theoretically whether $\hat{\psi}\dot{\xi}^T(t)Q\dot{\xi}(t)$ in (3.31) can introduce less conservatism than $\psi\dot{\xi}^T(t)Q\dot{\xi}(t)$ in (3.20) or not. Similar conclusions can be drawn for the inequalities in (3.21) and (3.23).

In conclusion, if the mutually exclusive distribution characteristic of the interval time-varying delay $\tau(t)$ is adopted to deal with integral inequalities for products of vectors, not a theoretical proof shows that dividing $[\tau_m,\ \eta)$ into $\rho+1$ equidistant time intervals can introduce less conservatism than dividing $[\tau_m,\ \eta)$ into ρ equidistant time intervals. However, the simulation results illustrate that dividing $[\tau_m,\ \eta)$ into $\rho+1$ equidistant time intervals can introduce less conservatism.

3.3 Data Reconstruction-Based FDF Design

This section is concerned with the data reconstruction-based FDF design for the closed-loop system (3.7). We state and establish the following result.

Theorem 3.3 *For given scalars τ_m, τ_M, δ, h, and ρ, the residual system (3.7) is asymptotically stable with an H_∞ norm bound γ and the FDF gains $L=V_2^TN^{-T}$ and $S=V_1^TN^{-T}$ if there exist symmetric positive definite matrices W_1, W_2, W_3, $\widetilde{Q}$, $\widetilde{M}$, $\widetilde{R}_1$, $\widetilde{R}_2$, $\widetilde{R}_3$, $\widetilde{R}_4$, $\widetilde{S}_1$, $\widetilde{S}_2$, $\widetilde{S}_3$, $\widetilde{S}_4$, and matrices V_1, V_2, N, such that (3.32) and (3.33) hold for every feasible value of λ_i and $\tilde{\lambda}_i$ $(i=1,\ 2,\ \ldots,\ \rho)$*

$$\begin{bmatrix} \bar{H}_{11} & \bar{H}_{12} \\ * & \bar{H}_{22} \end{bmatrix} < 0, \tag{3.32}$$

$$W_2 C^T = C^T N, \tag{3.33}$$

where

$$\bar{H}_{11} = \begin{bmatrix} \bar{\Pi}_{11} & \bar{\Pi}_{12} & \bar{\Pi}_{13} & 0 & \bar{\Pi}_{15} & \bar{\Pi}_{16} & 0 & \widetilde{D} \\ * & \bar{\Pi}_{22} & \bar{\Pi}_{23} & 0 & 0 & 0 & 0 & 0 \\ * & * & \bar{\Pi}_{33} & \bar{\Pi}_{34} & 0 & 0 & 0 & 0 \\ * & * & * & \bar{\Pi}_{44} & 0 & 0 & 0 & 0 \\ * & * & * & * & \bar{\Pi}_{55} & \bar{\Pi}_{56} & 0 & 0 \\ * & * & * & * & * & \bar{\Pi}_{66} & \bar{\Pi}_{67} & 0 \\ * & * & * & * & * & * & \bar{\Pi}_{77} & 0 \\ * & * & * & * & * & * & * & -\gamma I \end{bmatrix},$$

$$\bar{H}_{12} = \begin{bmatrix} \bar{\Phi}_1 & \bar{\Phi}_2 & \bar{\Phi}_2 & \bar{\Phi}_2 & \bar{\Phi}_2 & \bar{\Phi}_2 & \bar{\Phi}_2 \end{bmatrix},$$

$$\bar{H}_{22} = diag\{-\gamma I,\ \bar{\mathscr{X}}_1,\ \bar{\mathscr{X}}_2,\ \bar{\mathscr{X}}_3,\ \bar{\mathscr{X}}_4,\ \bar{\mathscr{X}}_5,\ \bar{\mathscr{X}}_6\},$$

$$\bar{\Pi}_{11} = \widetilde{A}W + W\widetilde{A}^T - \frac{\widetilde{S}_1}{\tau_m} - (\frac{\lambda_1}{\eta_1} + \frac{\lambda_2}{\eta_2} + \cdots + \frac{\lambda_\rho}{\eta})\widetilde{Q}$$

$$+ \widetilde{R}_1 + \widetilde{R}_3 - \frac{\widetilde{S}_3}{\tau_m + h} - (\frac{\tilde{\lambda}_1}{\tilde{\eta}_1} + \frac{\tilde{\lambda}_2}{\tilde{\eta}_2} + \cdots + \frac{\tilde{\lambda}_\rho}{\tilde{\eta}})\widetilde{M},$$

$$\bar{\Pi}_{12} = \frac{\widetilde{S}_1}{\tau_m},\ \bar{\Pi}_{13} = \bar{V}_2^T + (\frac{\lambda_1}{\eta_1} + \frac{\lambda_2}{\eta_2} + \cdots + \frac{\lambda_\rho}{\eta})\widetilde{Q},$$

$$\bar{\Pi}_{15} = \frac{\widetilde{S}_3}{\tau_m + h},\ \bar{\Pi}_{16} = \hat{V}_2^T + (\frac{\tilde{\lambda}_1}{\tilde{\eta}_1} + \frac{\tilde{\lambda}_2}{\tilde{\eta}_2} + \cdots + \frac{\tilde{\lambda}_\rho}{\tilde{\eta}})\widetilde{M},$$

$$\bar{\Pi}_{22} = \widetilde{R}_2 - \widetilde{R}_1 - \frac{\widetilde{S}_1}{\tau_m} - (\frac{\lambda_1}{\sigma_1} + \frac{\lambda_2}{2\sigma_1} + \cdots + \frac{\lambda_\rho}{\rho\sigma_1})\widetilde{S}_2,$$

$$\bar{\Pi}_{23} = (\frac{\lambda_1}{\sigma_1} + \frac{\lambda_2}{2\sigma_1} + \cdots + \frac{\lambda_\rho}{\rho\sigma_1})\widetilde{S}_2,$$

$$\bar{\Pi}_{33} = -(\frac{\lambda_1}{\eta_1} + \frac{\lambda_2}{\eta_2} + \cdots + \frac{\lambda_\rho}{\eta})\widetilde{Q} - (\frac{\lambda_1}{\sigma_1} + \frac{\lambda_2}{2\sigma_1} + \cdots + \frac{\lambda_\rho}{\rho\sigma_1})\widetilde{S}_2$$

$$-(\frac{\lambda_1}{\rho\sigma_1} + \frac{\lambda_2}{(\rho-1)\sigma_1} + \cdots + \frac{\lambda_\rho}{\sigma_1})\widetilde{S}_2,$$

$$\bar{\Pi}_{34} = (\frac{\lambda_1}{\rho\sigma_1} + \frac{\lambda_2}{(\rho-1)\sigma_1} + \cdots + \frac{\lambda_\rho}{\sigma_1})\widetilde{S}_2,$$

$$\bar{\Pi}_{44} = -\widetilde{R}_2 - (\frac{\lambda_1}{\rho\sigma_1} + \frac{\lambda_2}{(\rho-1)\sigma_1} + \cdots + \frac{\lambda_\rho}{\sigma_1})\widetilde{S}_2,$$

$$\bar{\Pi}_{55} = \widetilde{R}_4 - \widetilde{R}_3 - \frac{\widetilde{S}_3}{\tau_m + h} - (\frac{\tilde{\lambda}_1}{\sigma_3} + \frac{\tilde{\lambda}_2}{2\sigma_3} + \cdots + \frac{\tilde{\lambda}_\rho}{\rho\sigma_3})\widetilde{S}_4,$$

$$\bar{\Pi}_{56} = (\frac{\tilde{\lambda}_1}{\sigma_3} + \frac{\tilde{\lambda}_2}{2\sigma_3} + \cdots + \frac{\tilde{\lambda}_\rho}{\rho\sigma_3})\widetilde{S}_4,$$

$$\bar{\Pi}_{66} = -(\frac{\tilde{\lambda}_1}{\sigma_3} + \frac{\tilde{\lambda}_2}{2\sigma_3} + \cdots + \frac{\tilde{\lambda}_\rho}{\rho\sigma_3})\widetilde{S}_4 - (\frac{\tilde{\lambda}_1}{\rho\sigma_3} + \frac{\tilde{\lambda}_2}{(\rho-1)\sigma_3} + \cdots + \frac{\tilde{\lambda}_\rho}{\sigma_3})\widetilde{S}_4 - (\frac{\tilde{\lambda}_1}{\tilde{\eta}_1} + \frac{\tilde{\lambda}_2}{\tilde{\eta}_2} + \cdots + \frac{\tilde{\lambda}_\rho}{\tilde{\eta}})\widetilde{M},$$

$$\bar{\Pi}_{67} = (\frac{\tilde{\lambda}_1}{\rho\sigma_3} + \frac{\tilde{\lambda}_2}{(\rho-1)\sigma_3} + \cdots + \frac{\tilde{\lambda}_\rho}{\sigma_3})\widetilde{S}_4,$$

$$\bar{\Pi}_{77} = -\widetilde{R}_4 - (\frac{\tilde{\lambda}_1}{\rho\sigma_3} + \frac{\tilde{\lambda}_2}{(\rho-1)\sigma_3} + \cdots + \frac{\tilde{\lambda}_\rho}{\sigma_3})\widetilde{S}_4,$$

$$\bar{\Phi}_1 = \left[\widetilde{C}W \ 0 \ \bar{\Pi}_{39}^T \ 0 \ 0 \bar{\Pi}_{69}^T \ 0 \ \widetilde{F}\right]^T, \ \bar{\Phi}_2 = \left[\widetilde{A}W \ 0 \ \bar{\Pi}_{3,10}^T \ 0 \ 0 \ \bar{\Pi}_{6,10}^T \ 0 \ \widetilde{D}\right]^T,$$

$$\bar{\mathscr{X}}_1 = \psi^{-1}(\widetilde{Q} - 2W), \ \bar{\mathscr{X}}_2 = \tau_m^{-1}(\widetilde{S}_1 - 2W), \ \bar{\mathscr{X}}_3 = (\eta - \tau_m)^{-1}(\widetilde{S}_2 - 2W),$$

$$\bar{\mathscr{X}}_4 = \bar{\psi}^{-1}(\widetilde{M} - 2W), \ \bar{\mathscr{X}}_5 = (\tau_m + h)^{-1}(\widetilde{S}_3 - 2W),$$

$$\bar{\mathscr{X}}_6 = (\tilde{\eta} - \tau_m - h)^{-1}(\widetilde{S}_4 - 2W), \ W = \mathrm{diag}\{W_1, \ W_2, \ W_3\},$$

$$\psi = \lambda_1\rho\sigma_1 + \lambda_2(\rho-1)\sigma_1 + \cdots + \lambda_\rho\sigma_1,$$

$$\bar{\psi} = \tilde{\lambda}_1\rho\sigma_3 + \tilde{\lambda}_2(\rho-1)\sigma_3 + \cdots + \tilde{\lambda}_\rho\sigma_3,$$

$$\bar{\Pi}_{39} = \bar{V}_1, \bar{\Pi}_{3,10} = \bar{V}_2, \bar{\Pi}_{69} = \hat{V}_1, \bar{\Pi}_{6,10} = \hat{V}_2,$$

$$\bar{V}_1 = \begin{bmatrix} 0 \\ -\frac{a_1}{\bar{a}}C^T V_1 \\ 0 \end{bmatrix}, \ \bar{V}_2 = \begin{bmatrix} 0 & 0 & 0 \\ 0 & -\frac{a_1}{\bar{a}}C^T V_2 & 0 \\ 0 & 0 & 0 \end{bmatrix},$$

$$\hat{V}_1 = \begin{bmatrix} 0 \\ -\frac{a_2}{\bar{a}}C^T V_1 \\ 0 \end{bmatrix}, \ \hat{V}_2 = \begin{bmatrix} 0 & 0 & 0 \\ 0 & -\frac{a_2}{\bar{a}}C^T V_2 & 0 \\ 0 & 0 & 0 \end{bmatrix}.$$

Proof Construct the following Lyapunov functional

$$V(t, \xi_t) = \sum_{i=1}^{7} V_i(t, \xi_t), \tag{3.34}$$

where $V_l(t, \xi_t)$ $(l = 1, \ \ldots, \ 4)$ are the same as the corresponding items in (3.13), and

$$V_5(t, \xi_t) = (\tilde{\eta} - d(t))\int_{t-d(t)}^{t} \dot{\xi}^T(s)M\dot{\xi}(s)ds,$$

$$V_6(t, \xi_t) = \int_{t-\tau_m-h}^{t} \xi^T(s)R_3\xi(s)ds + \int_{t-\tilde{\eta}}^{t-\tau_m-h} \xi^T(s)R_4\xi(s)ds,$$

$$V_7(t, \xi_t) = \int_{-\tau_m-h}^{0}\int_{t+s}^{t} \dot{\xi}^T(\theta)S_3\dot{\xi}(\theta)d\theta ds + \int_{-\tilde{\eta}}^{-\tau_m-h}\int_{t+s}^{t} \dot{\xi}^T(\theta)S_4\dot{\xi}(\theta)d\theta ds,$$

M, R_3, R_4, S_3, and S_4 are symmetric positive definite matrices with appropriate dimensions.

The rest of the proof is similar to the proof of Theorem 3.1, here it is omitted. This completes the proof. □

Remark 3.6 To guarantee the asymptotic stability of the residual system (3.7), the FDF design criterion in Theorem 3.3 should hold for every feasible value of λ_i and $\tilde{\lambda}_i$ $(i = 1, 2, \ldots, \rho)$ with $\lambda_i = 1$ or $\lambda_i = 0$, and $\tilde{\lambda}_i = 1$ or $\tilde{\lambda}_i = 0$. Moreover, for any $\lambda_i = 1$, one has $\lambda_j = 0$, where $j = 1, 2, \ldots, \rho$, and $j \neq i$. Furthermore, for any $\tilde{\lambda}_i = 1$, one has $\tilde{\lambda}_j = 0$, where $j = 1, 2, \ldots, \rho$, and $j \neq i$.

For the data reconstruction-based FDF design criterion in Theorem 3.3, different data reconstruction parameters a_1 and a_2 presented in (3.6) lead to different fault detection performance. The following algorithm describes an approach for choosing appropriate a_1 and a_2.

Algorithm 3.1 Choose appropriate a_1 and a_2.

Step 1. For given scalars τ_m, τ_M, δ, h, and ρ, choose the initial values $a_{\epsilon,0} > 0$ ($\epsilon = 1$, 2, and $a_{1,0} < a_{2,0}$) and the final values $a_{\epsilon,ult} > 0$ ($a_{\epsilon,ult} < a_{\epsilon,0}$) for a_ϵ. Set appropriate step lengths $a_{\epsilon,dec} > 0$. Choose a large enough H_∞ norm bound γ_{opt} and set $a_{\epsilon,opt} = a_{\epsilon,0}$, $a_1 = a_{1,0}$.

Step 2. Set $a_2 = a_{2,0}$.

Step 3. Solve the FDF design criterion presented in Theorem 3.3. If $\gamma < \gamma_{opt}$, set $\gamma_{opt} = \gamma$, $a_{\epsilon,opt} = a_\epsilon$ and go to Step 4; otherwise, go to Step 4 directly.

Step 4. Set $a_2 = a_2 - a_{2,dec}$, if $a_2 \geq a_{2,ult}$, go to Step 3; otherwise, $a_1 = a_1 - a_{1,dec}$, if $a_1 \geq a_{1,ult}$, go to Step 2, otherwise, go to Step 5.

Step 5. Output the locally optimal $a_{\epsilon,opt}$ and γ_{opt}.

By using Algorithm 3.1, one can get the locally optimal $a_{\epsilon,opt}$ to guarantee better fault detection performance.

It should be mentioned that the mutually exclusive distribution characteristic of interval time-varying delays $\tau(t)$ and $d(t)$ is made full use in Theorem 3.3 to deal with integral inequalities for products of vectors. In fact, if such a mutually exclusive distribution characteristic is not considered, the FDF design method in Theorem 3.3 is still applicable, the corresponding result is omitted here for briefness.

Remark 3.7 In Theorem 3.1, Corollary 3.1, and Theorem 3.3, the FDF gains are described as $L = V_2^T N^{-T}$ and $S = V_1^T N^{-T}$. For a specific NCS, if the derived matrix N is not invertible, one should add a constraint such as $N + N^T < 5I$ for the FDF design criteria to get an invertible matrix N, and the corresponding FDF design in Theorem 3.1, Corollary 3.1, and Theorem 3.3 is a suboptimal solution. Moreover, by assuming that C^T is full column rank, where C is presented in (3.1), and adopting the method in Theorem 2 in [11], the equality constraint $W_2 C^T = C^T N$ in Theorem 3.1, Corollary 3.1, and Theorem 3.3 can be removed, and one can derive the FDF gains L and S which are independent of the matrix N. The detailed result is an extension of Theorem 2 in [11], here it is omitted for briefness.

Remark 3.8 Notice that both the robust design and the sensitivity constraint of an FDF are considered in [12]. From (29) in [12], one can see that when dealing with the robust design of the FDF, the fault signal is not considered. Similarly, from (50) in [12], one can see that external disturbances are not considered when dealing with the sensitivity constraint of the FDF. For the FDF design in this chapter, if the actuator fault signal $f(t)$ in (3.1) is assumed to be zero, the design criteria in Theorem 3.1, Corollary 3.1, and Theorem 3.3 reduce to criteria concerning the robustness of the residual signal to the disturbance and control input.

3.4 Performance Analysis and Discussion

In this section, we give an example to illustrate the merits and effectiveness of the proposed fault detection scheme.

Consider the two-tank benchmark. Suppose that the linearized continuous time model of the two-tank system is described as (see [13, 14])

$$\begin{cases} \dot{x}(t) = \begin{bmatrix} -3 & 3 \\ 3 & -6 \end{bmatrix} x(t) + \begin{bmatrix} -3 & 0 \\ 0 & 3 \end{bmatrix} u(t) + \begin{bmatrix} 0.7 \\ 0.3 \end{bmatrix} \omega(t) + \begin{bmatrix} 0 \\ 1 \end{bmatrix} f(t), \\ y(t) = \begin{bmatrix} 1 & 0 \\ 0 & 1 \end{bmatrix} x(t). \end{cases} \tag{3.35}$$

The reference model is given by

$$\begin{cases} \dot{e}_r(t) = \begin{bmatrix} -5 & 1 \\ 2 & -6 \end{bmatrix} e_r(t) + \begin{bmatrix} 0.2 \\ 1 \end{bmatrix} \omega(t) + \begin{bmatrix} 0.9 \\ -0.5 \end{bmatrix} f(t), \\ r_{ref}(t) = \begin{bmatrix} -5 & -2 \end{bmatrix} e_r(t) + 2\omega(t) - 0.8 f(t). \end{cases} \tag{3.36}$$

Firstly, we show the merits for considering the mutually exclusive distribution characteristic of interval time-varying delays. For Theorem 3.1 and Corollary 3.1, suppose that $\tau_m = 0.001$, $\tau_M = 0.05$, $\delta = 3$, $h = 0.05s$. To avoid that some elements of the obtained matrix S are close to zero, we assume that $N + N^T < 5I$ for Theorem 3.1, Corollary 3.1, and Theorem 3.3.

As discussed in Remark 3.4, the FDF design criterion in Theorem 3.1 should hold for every feasible value of λ_i $(i = 1, 2, \ldots, \rho)$ with $\lambda_i = 1$ or $\lambda_i = 0$. By solving the FDF design criterion in Theorem 3.1 for every feasible value of λ_i $(i = 1, 2, \ldots, \rho)$, one can obtain that the H_∞ norm bounds corresponding to $\rho = 2$ and $\rho = 3$ are $\gamma_1 = 2.8041$ and $\gamma_2 = 2.7993$, respectively. On the other hand, solving the FDF design criterion in Corollary 3.1, one can obtain the H_∞ norm bound $\gamma_3 = 2.8401$. From $\gamma_2 < \gamma_1$, one can see that if the mutually exclusive distribution characteristic of interval time-varying delays is adopted to deal with integral inequalities for products of vectors, the larger the number (that is ρ) of subintervals, the better the system performance. Moreover, $\gamma_1 < \gamma_3$ and $\gamma_2 < \gamma_3$ verify the fact that the FDF design scheme considering the mutually exclusive distribution characteristic of interval time-varying delays introduces less conservatism than the FDF design scheme without considering such a mutually exclusive distribution characteristic.

In the following, we illustrate the merits and effectiveness of the newly proposed fault detection scheme. For Theorems 3.1 and 3.3, suppose that $\delta = 0$, $\rho = 2$, while τ_m, τ_M and h are the same as the ones presented above. By solving the FDF design criterion in Theorem 3.1, one can get the FDF gain matrices derived by Theorem 3.1 as follows

$$L = \begin{bmatrix} 2.6508 & 2.0729 \\ 3.3788 & 1.7374 \end{bmatrix}, \quad S = \begin{bmatrix} 0.0283 & 0.0007 \end{bmatrix}.$$

For Theorem 3.3, suppose that $a_{1,0} = 1, a_{2,0} = 3, a_{1,ult} = 0.2, a_{2,ult} = 2, a_{1,dec} = 0.4, a_{2,dec} = 0.5$. By using Algorithm 3.1, one can get the locally optimal $a_{1,opt} = 1$, $a_{2,opt} = 3$, and the corresponding FDF gain matrices derived by Theorem 3.3 are

$$L = \begin{bmatrix} -0.2257 & 3.2462 \\ 2.7005 & -1.9388 \end{bmatrix}, \quad S = \begin{bmatrix} 0.0753 & -0.0393 \end{bmatrix}.$$

Suppose that the initial state of the augmented systems (3.4) and (3.7) is $\xi_0 = [0.1 \quad -0.1 \quad 0.05 \quad -0.05 \quad 0.2 \quad -0.2]^T$. The control input $u(t)$ is assumed to be zero. The disturbance input $\omega(t)$ is a normally (Gaussian) distributed random signal whose mean and variance are 0 and 0.01, respectively, and the curve of $\omega(t)$ is presented in Fig. 3.1. The fault signal $f(t)$ is described by

$$f(t) = \begin{cases} 1.8, & t \in [0.8s, \ 1.8s], \\ 0, & otherwise. \end{cases} \tag{3.37}$$

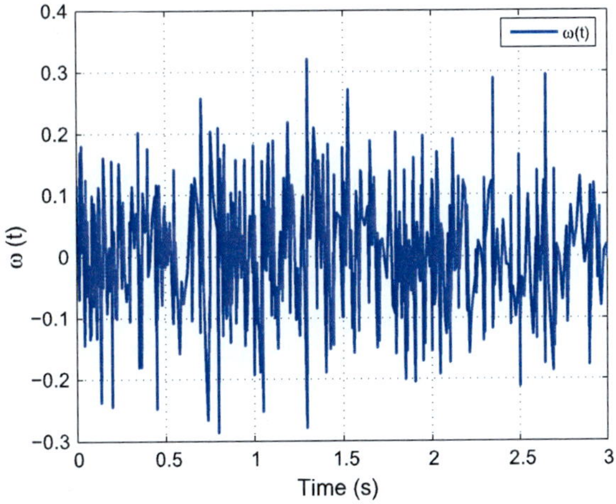

Fig. 3.1 The curve of $\omega(t)$

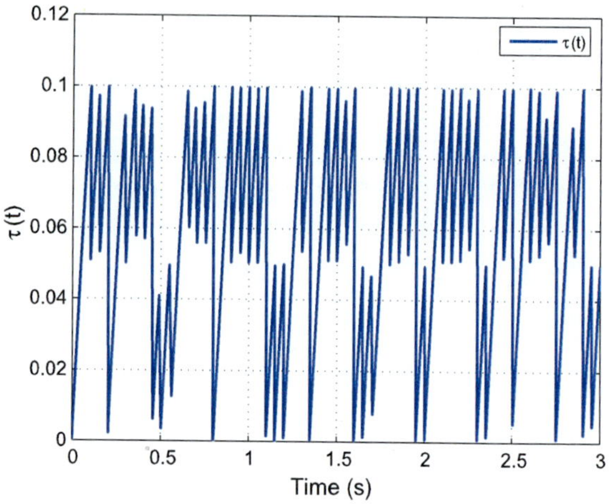

Fig. 3.2 The curve of $\tau(t)$

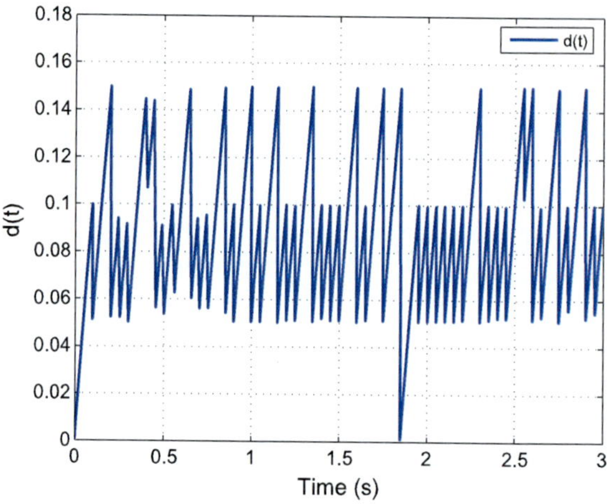

Fig. 3.3 The curve of $d(t)$

The interval time-varying delays $\tau(t)$ and $d(t)$ are given in Figs. 3.2 and 3.3, respectively. Then, Figs. 3.4 and 3.5 show the residual response $r_e(t)$ and the residual evaluation function response $J(t)$, respectively, for the system (3.4). Figures 3.6 and 3.7 show the residual response $r_e(t)$ and the residual evaluation function response $J(t)$, respectively, for the system (3.7) which considers the data reconstruction.

From Figs. 3.5 and 3.7, one can see that the newly proposed fault detection scheme can not only reflect the occurrence of faults in time (that is, the proposed fault detection scheme is sensitive to faults), but also recognize faults without confusing them with the disturbance $\omega(t)$.

To show the merits of the data reconstruction-based FDF design scheme, Fig. 3.8 presents the residual evaluation function response $J(t)$ for the system (3.4) without considering data reconstruction and the system (3.7) considering data reconstruction, where the fault signal is the same as the one in (3.37). From (3.9), Figs. 3.5 and

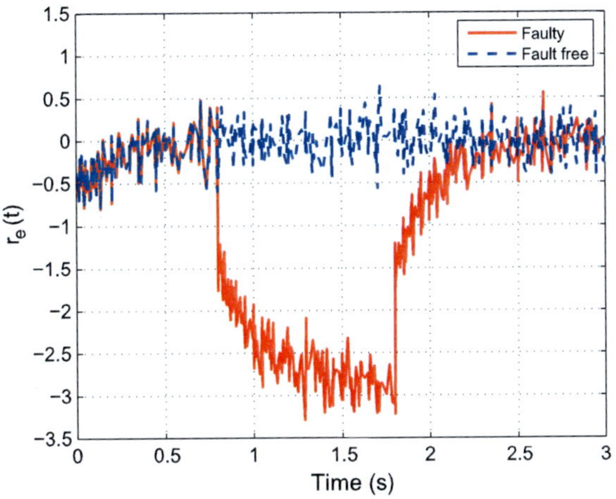

Fig. 3.4 The residual response $r_e(t)$ of the system (3.4)

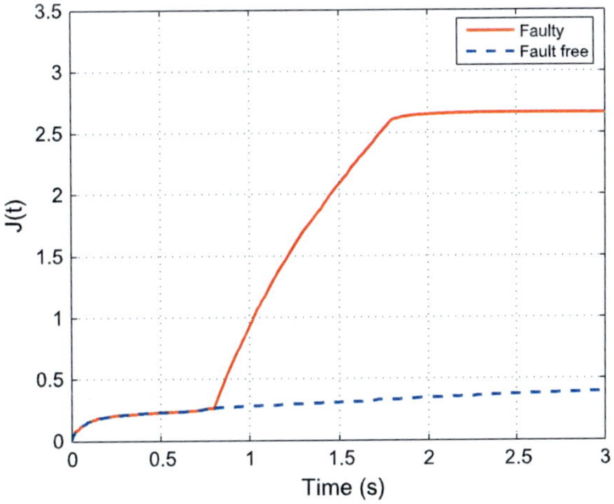

Fig. 3.5 The residual evaluation function response $J(t)$ of the system (3.4)

3.7, one can see that J_{th} for the system (3.4) and the system (3.7) are 0.3919 and 0.3878, respectively. Then, from Fig. 3.8, the fault detection logic presented in (3.10), and J_{th} presented above, one can see that the faults in the systems (3.4) and (3.7) are detected at the instants $0.8398s$ and $0.8357s$, respectively. For systems (3.4) and (3.7), the faults occur at the instant $0.8s$. Then, the fault detection time for the system (3.4) is $0.0398s$, while the fault detection time for the system (3.7) is $0.0357s$. Considering that $0.0357s$ is only 89.69% of $0.0398s$, one can conclude that the data reconstruction-based fault detection scheme provides a shorter fault detection time than the fault detection scheme without considering data reconstruction. This illustrates the merits of the data reconstruction-based fault detection scheme.

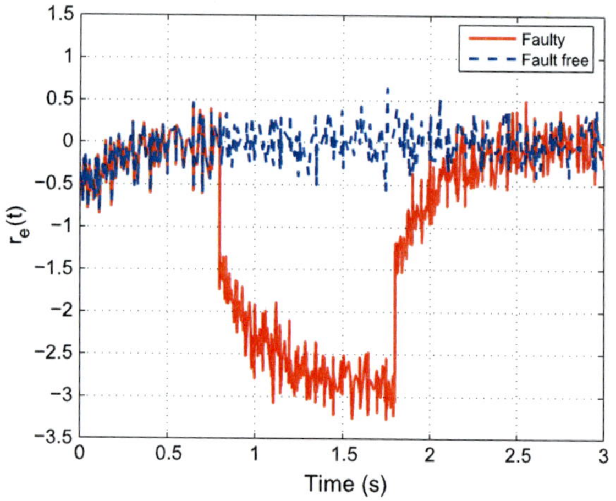

Fig. 3.6 The residual response $r_e(t)$ of the system (3.7)

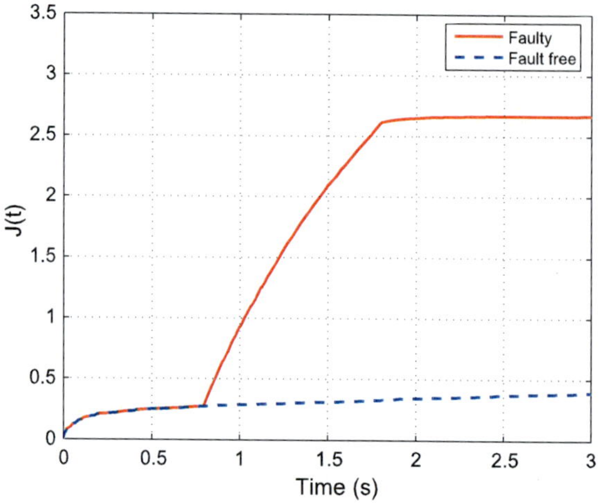

Fig. 3.7 The residual evaluation function response $J(t)$ of the system (3.7)

3.5 Conclusions

The FDF design for a continuous-time NCS considering packet dropouts, network-induced delays, and actuator faults has been investigated. The observer-based FDF and the data reconstruction scheme have been introduced to establish new closed-loop models. Based on the established models, FDF design criteria have been derived to asymptotically stabilize the residual systems. When dealing with integral inequalities for products of vectors, the mutually exclusive distribution characteristic of interval time-varying delays has been made full use to derive less conservative FDF design criteria. The designed FDFs can guarantee the sensitivity of the residual signal to faults, which has been verified by the benchmark example.

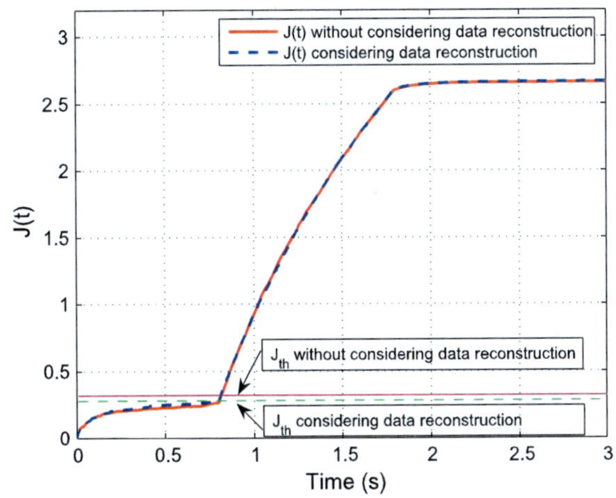

Fig. 3.8 The residual evaluation function response $J(t)$ of systems (3.4) and (3.7)

3.6 Notes

When dealing with the fault detection of NCSs, transmitting the data reconstruction-based measurement outputs to the FDF can provide satisfying fault detection performance. However, the data reconstruction scheme has attracted little attention in the literature. Moreover, making full use of the mutually exclusive distribution characteristic of interval time-varying delays can help to derive less conservative results. In this chapter, the data reconstruction scheme is proposed, and the mutually exclusive distribution characteristic of interval time-varying delays is made full use in FDF design. The designed FDFs can guarantee the sensitivity of the residual signal to faults. The results in this chapter are based mainly on [15]. For more results about fault detection of NCSs, one can also refer to [12, 16–18], etc. Combining the reciprocally convex approach [9, 10] and the mutually exclusive distribution characteristic-based approach is a future research topic.

References

1. Y. Wang, S.X. Ding, H. Ye, G. Wang, A new fault detection scheme for networked control systems subject to uncertain time-varying delay. IEEE Trans. Signal Process. **56**(10), 5258–5268 (2008)
2. Y. Zhao, J. Lam, H. Gao, Fault detection for fuzzy systems with intermittent measurements. IEEE Trans. Fuzzy Syst. **17**(2), 398–410 (2009)
3. P.M. Frank, X. Ding, Survey of robust residual generation and evaluation methods in observer-based fault detection systems. J. Process Control **7**(6), 403–424 (1997)
4. Q.-L. Han, A discrete delay decomposition approach to stability of linear retarded and neutral systems. Automatica **45**(2), 517–524 (2009)
5. K. Gu, V.L. Kharitonov, J. Chen, *Stability of Time-Delay Systems* (Birkhäuser, Boston, 2003)

6. X. Jia, D. Zhang, X. Hao, N. Zheng, Fuzzy H_∞ tracking control for nonlinear networked control systems in T-S fuzzy model. IEEE Trans. Syst. Man Cybern. Part B Cybern. **39**(4), 1073–1079 (2009)
7. X. Jiang, Q.-L. Han, S. Liu, A. Xue, A new H_∞ stabilization criterion for networked control systems. IEEE Trans. Autom. Control **53**(4), 1025–1032 (2008)
8. X.-L. Zhu, G.-H. Yang, Jensen integral inequality approach to stability analysis of continuous-time systems with time-varying delay. IET Control Theory Appl. **2**(6), 524–534 (2008)
9. W.I. Lee, P. Park, Second-order reciprocally convex approach to stability of systems with interval time-varying delays. Appl. Math. Comput. **229**, 245–253 (2014)
10. P. Park, J.W. Ko, C. Jeong, Reciprocally convex approach to stability of systems with time-varying delays. Automatica **47**(1), 235–238 (2011)
11. Y.-L. Wang, Q.-L. Han, Modelling and observer-based H_∞ controller design for networked control systems. IET Control Theory Appl. **8**(15), 1478–1486 (2014)
12. B. Liu, Y. Xia, Y. Yang, M. Fu, Robust fault detection of linear systems over networks with bounded packet loss. J. Frankl. Inst. **349**(7), 2480–2499 (2012)
13. Y.Q. Wang, S.X. Ding, H. Ye, L. Wei, P. Zhang, G.Z. Wang, Fault detection of networked control systems with packet based periodic communication. Int. J. Adapt. Control Signal Process. **23**(8), 682–698 (2009)
14. Y.Q. Wang, H. Ye, G.Z. Wang, Fault detection of NCS based on eigendecomposition, adaptive evaluation and adaptive threshold. Int. J. Control **80**(12), 1903–1911 (2007)
15. Y.-L. Wang, T.-B. Wang, Q.-L. Han, Fault detection filter design for data reconstruction-based continuous-time networked control systems. Inf. Sci. **328**, 577–594 (2016)
16. H. Dong, Z. Wang, J. Lam, H. Gao, Fuzzy-model-based robust fault detection with stochastic mixed time delays and successive packet dropouts. IEEE Trans. Syst. Man Cybern. Part B Cybern. **42**(2), 365–376 (2012)
17. X. He, Z. Wang, Y. Liu, D.H. Zhou, Least-squares fault detection and diagnosis for networked sensing systems using a direct state estimation approach. IEEE Trans. Ind. Inf. **9**(3), 1670–1679 (2013)
18. Y. Long, G.-H. Yang, Fault detection in finite frequency domain for networked control systems with missing measurements. J. Frankl. Inst. **350**(9), 2605–2626 (2013)

one can obtain from $\{t_k\}_{k\in\mathcal{I}_{\geq 0}}$ that $b_v = t_{\kappa_v}$ and $b_{v+1} = t_{\kappa_{v+1}}$ with $\kappa_v < \kappa_{v+1}$. Let MASPDs be the number of maximum allowable successive packet dropouts. That is, MASPDs $= \max\{\kappa_{v+1} - \kappa_v\} - 1$, which yields $\max\{b_{v+1} - b_v\} \leq (\text{MASPDs} + 1) \times \text{MATI}$. It implies that the input of controller is only updated at instants $\{b_v\}_{v\in\mathcal{I}_{\geq 0}}$. Under the IID scheduling protocol, the updating law of $\hat{y}_s(b_v)$ is expressed as

$$\hat{y}_s(b_v) = \begin{cases} y_s(b_v), & s = \sigma_{b_v}, \\ \hat{y}_s(b_{v-1}), & s \neq \sigma_{b_v}, \end{cases}$$

Then the input of the controller $\hat{y}(b_v)$ takes the form

$$\hat{y}(b_v) = \text{col}\{\hat{y}_1(b_{v-1}), \ldots, \hat{y}_{\sigma_{b_v}-1}(b_{v-1}), y_{\sigma_{b_v}}(b_v), \hat{y}_{\sigma_{b_v}+1}(b_{v-1}), \ldots, \hat{y}_N(b_{v-1})\}. \tag{4.3}$$

By using the hold input mechanism, the controller maintains its input as a constant until new signals arrive at the controller i.e., $\hat{y}(t) = \hat{y}(b_v)$ for $t \in [b_v, b_{v+1})$.

(2) *Network-induced delays.* The effect of network-induced delays is described as η_k with $\eta_k \in [\tau_1, \text{MAD}]$, $k \in \mathcal{I}_{\geq 0}$, where τ_1 and MAD denote the minimum and maximum allowable delays [4]. Connecting to the time-stamp set $\{b_v\}_{v\in\mathcal{I}_{\geq 0}}$ and taking into account the updating law of $\hat{y}_i(b_v)$, one has

$$\hat{y}(t) = \text{col}\{\hat{y}_1(b_{v-1}), \ldots, \hat{y}_{\sigma_{b_v}-1}(b_{v-1}), y_{\sigma_{b_v}}(b_v), \hat{y}_{\sigma_{b_v}+1}(b_{v-1}), \ldots, \hat{y}_N(b_{v-1})\}, \tag{4.4}$$

where $t \in [d_v, d_{v+1})$ with $d_v = b_v + \eta_{b_v}$.

Remark 4.1 Different from some existing works (see, e.g. [5, 6]) where a stochastic variable is introduced to describe the packet dropouts in the communication, a time-stamp technique is used in this chapter to define the available packets [7]. It is convenient for us to take into account the scheduling protocol and non-ideal quality of services (QoS) in a unified framework.

4.1.3 Description of the Output Feedback Controller

Based on the fact that $\hat{y}(t)$ is available for $t \in [d_v, d_{v+1})$, a DOFC is formulated as follows

$$\begin{cases} \dot{x}_c(t) = A_c x_c(t) + B_c x_c(b_v) + C_c \hat{y}(t), \\ u(t) = D_c x_c(t), \ d_v \leq t < d_{v+1}, \end{cases} \tag{4.5}$$

where A_c, B_c, C_c and D_c are the controller gains to be designed with the gain C_c being described as $C_c = [C_{c1}, C_{c2}, \ldots, C_{cN}]$.

Remark 4.2 For the controller state vector $x_c(t)$, we consider several items, i.e., $x_c(t)$, $x_c(b_v)$ and $\hat{y}(t)$. Compared with [2], the item $x_c(b_v)$ is additionally introduced in the design of DOFC. The reason for including the item $x_c(b_v)$ lies in that it makes the controller synthesis tractable in terms of linear matrix inequalities (LMIs). Note that the time-stamp of the item $x_c(b_v)$ is the same as the one of $y_{\sigma_{b_v}}(b_v)$, which can be easily realized by adding a time-stamp to the data at the sampler. Therefore, the controller (4.5) is online implementable [8, 9] under the supposition that the controller matrices are pre-known. If the controller matrices are unknown, we can apply the stabilization criterion given in the next section to design them offline.

For further developments, define scheduling errors over $t \in [d_v, \ d_{v+1})$ as

$$e_s(t) = \hat{y}_s(b_{v-1}) - y_s(b_v), \ s = 1, \ \ldots, \ N, \tag{4.6}$$

from which one can obtain that for $s = \sigma_{b_v}$,

$$e_s(d_{v+1}) = y_s(b_v) - y_s(b_{v+1}) = C_s(x(b_v) - x(b_{v+1})),$$

and for $s = 1, \ \ldots, \ \sigma_{b_v} - 1, \ \sigma_{b_v} + 1, \ \ldots, \ N$,

$$e_s(d_{v+1}) = \hat{y}_s(b_{v-1}) - y_s(b_{v+1}) = e_s(d_v) + C_s(x(b_v) - x(b_{v+1})).$$

It follows from (4.4) and (4.6) that

$$\hat{y}(t) = y(b_v) + \sum_{s=1, \ s \neq \sigma_{b_v}}^{N} \mathcal{J}_s e_s(t), \ d_v \leq t < d_{v+1}, \tag{4.7}$$

where $\mathcal{J}_s = \mathrm{col}\{\underbrace{0, \ \ldots, \ 0}_{s-1}, I, \underbrace{0, \ \ldots, \ 0}_{N-s}\}$. Then, one gets a new expression of $\hat{y}(t)$, which is related to the scheduling error $e_s(t)$ and the measurement output $y(t)$ sampled at the instant $\{b_v\}_{v \in \mathcal{I}_{\geq 0}}$. The controller given in (4.5) can be rewritten as

$$\begin{cases} \dot{x}_c(t) = A_c x_c(t) + B_c x_c(b_v) + C_c C x(b_v) + \sum\limits_{s=1, \ s \neq \sigma_{b_v}}^{N} C_{cs} e_s(t), \\ u(t) = D_c x_c(t), \ d_v \leq t < d_{v+1}. \end{cases} \tag{4.8}$$

4.1.4 Stochastic Impulsive System Modeling

By setting $\zeta(t) = \mathrm{col}\{x(t), \ x_c(t)\}$, one can see that the connection of the controller (4.8) with the system (4.1) produces a closed-loop system of the form

$$\dot{\zeta}(t) = \mathscr{A}\zeta(t) + \mathscr{B}\zeta(b_v) + \sum_{s=1, s\neq\sigma_{b_v}}^{N} \mathscr{C}_s e_s(t),\ d_v \leq t < d_{v+1},$$

where

$$\mathscr{A} = \begin{bmatrix} A & BD_c \\ 0 & A_c \end{bmatrix},\ \mathscr{B} = \begin{bmatrix} 0 & 0 \\ C_c C & B_c \end{bmatrix},\ \mathscr{C}_s = \begin{bmatrix} 0 \\ C_{cs} \end{bmatrix}.$$

Then following the time-delay system approach [10], we define an artificial delay $\tau(t) = t - b_v$ for $t \in [d_v,\ d_{v+1})$, from which one obtains

$$\tau_1 \leq \eta_{b_v} \leq \tau(t) \leq b_{v+1} - b_v + \eta_{b_{v+1}} \leq \tau_2,$$

and $\dot{\tau}(t) = 1$ with $\tau_2 = (\text{MASPDs} + 1) \times \text{MATI} + \text{MAD}$. Moreover, we obtain from $\{b_v\}_{v\in\mathcal{I}_{\geq 0}} \subseteq \{t_k\}_{k\in\mathcal{I}_{\geq 0}}$ that $\text{Prob}\{\sigma_{b_v} = s\}$ is with the property as given in (4.2). The following indicator functions are introduced

$$\varepsilon\{\sigma_{b_v},\ s\} = \begin{cases} 1, & \sigma_{b_v} = s, \\ 0, & \sigma_{b_v} \neq s, \end{cases}$$

for $s = 1,\ \ldots,\ N$. Based on (4.2), the mathematical expectations and covariance of $\varepsilon_{\sigma_{b_v},s}$ are obtained as $E\{\varepsilon_{\sigma_{b_v},s}\} = E\{\varepsilon^2_{\sigma_{b_v},s}\} = \text{Prob}\{\sigma_{b_v} = s\} = \beta_s$, while $E\{(\varepsilon_{\sigma_{b_v},s} - \beta_s)(\varepsilon_{\sigma_{b_v},m} - \beta_m)\}$ is equal to $-\beta_s\beta_m$ if $s \neq m$, and $\beta_s(1 - \beta_s)$ otherwise.

By utilizing the proposed indicator function, the closed-loop system can be further written as the following stochastic impulsive system

$$\dot{\zeta}(t) = \mathscr{A}\zeta(t) + \mathscr{B}\zeta(t - \tau(t)) + \sum_{s=1}^{N}(1 - \varepsilon_{\sigma_{b_v},s})\mathscr{C}_s e_s(t), \tag{4.9}$$

for $t \in [d_v,\ d_{v+1})$. The overall updating law of scheduling errors is described by

$$e_s(d_{v+1}) = (1 - \varepsilon_{\sigma_{b_v},s})e_s(d_v) + C_s H(\zeta(b_v) - \zeta(b_{v+1})),$$

with $H = [I,\ 0]$. The initial value of state $\zeta(t)$ is supplemented as $\zeta(t) = \text{col}\{\phi(t),\ 0\}$, $t \in [d_0 - \tau_2,\ d_0]$, $\phi(0) = x_0$, and $e_s(d_0) = -C_s(d_0 - \eta_{b_0}) = -C_s x_0$, where $\phi(t)$ is a continuous function on $t \in [d_0 - \tau_2,\ d_0]$.

Definition 4.1 The closed-loop system (4.9) is said to be asymptotically mean-square stable if there exists a constant $b > 0$ such that the following bound holds

$$E\{\|\zeta(t)\|^2\} \leq bE\{\|\zeta_{d_0}\|^2_w + \|e(d_0)\|^2\},\ t \geq d_0, \tag{4.10}$$

for the solutions of the system (4.9) initialized with $e(d_0)$ and $\zeta(t) = \text{col}\{\phi(t),\ 0\}$, $t \in [d_0 - \tau_2,\ d_0]$.

The objective of this chapter is to derive sufficient conditions for stabilization of the studied system. That is, for prescribed scalars τ_1 and τ_2, design the controller gains A_c, B_c, C_{cs} $(s = 1, \ldots, N)$, and D_c such that the closed-loop system (4.9) subject to the updating law of scheduling errors is asymptotically stable in the mean-square sense.

4.2 Stability Analysis and Stabilization

In this section, we present a stability criterion guaranteeing the asymptotic stability of the system (4.9) in the mean-square sense. For this purpose, we choose a Lyapunov functional candidate as

$$V(t) = V_1(t) + V_2(t) + \sum_{s=1}^{N} e_s^T(d_v) S_s e_s(d_v), \tag{4.11}$$

where

$$\begin{aligned}
V_1(t) =& \zeta^T(t) P \zeta(t) + \int_{t-\tau_1}^{t} \zeta^T(\psi) Q_1 \zeta(\psi) d\psi + \int_{t-\tau_2}^{t-\tau_1} \zeta^T(\psi) Q_2 \zeta(\psi) d\psi \\
&+ \int_{-\tau_1}^{0} \int_{t+\psi}^{t} \tau_1 \dot{\zeta}^T(\rho) R_1 \dot{\zeta}(\rho) d\rho d\psi \\
&+ \int_{-\tau_2}^{-\tau_1} \int_{t+\psi}^{t} (\tau_2 - \tau_1) \dot{\zeta}^T(\rho) R_2 \dot{\zeta}(\rho) d\rho d\psi, \\
V_2(t) =& \sum_{s=1}^{N} \int_{b_v}^{t} (\tau_2 - \tau_1) \| \sqrt{G_s} C_s H \dot{\zeta}(\psi) \|^2 d\psi,
\end{aligned}$$

for $t \in [d_v, d_{v+1})$, $P > 0$, $Q_1 > 0$, $Q_2 > 0$, $R_1 > 0$, $R_2 > 0$, $G_s > 0$ and $S_s > 0$ $(s = 1, \ldots, N)$.

Remark 4.3 Note that the constructed Lyapunov functional (4.11) demonstrates the following three characteristics: (a) The item $V_1(t)$ is employed to fully utilize the lower and upper bounds of artificial delay $\tau(t)$; (b) The item $V_2(t)$ does not grow in the jumps when $t = d_{v+1}$. The reason lies in that the difference $\Delta V_2(d_{v+1}) = V_2(d_{v+1}) - V_2(d_{v+1}^-)$ is negative, where $V_2(d_{v+1})$, and $V_2(d_{v+1}^-)$ are given by

$$\begin{aligned}
V_2(d_{v+1}) &= \sum_{s=1}^{N} \int_{b_{v+1}}^{d_{v+1}} (\tau_2 - \tau_1) \| \sqrt{G_s} C_s H \dot{\zeta}(\psi) \|^2 d\psi, \\
V_2(d_{v+1}^-) &= \sum_{s=1}^{N} \int_{b_v}^{d_{v+1}^-} (\tau_2 - \tau_1) \| \sqrt{G_s} C_s H \dot{\zeta}(\psi) \|^2 d\psi,
\end{aligned}$$

respectively. Taking into account $d_{v+1} = d_{v+1}^-$, one obtains that

$$\Delta V_2(d_{v+1}) = -\sum_{s=1}^{N}\int_{b_v}^{b_{v+1}}(\tau_2-\tau_1)\|\sqrt{G_s}C_s H\dot{\zeta}(\psi)\|^2 d\psi \\ \leq -\sum_{s=1}^{N}\|\sqrt{G_s}C_s(x(b_v)-x(b_{v+1}))\|^2, \tag{4.12}$$

where Jensen inequality is adopted here; and (c) The constant item $\sum_{s=1}^{N} e_s^T(d_v)S_s e_s(d_v)$ together with the item $V_2(t)$ are utilized to cope with the updating law of scheduling errors.

Based on (4.11) and (4.12), one can get the following theorem.

Theorem 4.1 *For given scalars τ_1, τ_2, β_s, and matrices $\mathscr{A}$, $\mathscr{B}$, and $\mathscr{C}_s$ $(s = 1, \ldots, N)$, if there exist real matrices $P > 0$, $Q_1 > 0$, $Q_2 > 0$, $R_1 > 0$, $R_2 > 0$, $G_s > 0$, $S_s > 0$, $T_s > 0$ $(s = 1, \ldots, N)$, and U with appropriate dimensions such that*

$$\begin{bmatrix} \Xi_{11} & * & * \\ \Xi_{21} & \Xi_{22} & * \\ \Xi_{31} & 0 & \Xi_{33} \end{bmatrix} < 0, \tag{4.13}$$

$$\begin{bmatrix} R_2 & * \\ U & R_2 \end{bmatrix} \geq 0, \tag{4.14}$$

$$\Omega_s < 0,\ s = 1, \ldots, N, \tag{4.15}$$

where

$$\Xi_{11} = \begin{bmatrix} \Gamma_{11} & * & * & * & * \\ \Gamma_{21} & \Gamma_{22} & * & * & * \\ \Gamma_{31} & \Gamma_{32} & \Gamma_{33} & * & * \\ 0 & \Gamma_{42} & \Gamma_{43} & \Gamma_{44} & * \\ \Gamma_{51} & 0 & 0 & 0 & \Gamma_{55} \end{bmatrix},\ \Xi_{21} = \mathrm{col}\{\tau_1 R_1 F_1, (\tau_2-\tau_1)R_2 F_1, \Sigma_1, \ldots, \Sigma_N\},$$

$$\Xi_{22} = \beta_s\Psi_s = \mathrm{diag}\{-R_1, -R_2, -G_1, \ldots, -G_N\},\ \Xi_{31} = \mathrm{col}\{\Upsilon_1, \ldots, \Upsilon_N\},$$

$$\Xi_{33} = \mathrm{diag}\{\Psi_1, \ldots, \Psi_N\},\ \Upsilon_s = \mathrm{col}\{\tau_1 R_1 F_{2s}, (\tau_2-\tau_1)R_2 F_{2s}, \Delta_{s1}, \ldots, \Delta_{sN}\},$$

$$\Omega_s = \begin{bmatrix} (\tau_2-\tau_1)T_s - \beta_s S_s & * \\ (1-\beta_s)S_s & S_s - G_s \end{bmatrix},\ \Gamma_{11} = \mathscr{A}^T P + P\mathscr{A} + Q_1 - R_1,\ \Gamma_{21} = R_1,$$

$$\Gamma_{22} = -Q_1 + Q_2 - R_1 - R_2,\ \Gamma_{31} = \mathscr{B}^T P,\ \Gamma_{32} = R_2 - U,\ \Gamma_{33} = -2R_2 + U + U^T,$$

$$\Gamma_{42} = U,\ \Gamma_{43} = R_2 - U,\ \Gamma_{44} = -Q_2 - R_2,\ \Gamma_{55} = \mathrm{diag}\{-T_1, \ldots, -T_N\},$$

$$\Gamma_{51} = \mathrm{col}\{(1-\beta_1)\mathscr{C}_1^T P, \ldots, (1-\beta_N)\mathscr{C}_N^T P\},$$

$$\Sigma_s = \sqrt{\tau_2 - \tau_1} G_s C_s H \digamma_1, \ s = 1, \ \ldots, \ N,$$
$$\Delta_{sl} = \sqrt{\tau_2 - \tau_1} G_l C_l H \digamma_{2s}, \ s, \ l = 1, \ \ldots, \ N,$$
$$\digamma_1 = [\mathscr{A}, 0, \mathscr{B}, 0, (1-\beta_1)\mathscr{C}_1, \ldots, (1-\beta_N)\mathscr{C}_N],$$
$$\digamma_{2s} = [\underbrace{0, \ldots, 0}_{s+3}, \mathscr{C}_s, \underbrace{0, \ldots, 0}_{N-s}], \ s = 1, \ \ldots, \ N,$$

then the system (4.9) is asymptotically stable in the mean-square sense.

Proof For convenience of presentation, define

$$\xi_t = \mathrm{col}\{\zeta(t), \zeta(t-\tau_1), \zeta(t-\tau(t)), \zeta(t-\tau_2), e_1(b_v), \ldots, e_N(b_v)\},$$

Then one has

$$\dot{\zeta}(t) = (\digamma_1 + \sum_{s=1}^{N} (\beta_s - \varepsilon_{\sigma_{b_v}, s}) \digamma_{2s}) \xi_t,$$

where $\digamma_1$, $\digamma_{2s}$ are defined in Theorem 4.1. Utilizing the characteristic of $\varepsilon\{\sigma_{b_v}, s\}$ with $\Theta = \tau_1^2 R_1 + (\tau_2 - \tau_1)^2 R_2 + \sum_{l=1}^{N} (\tau_2 - \tau_1) H^T C_l^T G_l C_l H$ yields

$$E\{\dot{\zeta}^T(t) \Theta \dot{\zeta}(t)\} \leq \xi_t^T (\digamma_1^T \Theta \digamma_1 + \sum_{s=1}^{N} \beta_s \digamma_{2s}^T \Theta \digamma_{2s}) \xi_t.$$

Next, applying the infinitesimal operator $\mathscr{L}$ for (4.11) yields

$$\begin{aligned} \mathscr{L}V(t) = {} & 2\zeta^T(t) P \dot{\zeta}(t) + \zeta^T(t) Q_1 \zeta(t) - \zeta^T(t-\tau_1)(Q_1 - Q_2)\zeta(t-\tau_1) \\ & - \zeta^T(t-\tau_2) Q_2 \zeta(t-\tau_2) + \dot{\zeta}^T(t) \Theta \dot{\zeta}(t) - \int_{t-\tau_1}^{t} \tau_1 \dot{\zeta}^T(\rho) R_1 \dot{\zeta}(\rho) d\rho \\ & - \int_{t-\tau_2}^{t-\tau_1} (\tau_2 - \tau_1) \dot{\zeta}^T(\rho) R_2 \dot{\zeta}(\rho) d\rho, \end{aligned} \quad (4.16)$$

By using the Jensen inequality [11] and reciprocally convex approach [12] with (4.14), one has

$$\int_{t-\tau_1}^{t} \tau_1 \dot{\zeta}^T(\rho) R_1 \dot{\zeta}(\rho) d\rho \geq \begin{bmatrix} \vartheta_1 \\ \vartheta_2 \end{bmatrix}^T \begin{bmatrix} R_1 & * \\ -R_1 & R_1 \end{bmatrix} \begin{bmatrix} \vartheta_1 \\ \vartheta_2 \end{bmatrix},$$
$$\int_{t-\tau_2}^{t-\tau_1} (\tau_2 - \tau_1) \dot{\zeta}^T(\rho) R_2 \dot{\zeta}(\rho) d\rho \geq \begin{bmatrix} \vartheta_3 \\ \vartheta_4 \end{bmatrix}^T \begin{bmatrix} R_2 & * \\ U & R_2 \end{bmatrix} \begin{bmatrix} \vartheta_3 \\ \vartheta_4 \end{bmatrix},$$

where $\vartheta_1 = \zeta(t)$, $\vartheta_2 = \zeta(t-\tau_1)$, $\vartheta_3 = \zeta(t-\tau_1) - \zeta(t-\tau(t))$, and $\vartheta_4 = \zeta(t-\tau(t)) - \zeta(t-\tau_2)$. Then one has

$$\mathscr{L}V(t) \leq \dot{\zeta}^T(t)\Theta\dot{\zeta}(t) + \sum_{s=1}^{N} e_s^T(d_v)T_s e_s(d_v) + \xi_t^T \Xi_{11}\xi_t$$
$$+ \sum_{s=1}^{N} 2x^T(t)P(\beta_s - \varepsilon_{\sigma_{b_v},s})F_{2s}\xi_t. \tag{4.17}$$

Taking the mathematical expectation of (4.17), one obtains

$$E\{\mathscr{L}V(t)\} \leq E\{\sum_{s=1}^{N} e_s^T(d_v)T_s e_s(d_v)\} + \xi_t^T(\Xi_{11} + F_1^T\Theta F_1 + \sum_{s=1}^{N}\beta_s F_{2s}^T\Theta F_{2s})\xi_t.$$

By using the Schur complement, it follows from (4.13) that

$$\Xi_{11} + F_1^T\Theta F_1 + \sum_{s=1}^{N}\beta_s F_{2s}^T\Theta F_{2s} < 0,$$

from which one obtains

$$E\{\mathscr{L}V(t)\} \leq E\{\sum_{s=1}^{N} e_s^T(d_v)T_s e_s(d_v)\}. \tag{4.18}$$

Since $V(t)$ is continuous on $[d_v,\ d_{v+1})$, the integration on both sides of (4.18) yields

$$E\{V(t) - V(d_v)\} \leq \int_{d_v}^{t} E\{\sum_{s=1}^{N} e_s^T(d_v)T_s e_s(d_v)\}d\psi, \tag{4.19}$$

for $t \in [d_v,\ d_{v+1})$. The inequality (4.19) with $t = d_{v+1}^-$ and $d_{v+1}^- - d_v \leq \tau_2 - \tau_1$ implies that

$$E\{V(d_{v+1}^-) - V(d_v)\} \leq (\tau_2 - \tau_1)E\{\sum_{s=1}^{N} e_s^T(d_v)T_s e_s(d_v)\},$$

Then one has

$$E\{V(d_{v+1})\} \leq E\{V(d_{v+1})\} - E\{V(d_{v+1}^-)\}$$
$$+ E\{V(d_v)\} + (\tau_2 - \tau_1)E\{\sum_{s=1}^{N} e_s^T(d_v)T_s e_s(d_v)\}.$$

It follows from (4.12) that

$$E\{V(d_{v+1})\} \leq E\{V(d_v)\} + \sum_{s=1}^{N} \Phi_s, \tag{4.20}$$

where

$$\begin{aligned}\Phi_s &= E\{e_s^T(d_{v+1})S_s e_s(d_{v+1})\} + E\{e_s^T(d_v)((\tau_2 - \tau_1)T_s - S_s)e_s(d_v)\}\\ &\quad - E\{(x(b_v) - x(b_{v+1}))^T C_s^T G_s C_s(x(b_v) - x(b_{v+1}))\}.\end{aligned}$$

Noting that the following equality can be verified

$$\begin{aligned}E\{e_s^T(d_{v+1})S_s e_s(d_{v+1})\} = &(1 - \beta_s)E\{e_s^T(d_v)S_s e_s(d_v)\}\\ &+ (1 - \beta_s)E\{e_s^T(d_v)S_s C_s(x(b_v) - x(b_{v+1}))\}\\ &+ E\{(x(b_v) - x(b_{v+1}))^T C_s^T S_s C_s(x(b_v) - x(b_{v+1}))\},\end{aligned}$$

From (4.15), one can obtain that

$$\Phi_s = E\{\kappa_s^T(v)\Omega_s\kappa_s(v)\} < 0, \ s = 1, \ \ldots, \ N, \tag{4.21}$$

with $\kappa_s(v) = \text{col}\{e_s(d_v), \ C_s(x(b_v) - x(b_{v+1}))\}$. Substituting (4.21) into (4.20) yields $E\{V(d_{v+1})\} \leq E\{V(d_v)\} \leq \cdots \leq E\{V(d_0)\}$, one can conclude from (4.19) that

$$E\{V(t)\} \leq E\{V(d_0)\} + (\tau_2 - \tau_1)E\{\sum_{s=1}^{N} e_s^T(d_v)T_s e_s(d_v)\}, \ t \in [d_v, \ d_{v+1}).$$

From (4.11), one obtains

$$E\{V(t)\} \geq \lambda_{\min}\{P\}E\{\|\zeta(t)\|^2\} + E\{\sum_{s=1}^{N} e_s^T(d_v)S_s e_s(d_v)\}.$$

That is,

$$\lambda_{\min}\{P\}E\{\|\zeta(t)\|^2\} \leq E\{V(d_0)\} + E\{\sum_{s=1}^{N} e_s^T(d_v)((\tau_2 - \tau_1)T_s - S_s)e_s(d_v)\}.$$

Then it follows from $\Omega_s \leq 0$ that $(\tau_2 - \tau_1)T_s \leq \beta_s S_s \leq S_s$, for $s = 1, \ \ldots, \ N$. Then one gets $\lambda_{\min}\{P\}E\{\|\zeta(t)\|^2\} \leq E\{V(d_0)\}$, which implies that the mean-square stability of the system (4.9) can be guaranteed by applying Definition 4.1 and $E\{V(d_0)\} \leq \mu E\{\|\zeta(d_0)\|_w^2 + \|e(d_0)\|^2\}$ with $\mu > 0$. This completes the proof. □

Remark 4.4 For a special case of non-scheduling protocol, i.e., $\beta_s = 1$ $(s = 1, \ldots, N)$, Theorem 4.1 with $\tau_1 = 0$, $G_s = 0$, and $S_s = 0$ will lead to $\tau_2 = 1.07$, where the matrices $\mathscr{A}$ and $\mathscr{B}$ in system (4.9) are given as $\mathscr{A} = \begin{bmatrix} 0 & 1 \\ 0 & -0.1 \end{bmatrix}$ and $\mathscr{B} = \begin{bmatrix} 0 & 0 \\ -0.375 & -1.15 \end{bmatrix}$, respectively. Note that the obtained result coincides with that in [4], which is larger than the value 1.04 in [7]. Moreover, the time-delay system approach used in this chapter can lead to less conservative results than the hybrid system approach, which will be shown in Sect. 4.4.

Based on Theorem 4.1, we can not obtain the controller gains A_c, B_c, C_{cs} $(s = 1, \ldots, N)$, and D_c directly since they are coupled with matrix P and others. In order to derive the controller gain matrices, we now state and establish the following theorem.

Theorem 4.2 *For given scalars τ_1, τ_2, and β_s $(s = 1, \ldots, N)$, if there exist real matrices $X > 0$, $Y > 0$ $\tilde{Q}_1 > 0$, $\tilde{Q}_2 > 0$, $\tilde{R}_1 > 0$, $\tilde{R}_2 > 0$, $G_s > 0$, $S_s > 0$, $T_s > 0$, $\tilde{U}$, W_1, W_2, W_{3s} $(s = 1, \ldots, N)$, and W_4 with appropriate dimensions, such that (4.15) and*

$$\begin{bmatrix} \tilde{\Xi}_{11} & * & * \\ \tilde{\Xi}_{21} & \tilde{\Xi}_{22} & * \\ \tilde{\Xi}_{31} & 0 & \tilde{\Xi}_{33} \end{bmatrix} < 0, \tag{4.22}$$

$$\begin{bmatrix} \tilde{R}_2 & * \\ \tilde{U} & \tilde{R}_2 \end{bmatrix} \geq 0, \tag{4.23}$$

$$\mathcal{J} = \begin{bmatrix} X & * \\ I & Y \end{bmatrix} > 0, \tag{4.24}$$

where

$$\tilde{\Xi}_{11} = \begin{bmatrix} \tilde{\Gamma}_{11} & * & * & * & * \\ \tilde{\Gamma}_{21} & \tilde{\Gamma}_{22} & * & * & * \\ \tilde{\Gamma}_{31} & \tilde{\Gamma}_{32} & \tilde{\Gamma}_{33} & * & * \\ 0 & \tilde{\Gamma}_{42} & \tilde{\Gamma}_{43} & \tilde{\Gamma}_{44} & * \\ \tilde{\Gamma}_{51} & 0 & 0 & 0 & \tilde{\Gamma}_{55} \end{bmatrix}, \ \tilde{\Xi}_{21} = \mathrm{col}\{\tau_1 \tilde{F}_1, \ (\tau_2 - \tau_1)\tilde{F}_1, \ \tilde{\Sigma}_1, \ \ldots, \ \tilde{\Sigma}_N\},$$

$$\tilde{\Xi}_{22} = -\mathrm{diag}\{\mathcal{J}^T \tilde{R}_1^{-1} \mathcal{J}, \mathcal{J}^T \tilde{R}_i^{-1} \mathcal{J}, G_1^{-1}, \ldots, G_N^{-1}\},$$

$$\tilde{\Xi}_{31} = \mathrm{col}\{\tilde{\Upsilon}_1, \ldots, \tilde{\Upsilon}_N\}, \ \tilde{\Xi}_{33} = \mathrm{diag}\{\frac{\tilde{\Xi}_{22}}{\beta_1}, \ldots, \frac{\tilde{\Xi}_{22}}{\beta_N}\},$$

$$\tilde{\Upsilon}_s = \mathrm{col}\{\tau_1 \tilde{F}_{2s}, (\tau_2 - \tau_1)\tilde{F}_{2s}, \tilde{\Delta}_{s1}, \ldots, \tilde{\Delta}_{sN}\},$$

$$\tilde{\Gamma}_{11} = \Pi_1 + \Pi_1^T + \tilde{Q}_1 - \tilde{R}_1, \ \tilde{\Gamma}_{21} = \tilde{R}_1, \ \tilde{\Gamma}_{22} = -\tilde{Q}_1 + \tilde{Q}_2 - \tilde{R}_1 - \tilde{R}_2, \ \tilde{\Gamma}_{31} = \Pi_2^T,$$

$$\tilde{\Gamma}_{32} = \tilde{R}_2 - \tilde{U},\ \tilde{\Gamma}_{33} = -2\tilde{R}_2 + \tilde{U} + \tilde{U}^T,\ \tilde{\Gamma}_{42} = \tilde{U},\ \tilde{\Gamma}_{43} = \tilde{R}_2 - \tilde{U},\ \tilde{\Gamma}_{44} = -\tilde{Q}_2 - \tilde{R}_2,$$

$$\tilde{\Gamma}_{51} = \mathrm{col}\{(1-\beta_1)\Pi_{31}^T, \ldots, (1-\beta_N)\Pi_{3N}^T\},\ \tilde{\Gamma}_{55} = \mathrm{diag}\{-T_1, \ldots, -T_N\},$$

$$\tilde{\Sigma}_s = \sqrt{\tau_2 - \tau_1} C_s H \tilde{F}_1,\ s = 1,\ \ldots,\ N,$$

$$\tilde{\Delta}_{sl} = \sqrt{\tau_2 - \tau_1} C_l H \tilde{F}_{2s},\ s,\ l = 1,\ \ldots,\ N,$$

$$\tilde{F}_1 = [\Pi_1, 0, \Pi_2, 0, (1-\beta_1)\Pi_{31}, \ldots, (1-\beta_N)\Pi_{3N}],$$

$$\tilde{F}_{2s} = [\underbrace{0, \ldots, 0}_{s+3}, \Pi_{3s}, \underbrace{0, \ldots, 0}_{N-s}],\ s = 1,\ \ldots,\ N,$$

$$\Pi_1 = \begin{bmatrix} AX + BW_1 & A \\ W_2 & YA \end{bmatrix},\ \Pi_2 = \begin{bmatrix} 0 & 0 \\ W_4 & \sum_{s=1}^N W_{3s} \mathcal{J}_s^T C \end{bmatrix},$$

$$\Pi_{3s} = \begin{bmatrix} 0 \\ W_{3s} \end{bmatrix},\ s = 1,\ \ldots,\ N,$$

then the system (4.9) is asymptotically stable in mean-square sense.

Proof Suppose that there exist real matrices X, Y, $\tilde{Q}_1$, $\tilde{Q}_2$, $\tilde{R}_1$, $\tilde{R}_2$, G_s, S_s, T_s, $\tilde{U}$, W_1, W_2, W_{3s} ($s = 1,\ \ldots,\ N$), and W_4 such that (4.15), (4.22), (4.23), (4.24) are satisfied. Utilizing the Schur complement with (4.24) implies that $Y - X^{-1} > 0$. Hence, the matrix $I - YX$ is nonsingular, and there always exists a nonsingular matrix Z satisfying $ZZ^{-1}(I - YX) = I - YX$. Then, two nonsingular matrices are introduced as follows

$$J_1 = \begin{bmatrix} X & I \\ Z^{-1}(I - YX) & 0 \end{bmatrix},\ J_2 = \begin{bmatrix} I & Y \\ 0 & Z^T \end{bmatrix},$$

Set $P = J_2 J_1^{-1}$, $Q_1 = J_1^{-T} \tilde{Q}_1 J_1^{-1}$, $Q_2 = J_1^{-T} \tilde{Q}_2 J_1^{-1}$, $R_1 = J_1^{-T} \tilde{R}_1 J_1^{-1}$, $R_2 = J_1^{-T} \tilde{R}_2 J_1^{-1}$, $U = J_1^{-T} \tilde{U} J_1^{-1}$, and $\Theta = J_1^{-T} \tilde{\Theta} J_1^{-1}$. One can verify that

$$P = \begin{bmatrix} Y & Z \\ Z^T & Z^T(Y - X^{-1})^{-1} Z \end{bmatrix} > 0,$$

since the following inequalities hold

$$\begin{cases} Z^T(Y - X^{-1})^{-1} Z > 0, \\ Y - Z(Z^T(Y - X^{-1})^{-1} Z)^{-1} Z^T = X^{-1} > 0. \end{cases}$$

Define

$$\mathcal{J}_1 = \mathrm{diag}\{J_1^{-1}, J_1^{-1}, J_1^{-1}, J_1^{-1}, \underbrace{I, \ldots, I}_{N}\},$$

$$\mathcal{J}_2 = \mathrm{diag}\{R_1 J_2^{-1}, R_2 J_2^{-1}, G_1, \ldots, G_N\},$$

and

$$\begin{cases} W_1 = D_c Z^{-1}(I - YX), \\ W_2 = YAX + YBW_1 + ZA_c Z^{-1}(I - YX), \\ W_{3s} = ZC_{cs},\ s = 1,\ \ldots,\ N, \\ W_4 = \sum_{s=1}^{N} W_{3s} \mathcal{J}_s^T CX + ZB_c Z^{-1}(I - YX). \end{cases}$$

Pre- and post-multiply (4.22), (4.23) with

$$\text{diag}\{\underbrace{\mathcal{J}_1^T, \mathcal{J}_2^T, \ldots, \mathcal{J}_2^T}_{N+1}\},\ \text{diag}\{J_1^{-T}, J_1^{-T}\},$$

and their transposes, respectively, we arrive at (4.13), (4.14). Then, one can conclude from Theorem 4.1 that if (4.15), (4.22), (4.23), (4.24) are satisfied, the system (4.9) is asymptotically stable in mean-square sense. This completes the proof. □

As one can see, Theorem 4.2 presents an explicit expression on the solution of matrices X, Y, W_1, W_2, W_{3s}, and W_4. However, it is difficult to solve them by using the *Matlab LMI toolbox* since there exist nonlinear items $-\mathcal{J}^T \tilde{R}_i^{-1} \mathcal{J}$ $(i = 1, 2)$ and $-G_s^{-1}$ $(s = 1, \ldots, N)$. Based on the fact that $(\rho_i \tilde{R}_i - \mathcal{J})^T \tilde{R}_i^{-1}(\rho_i \tilde{R}_i - \mathcal{J}) \geq 0$ with matrices $\tilde{R}_i > 0$ and tuning parameters ρ_i $(i = 1, 2)$, one obtains

$$-\mathcal{J}^T \tilde{R}_i^{-1} \mathcal{J} \leq \rho_i^2 \tilde{R}_i - 2\rho_i \mathcal{J},\ i = 1, 2. \tag{4.25}$$

Note that in [8, 13], scalars ρ_i are given as $\rho_i = 1$ for $i = 1, 2$. Such a value assignment for ρ_i leads to conservative results, whereas the results derived with an adjustable value of ρ_i in this chapter can reduce the conservatism of the results in [8, 13]. Moreover, define matrices $\mathcal{G}_s$ satisfying $G_s^{-1}\mathcal{G}_s = I$ $(s = 1, \ldots, N)$. Using the cone complementary linearization algorithm [14], one arrives at

$$\begin{bmatrix} \mathcal{G}_s & * \\ I & G_s \end{bmatrix} > 0,\ s = 1, \ldots, N. \tag{4.26}$$

Then the corresponding minimization problem is formulated as

$$\begin{aligned} &\min\ \text{tr}\left\{\sum_{s=1}^{N} G_s \mathcal{G}_s\right\}, \\ &\text{subject to :(4.15), (4.22)*, (4.23), (4.24), (4.26),} \end{aligned} \tag{4.27}$$

where (4.22)* is derived from (4.22) by replacing $-\mathcal{J}^T \tilde{R}_i^{-1} \mathcal{J}$ and $-G_s^{-1}$ with $\rho_i^2 \tilde{R}_i - 2\rho_i \mathcal{J}$ and $-\mathcal{G}_s$, respectively. For more details about solving the minimization problem (4.27), one can refer to [15]. Note that if $\text{tr}\left\{\sum_{s=1}^{N} G_s \mathcal{G}_s - n_y\right\} < \epsilon$ for a sufficiently small scalar $\epsilon > 0$, then one can calculate $X, Y, W_1, W_2, W_{3s}, W_4$, and exit. Otherwise, continue the iteration process and set $k = k + 1$ where k indicates the iteration number. If $k > \mathcal{K}$ ($\mathcal{K}$ denotes the allowed maximum number of iterations), then exit. We now obtain the matrices X, Y, W_1, W_2, W_{3s}, and W_4 by solving this minimization problem. Then the controller gains are derived as

$$\begin{cases} A_c = Z^{-1}(W_2 - YAX - YBW_1)(I - YX)^{-1}Z, \\ B_c = Z^{-1}(W_4 - \sum_{s=1}^{N} W_{3s}\mathcal{I}_s^T CX)(I - YX)^{-1}Z, \\ C_{cs} = Z^{-1}W_{3s}, \ s = 1, \ \ldots, \ N, \\ D_c = W_1(I - YX)^{-1}Z. \end{cases}$$

It should be mentioned that matrix Z is unknown and independent on Theorem 4.2. For this reason, one performs an irreducible transformation as $x_c(t) = Z^{-1}\hat{x}_c(t)$. That is, the DOFC (4.5) with (A_c, B_c, C_{cs}, D_c) is algebraically equivalent to the one with $(\mathscr{A}_c, \mathscr{B}_c, \mathscr{C}_{cs}, \mathscr{D}_c) = (Z^{-1}A_cZ, Z^{-1}B_cZ, Z^{-1}C_{cs}, D_cZ^{-1})$.

To summarize, we present the following controller design criterion.

Theorem 4.3 *For given scalars τ_1, τ_2, and β_s $(s = 1, \ \ldots, \ N)$, if there exist real matrices $X > 0$, $Y > 0$, $\tilde{Q}_1 > 0$, $\tilde{Q}_2 > 0$, $\tilde{R}_1 > 0$, $\tilde{R}_2 > 0$, $G_s > 0$, $S_s > 0$, $T_s > 0$, $\tilde{U}$, W_1, W_2, W_{3s} $(s = 1, \ \ldots, \ N)$, and W_4 with appropriate dimensions such that the minimization problem (4.27) is solvable, then the system (4.9) is asymptotically stable in mean-square sense and the controller gain matrices are given by*

$$\begin{cases} \mathscr{A}_c = (W_2 - YAX - YBW_1)(I - YX)^{-1}, \\ \mathscr{B}_c = (W_4 - \sum_{s=1}^{N} W_{3s}\mathcal{I}_s^T CX)(I - YX)^{-1}, \\ \mathscr{C}_{cs} = W_{3s}, \ s = 1, \ \ldots, \ N, \\ \mathscr{D}_c = W_1(I - YX)^{-1}. \end{cases} \tag{4.28}$$

Remark 4.5 If the item $x_c(b_v)$ is discarded (that is, $B_c = \mathscr{B}_c = 0$), then one obtains from (4.28) that $W_4 = \sum_{s=1}^{N} W_{3s}\mathcal{I}_s^T CX$. However, the newly obtained conditions can not be solved since some real matrices W_{3s} $(s = 1, \ \ldots, \ N)$ and X to be determined are coupled in W_4. Different from [2], by introducing the additional item $x_c(b_v)$, the dynamic output feedback controller can be designed readily in this chapter.

4.3 The IID Protocol Optimization Algorithm

Based on Theorem 4.3, one can derive the upper bound of τ_2, i.e., $\max\{\tau_2\}$ and controller gain matrices $(\mathscr{A}_c, \mathscr{B}_c, \mathscr{C}_{cs}, \mathscr{D}_c)$ by solving the minimization problem (4.27). Consequently, the value of $\max\{\tau_2\}$ is affected by the selection of scalars β_s $(s = 1, \ \ldots, \ N)$. Since β_s denote the probabilities of the events given in (4.2), a non-optimum IID protocol can not lead to an optimal value of τ_2. For enlarging $\max\{\tau_2\}$, an IID protocol optimization algorithm subject to the constraint $\sum_{s=1}^{N} \beta_s = 1$ is given as follows.

Algorithm 4.1 Find the optimal value of τ_2, the optimal protocol parameters β_s, and controller gain matrices $(\mathscr{A}_c, \mathscr{B}_c, \mathscr{C}_{cs}, \mathscr{D}_c)$.

Step 1. Choose step increments θ_s, τ_1, small enough $\tau_2^0 > \tau_1$, β_s^0 and $(\mathscr{A}_c^0, \mathscr{B}_c^0, \mathscr{C}_{cs}^0, \mathscr{D}_c^0)$, $s = 1, \ldots, N$. For $\beta_1 = \theta_1 : \theta_1 : 1, \ldots, \beta_N = \theta_N : \theta_N : 1$, repeat Step 2 and Step 3.
Step 2. If $\sum_{s=1}^{N} \beta_s = 1$, find $\max\{\tau_2\}$ by solving (4.27); calculate $(\mathscr{A}_c, \mathscr{B}_c, \mathscr{C}_{cs}, \mathscr{D}_c)$, $s = 1, \ldots, N$; else, break.
Step 3. If $\max\{\tau_2\} > \tau_2^0$, update $\tau_2^0 = \max\{\tau_2\}$, $\beta_s^0 = \beta_s$, $(\mathscr{A}_c^0, \mathscr{B}_c^0, \mathscr{C}_{cs}^0, \mathscr{D}_c^0) = (\mathscr{A}_c, \mathscr{B}_c, \mathscr{C}_{cs}, \mathscr{D}_c)$, $s = 1, \ldots, N$.
Step 4. Return τ_2^0, β_s^0, $(\mathscr{A}_c^0, \mathscr{B}_c^0, \mathscr{C}_{cs}^0, \mathscr{D}_c^0)$, $s = 1, \ldots, N$.

Remark 4.6 Note that in each iteration, β_s take some values, i.e, $\beta_s \in [\theta_s, 1]$ and satisfy $\sum_{i=1}^{N} \beta_s = 1$. With those values of β_s, the upper bound of τ_2 can be obtained. If $\max\{\tau_2\} > \tau_2^0$, they will be updated and saved. It is clear that the value of $\max\{\tau_2\}$ and the controller matrices are obtained simultaneously by applying Algorithm 4.1.

4.4 Performance Analysis and Discussion

This section gives two examples to demonstrate the effectiveness of the proposed method.

Example 4.1 Consider the benchmark example of a batch reactor [16] with the form (4.9) under dynamic output control, where $N = 2$, $\mathcal{A} = \begin{bmatrix} \mathcal{A}_1 & \mathcal{A}_2 \\ 0 & 0 \end{bmatrix}$, $\mathcal{B}$, $\mathcal{C}_1$, and $\mathcal{C}_2$ are given as

$$\mathcal{A}_1 = \begin{bmatrix} 1.380 & -0.208 & 6.715 & -5.676 \\ -0.581 & -4.2902 & 0 & 0.675 \\ 1.067 & 4.273 & -6.654 & 5.893 \\ 0.048 & 4.273 & 1.343 & -2.104 \end{bmatrix},$$

$$\mathcal{A}_2^T = \begin{bmatrix} 0 & -11.358 & -2.272 & -2.272 \\ 0 & 0 & 25.168 & 0 \end{bmatrix},$$

$$\mathcal{B}_1^T = \begin{bmatrix} 0 & 0 & -15.73 & 0 & 0 & 1 \\ 0 & -11.3580 & -2.2720 & -2.2720 & 1 & 0 \end{bmatrix},$$

$$\mathcal{C}_1 = \begin{bmatrix} 0 & 0 & 0 & 0 \\ 1 & 0 & 1 & -1 \end{bmatrix}, \mathcal{C}_2 = \begin{bmatrix} 0 & 0 & 0 & 0 \\ 0 & 1 & 0 & 0 \end{bmatrix}.$$

Applying Theorem 4.1 with $\tau_1 = 0$, $\beta_1 = 0.6$, and $\beta_2 = 0.4$, one can see that the value of $\max\{\tau_2\}$ is 0.022, which is larger than 0.0088 (under RR protocol) and 0.0108 (under TOD protocol) in [16]. It is clear that the stability criterion derived in this chapter significantly improves the results in [16].

Table 4.1 Maximum values of τ_2

ρ	0.01	0.1	0.2	0.5	1	2
$\max\{\tau_2\}$	0.17	0.25	0.25	0.21	–	–

Example 4.2 Consider the satellite system under the dynamic output feedback control [8]. The parameters of system (4.1) are described by

$$A = \begin{bmatrix} 0 & 0 & 1 & 0 \\ 0 & 0 & 0 & 1 \\ -0.09 & 0.09 & -0.04 & 0.04 \\ 0.09 & -0.09 & 0.04 & -0.04 \end{bmatrix}, \quad B = \begin{bmatrix} 0 \\ 0 \\ 1 \\ 0 \end{bmatrix},$$

and three sensors are assembled to measure the system states with $C_1 = [0, 1, 1, 0]$, $C_2 = [1, 0, 1, 1]$, and $C_3 = [0, 1, 0, 1]$. The initial condition is given as $x(0) = \text{col}[0.10, -0.18, -0.07, 0.04]$.

By employing Theorem 4.3 with $\tau_1 = 0.05$, $\beta_1 = 0.3$, $\beta_2 = 0.3$, $\beta_3 = 0.4$, $\epsilon = 0.0001$, and $\mathcal{K} = 5$, the obtained values of $\max\{\tau_2\}$ are listed in Table 4.1 for different $\rho = \rho_1 = \rho_2$. It is clear that when $\rho = 1,\ 2$, no feasible result can be obtained by Theorem 4.3. Thus, the adjustable scalars ρ_i $(i = 1,\ 2)$ can provide a relaxed result.

Now, we apply Algorithm 4.1 to find the optimal value of τ_2 by searching the optimal values of protocol parameters β_i $(i = 1,\ 2,\ 3)$. Set $\tau_1 = 0.02$, $\rho_1 = \rho_2 = 0.2$, $\theta_i = 0.05$ $(i = 1,\ 2,\ 3)$, $\epsilon = 0.0001$, and $\mathcal{K} = 5$. By using Algorithm 4.1, one can obtain that the optimal value of τ_2 is 0.3038. The corresponding parameters of scheduling protocol are $\beta_1 = 0.05$, $\beta_2 = 0.5$, and $\beta_3 = 0.45$, and the controller gain matrices $(\mathscr{A}_c, \mathscr{B}_c, \mathscr{C}_{cs}, \mathscr{D}_c)$ are obtained as

$$\begin{cases} \mathscr{A}_c = \begin{bmatrix} -0.7047 & 0.3734 & -0.0129 & -0.0378 \\ -0.3883 & 0.2130 & -0.1397 & 0.1858 \\ -3.0801 & 2.4094 & -2.9646 & 0.2816 \\ 0.9526 & -0.5428 & 0.9152 & -1.4213 \end{bmatrix}, \\ \mathscr{B}_c = \begin{bmatrix} -0.7505 & 0.0760 & -0.3038 & -0.0374 \\ -0.1321 & -0.5494 & -0.1468 & -0.0776 \\ -0.9615 & 0.2338 & -0.3496 & -0.0423 \\ 0.1771 & -1.2372 & -0.0676 & 0.1007 \end{bmatrix}, \end{cases}$$

$$\begin{cases} \mathscr{C}_{c1} = \begin{bmatrix} -0.0001 & -0.0012 & -0.0017 & 0.0005 \end{bmatrix}^T, \\ \mathscr{C}_{c2} = \begin{bmatrix} -0.3216 & -0.0989 & -0.3924 & -0.1641 \end{bmatrix}^T, \\ \mathscr{C}_{c3} = \begin{bmatrix} -0.0294 & -0.4817 & 0.0731 & -0.8323 \end{bmatrix}^T, \\ \mathscr{D}_c = \begin{bmatrix} 2.4792 & -2.1987 & 3.1792 & -0.0352 \end{bmatrix}. \end{cases}$$

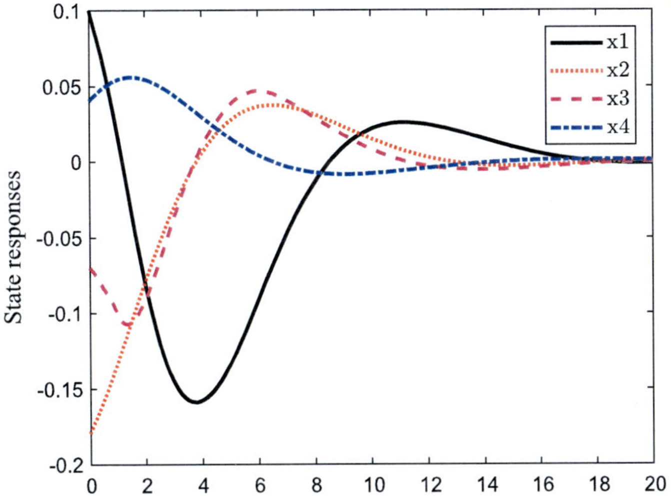

Fig. 4.2 The state responses

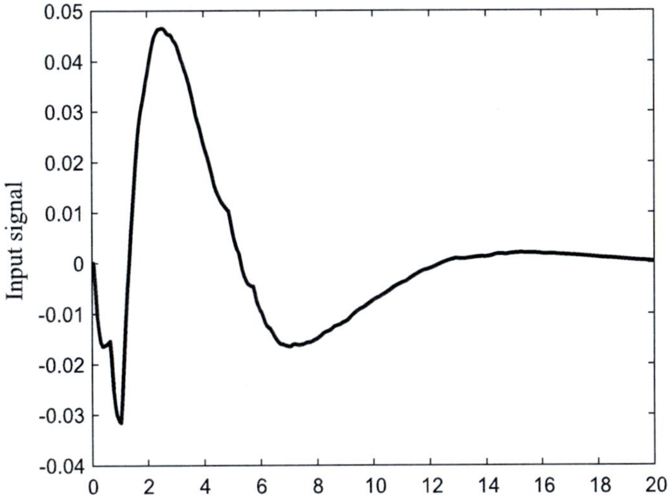

Fig. 4.3 The input signal

For this example, aperiodic sampling technique is employed where $t_{k+1} - t_k \in [0.055,\ 0.07]$. Suppose that MASPDs = 3, one gets MAD = 0.0238 from the definition of τ_2. Then Figs. 4.2, 4.3, and 4.4 depict the state responses, the input signal, and the active node sequence and data transmission, respectively. From these figures, one can see that the designed DOFC can stabilize the plant in the presence of the IID scheduling protocol, network-induced delays and packet dropouts.

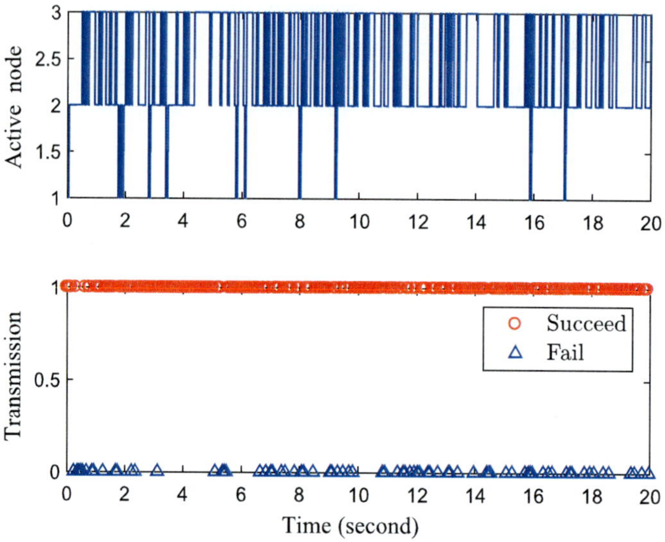

Fig. 4.4 The active node sequence and data transmission (○: succeed; △: fail)

4.5 Conclusions

The dynamic output feedback control of NCSs with a scheduling protocol, network-induced delays and packet dropouts has been addressed. To tackle with limited network bandwidth in the communication network, an IID scheduling protocol has been introduced to choose the node which can access the communication network. A stochastic impulsive system model has been established for the studied system. Stability and stabilization criteria have been derived to guarantee the stability of the stochastic impulsive system in mean-square sense. An optimization algorithm has been presented to achieve the parameters of DOFC and optimal IID protocol simultaneously. Benchmark examples have been provided to demonstrate the validity of the proposed method.

4.6 Notes

To deal with the node collision problem of NCSs with limited network bandwidth, some scheduling protocols, such as TOD protocol [17], RR protocol [18, 19], and stochastic protocol [3, 20], are proposed. The time-delay system approach for NCSs under scheduling protocols is developed in [2, 10, 17, 21]. Different from aforementioned results, the dynamic output feedback control of NCSs under an IID scheduling protocol and nonideal network quality of services (network-induced delays and packet dropouts) is investigated in this chapter. The results in this chapter are based mainly on [22]. For more results about the dynamic output feedback control and scheduling protocols of NCSs, refer to [3, 13, 19, 23], etc. Future works will involve the consideration of other protocols, e.g., TOD protocol and Markovian protocol, and the implementation of scheduling protocols in a real wireless communication network.

References

1. L. Zou, Z. Wang, H. Gao, Set-membership filtering for time-varying systems with mixed time-delays under Round-Robin and weighted Try-Once-Discard protocols. Automatica **74**, 341–348 (2016)
2. K. Liu, E. Fridman, K.H. Johansson, Networked control with stochastic scheduling. IEEE Trans. Autom. Control **60**(11), 3071–3076 (2015)
3. J. Zhang, C. Peng, M.-R. Fei, Y.-C. Tian, Output feedback control of networked systems with a stochastic communication protocol. J. Frankl. Inst. **354**(9), 3838–3853 (2017)
4. C. Peng, T.C. Yang, Event-triggered communication and H_∞ control co-design for networked control systems. Automatica **49**(5), 1326–1332 (2013)
5. F. Yang, Z. Wang, D.W.C. Ho, M. Gani, Robust H_∞ control with missing measurements and time delays. IEEE Trans. Autom. Control **52**(9), 1666–1672 (2007)
6. W.-A. Zhang, L. Yu, Modelling and control of networked control systems with both network-induced delay and packet-dropout. Automatica **44**(12), 3206–3210 (2008)
7. D. Yue, Q.-L. Han, J. Lam, Network-based robust H_∞ control of systems with uncertainty. Automatica **41**(6), 999–1007 (2005)
8. H. Gao, X. Meng, T. Chen, J. Lam, Stabilization of networked control systems via dynamic output-feedback controllers. SIAM J. Control Optim. **48**(5), 3643–3658 (2010)
9. Y.-L. Wang, Q.-L. Han, Network-based modelling and dynamic output feedback control for unmanned marine vehicles in network environments. Automatica **91**, 43–53 (2018)
10. E. Fridman, *Introduction to Time-Delay Systems: Analysis and Control* (Birkhäuser, Basel, 2014)
11. K. Gu, V.L. Kharitonov, J. Chen, *Stability of Time-Delay Systems* (Birkhäuser, Boston, 2003)
12. P. Park, J.W. Ko, C. Jeong, Reciprocally convex approach to stability of systems with time-varying delays. Automatica **47**(1), 235–238 (2011)
13. X.-M. Zhang, Q.-L. Han, Event-triggered dynamic output feedback control for networked control systems. IET Control Theory Appl. **8**(4), 226–234 (2014)
14. L. El Ghaoui, F. Oustry, M. AitRami, A cone complementarity linearization algorithm for static output-feedback and related problems. IEEE Trans. Autom. Control **42**(8), 1171–1176 (1997)
15. S. Hu, D. Yue, X. Xie, Z. Du, Event-triggered H_∞ stabilization for networked stochastic systems with multiplicative noise and network-induced delays. Inf. Sci. **299**, 178–197 (2015)
16. W.P.M.H. Heemels, A.R. Teel, N. van de Wouw, D. Nešić, Networked control systems with communication constraints: tradeoffs between transmission intervals, delays and performance. IEEE Trans. Autom. Control **55**(8), 1781–1796 (2010)
17. D. Freirich, E. Fridman, Decentralized networked control of systems with local networks: a time-delay approach. Automatica **69**, 201–209 (2016)
18. K. Liu, E. Fridman, L. Hetel, Stability and L_2-gain analysis of networked control systems under round-robin scheduling: a time-delay approach. Syst. Control Lett. **61**(5), 666–675 (2012)
19. D. Ding, Z. Wang, Q.-L. Han, G. Wei, Neural-network-based output-feedback control under round-robin scheduling protocols. IEEE Trans. Cybern. **49**(6), 2372–2384 (2019)

20. V. Gupta, T.H. Chung, B. Hassibi, R.M. Murray, On a stochastic sensor selection algorithm with applications in sensor scheduling and sensor coverage. Automatica **42**(2), 251–260 (2006)
21. G. Wen, Y. Wan, J. Cao, T. Huang, W. Yu, Master-slave synchronization of heterogeneous systems under scheduling communication. IEEE Trans. Syst. Man Cybern. Syst. **48**(3), 473–484 (2018)
22. J. Zhang, C. Peng, X. Xie, D. Yue, Output feedback stabilization of networked control systems under a stochastic scheduling protocol. IEEE Trans. Cybern. **50**(6), 2851–2860 (2020)
23. M. Jungers, E.B. Castelan, V.M. Moraes, U.F. Moreno, A dynamic output feedback controller for NCS based on delay estimates. Automatica **49**(3), 788–792 (2013)

Chapter 5
Network-Based Heading Control of UMVs

This chapter addresses network-based heading control and rudder oscillation reduction for a UMV equipped with single rudder in network environments. A novel network-based model is first established by constructing a heading control error system and purposely dropping some control input packets, which are received by a steering machine. Then a stabilization criterion is derived to guarantee the heading angle tracking performance and to reduce the oscillation of the rudder angle. Some algorithms for selecting the number of purposely dropped control input packets are presented. It is shown through heading control and rudder oscillation reduction performance analysis that compared with the controller design without dropping control input packets purposely, the proposed intentional packet dropouts-based controller design is more effective in improving the control performance of the UMV.

5.1 Model Transformation

The motion of a marine vehicle in 6 degrees of freedom includes sway, yaw, roll, surge, heave, and pitch [1–3]. The main concern of this chapter is the motion in sway, yaw, and roll. The influence of surge, heave, and pitch is treated as a disturbance. Applying Newton's laws in a space-fixed coordinate system, one can get the equations for sway, yaw, and roll as

$$\begin{cases} m_{y_a}\dfrac{d^2 y_a}{dt^2} = F_{y_a}, & \text{sway} \\ I_{zz_a}\dfrac{d^2 h_a}{dt^2} = N_a, & \text{yaw} \\ I_{xx_a}\dfrac{d^2 f_a}{dt^2} = K_a, & \text{roll} \end{cases} \tag{5.1}$$

Y.-L. Wang et al., *Network-Based Control of Unmanned Marine Vehicles*,
https://doi.org/10.1007/978-3-031-28605-6_5

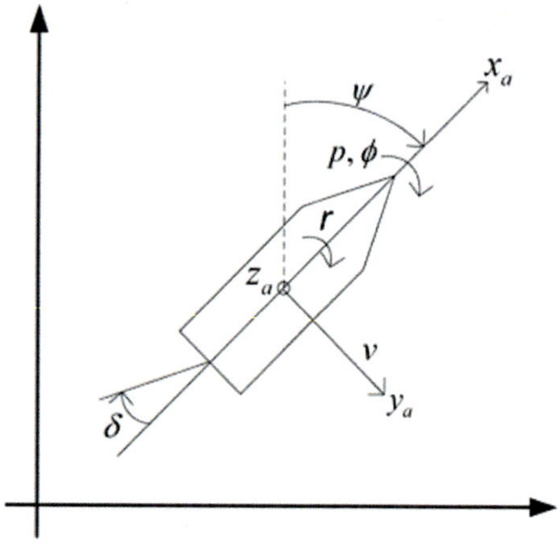

Fig. 5.1 The motion coordinate system for a marine vehicle

where x_a, y_a, and z_a denote the longitudinal axis, transverse axis, and normal axis, respectively; m_{y_a} and F_{y_a} denote the effective mass and the force of the marine vehicle in the y_a direction, respectively; I_{zz_a} and I_{xx_a} denote moments of inertia with respect to the z_a and x_a axes, respectively; h_a and f_a denote the heading angle and the roll angle, respectively; N_a and K_a denote moments with respect to the z_a and x_a axes, respectively.

By translating equations in system (5.1) to the motion coordinate system in Fig. 5.1, utilizing Taylor series expansions, Laplace transformation, and a model simplification, one can obtain the following state-space model, whose block diagram is presented in Fig. 5.2, for the sway-yaw and roll subsystems.

$$\begin{cases} \dot{x}(t) = Ax(t) + B_1\delta(t) + B_2\omega(t), \\ x(t_0) = x_0, \end{cases} \tag{5.2}$$

where $x(t) = [v(t),\ r(t),\ \psi(t),\ p(t),\ \phi(t)]^T \in \mathbb{R}^n$ with $v(t)$, $r(t)$, $\psi(t)$, $p(t)$, and $\phi(t)$ denoting the sway velocity caused by the rudder motion alone, the yaw velocity, the heading angle, the roll velocity, and the roll angle, respectively (see Fig. 5.1, where x_a, y_a, and z_a denote the directions of 3 coordinates); $\delta(t)$ represents the rudder angle; $\omega(t) = [\omega_\psi(t),\ \omega_\phi(t)]^T$ denotes the wave-induced disturbances, and $\omega(t) \in L_2[t_0, \infty)$ with t_0 denoting the initial instant; $x_0 \in \mathbb{R}^n$ denotes the initial condition; A, B_1, and B_2 are given by

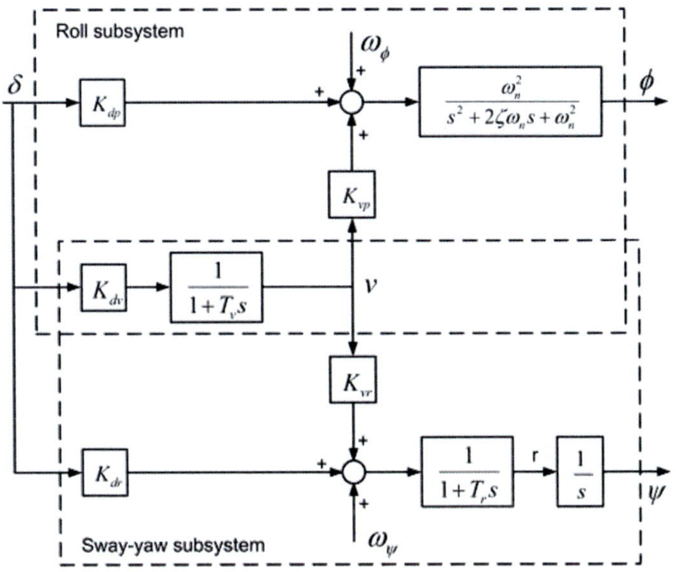

Fig. 5.2 The block diagram of the marine vehicle model

$$A = \begin{bmatrix} -\frac{1}{T_v} & 0 & 0 & 0 & 0 \\ \frac{K_{vr}}{T_r} & -\frac{1}{T_r} & 0 & 0 & 0 \\ 0 & 1 & 0 & 0 & 0 \\ \omega_n^2 K_{vp} & 0 & 0 & -2\zeta\omega_n & -\omega_n^2 \\ 0 & 0 & 0 & 1 & 0 \end{bmatrix},$$

$$B_1 = \begin{bmatrix} \frac{K_{dv}}{T_v} \\ \frac{K_{dr}}{T_r} \\ 0 \\ \omega_n^2 K_{dp} \\ 0 \end{bmatrix}, \quad B_2 = \begin{bmatrix} 0 & 0 \\ \frac{1}{T_r} & 0 \\ 0 & 0 \\ 0 & \omega_n^2 \\ 0 & 0 \end{bmatrix},$$

where T_v and T_r denote time constants of transfer functions; ζ and ω_n denote the damping ratio and the natural frequency under no damping, respectively; K_{vr}, K_{vp}, K_{dv}, K_{dr}, and K_{dp} denote given gains.

The model of the sway-yaw and roll subsystems in system (5.2) can be adopted to describe the dynamics of a UMV. The design, application and heading control of the UMV have received considerable attention and a number of interesting results have been obtained [4–7]. It should be pointed out that if no control is applied to the UMV, the actual heading angle will deviate inevitably from the desired heading angle. For example, choose the parameters in system matrices A, B_1, and B_2 in system (5.2) as

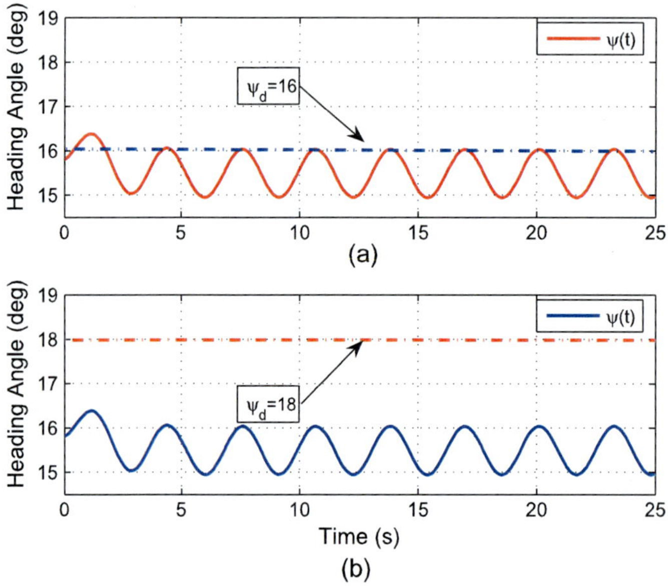

Fig. 5.3 Curves of heading angle for system (5.2) without control

$$
\begin{aligned}
U &= 7.8\ \text{(m/s)},\ T_v = 8/U,\ T_r = 6/U,\ K_{dv} = 0.01U,\\
K_{dr} &= -0.6027U,\ K_{dp} = -0.0014U^2,\ K_{vr} = -0.46\ \text{(m/s)},\\
K_{vp} &= 0.21U,\ \omega_n = 0.63\ \text{(rad/s)},\ \zeta = 0.86 + 0.0038U,
\end{aligned}
\tag{5.3}
$$

where U is the forward speed of the UMV. Suppose that the initial state of the system (5.2) is $x_0 = [1\ \ 0.2\ \ 15.8\ \ 0.1\ \ 1]^T$. The external wave-induced disturbances are given by

$$
\begin{cases}
\omega_\psi(t) = 2cos(2t), & 0s \le t \le 25s,\\
\omega_\phi(t) = -cos(t), & 0s \le t \le 25s.
\end{cases}
\tag{5.4}
$$

When no control is applied to the model of the sway-yaw and roll subsystems in system (5.2), the curves of heading angle $\psi(t)$ corresponding to the desired heading angle $\psi_d = 16$ and $\psi_d = 18$ are plotted in Fig. 5.3a, b, respectively; the curves of heading angle deviation, that is $\psi(t) - \psi_d$, corresponding to the desired heading angle $\psi_d = 16$ and $\psi_d = 18$ are plotted in Fig. 5.4a, b, respectively, from which one can see clearly that the uncontrolled heading angle and heading angle deviation indeed oscillate. Therefore, the system of the UMV subject to the heading angle oscillation and the heading angle deviation oscillation is a damping-like system. For the UMV, it is practically valuable to reduce the heading angle deviation. Thus, it is of paramount importance to study the problem of heading control for the UMV. The heading control for the UMV is usually based on a remote land-based/mother ship-based control station in network environments. In such a system, the sampled

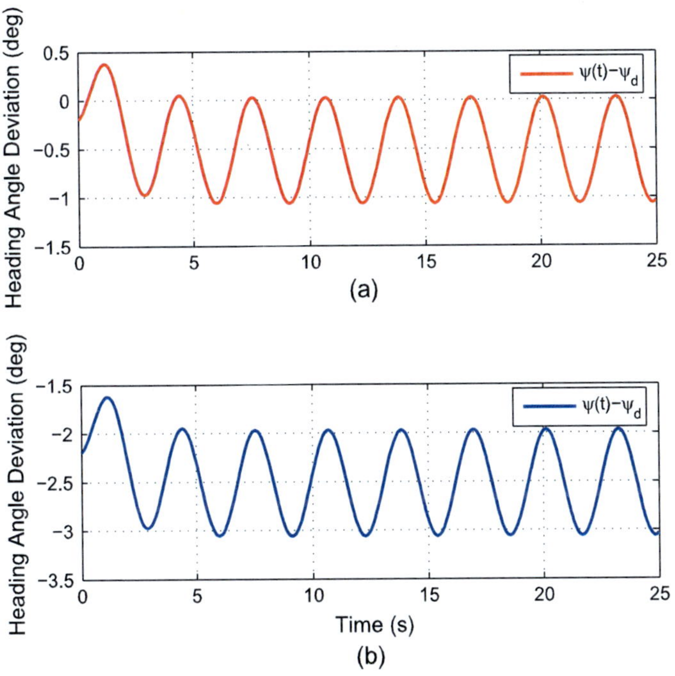

Fig. 5.4 Curves of heading angle deviation for system (5.2) without control

state data such as yaw velocity and heading angle are transmitted to the control station. Then the control station constructs and transmits control inputs to the steering machine in the UMV. The communication between the UMV and the control station is completed through communication networks. Thus, a network-based control system results from the UMV, the remote land-based/mother ship-based control station, and communication networks.

5.2 Network-Based Modeling

We first consider the network-based modeling for the UMV without dropping control input packets purposely. Throughout this chapter, we consider the case where the UMV is controlled by a land-based control station.

The roll angle $\phi(t)$ and heading angle $\psi(t)$ are measured by gyros and their derivatives; $p(t)$ and $r(t)$ are measured by rate gyros; the sway velocity $v(t)$ is obtained by a state estimator. The sampled state $x(t)$ with $x(t) = [v(t),\ r(t),\ \psi(t),\ p(t),\ \phi(t)]^T$ is transmitted to the land-based control station. The control station is used to construct and transmit the control signal (input) to the UMV, and the communication between the UMV and the control station is completed through communication networks. The actuator is chosen as a zero order holder which is connected to the steering machine. In this chapter, we assume that communication networks are reliable and there do not exist network-induced packet dropouts; the gyros and state estimator are time-driven, while the controller and actuator are event-driven.

The control objective of heading control for the UMV is to make $\psi(t) \rightarrow \psi_d$, where ψ_d is the desired heading angle, which is assumed to be constant. Then, the desired state of the UMV is $x_d = [0,\ 0,\ \psi_d,\ 0,\ 0]^T$. To achieve the control objective of heading control, one can define the following state error,

$$e(t) = x(t) - x_d = [v(t),\ r(t),\ \psi(t) - \psi_d,\ p(t),\ \phi(t)]^T. \tag{5.5}$$

From (5.2) and (5.5), one has

$$\dot{e}(t) = Ae(t) + B_1\delta(t) + B_2\omega(t) + Ax_d. \tag{5.6}$$

Since the elements in the third column of the matrix A are all equal to zero, Ax_d is a zero vector. Then (5.6) is converted to

$$\dot{e}(t) = Ae(t) + B_1\delta(t) + B_2\omega(t).$$

For the UMV without considering packet dropouts, suppose that $t_k, t_{k+1}, \ldots$ $(k = 0, 1, 2, \ldots)$ denote the instants that the gyros samples data from the UMV, and $t_{k+1} - t_k = h$ with h denoting the length of the sampling period. Then, for $t \in [t_k, t_{k+1})$, the control input used by the UMV with single rudder is given by

$$\delta(t) = K(x(t_k) - x_d) = Ke(t_k), k = 0, 1, 2, \ldots, \tag{5.7}$$

where K is a state feedback controller gain to be designed.

Remark 5.1 It should be mentioned that $\delta(t)$ in (5.7) is a piecewise constant function. Note that the transformation of $\delta(t)$ from $Ke(t_k)$ to $Ke(t_{k+1})$ is not instantaneous, and a rudder angle transformation delay is inevitable. The length of the rudder angle transformation delay is determined by the maximum rudder speed and the transformation value of the rudder angle. For a specific maximum rudder speed, if the transformation value of the rudder angle is small, the rudder angle transformation delay is negligible.

Define $\tau(t) = t - t_k$. Then one has $\tau(t) \in [0,\ h)$. The closed-loop system is described as

$$\dot{e}(t) = Ae(t) + B_1Ke(t - \tau(t)) + B_2\omega(t), \tag{5.8}$$

where $t \in [t_k,\ t_{k+1})$.

For the system (5.8), the initial condition of the state $e(t)$ on $[t_0 - h,\ t_0]$ is supplemented as

$$e(\varsigma) = \vartheta(\varsigma), \ \varsigma \in [t_0 - h, \ t_0], \tag{5.9}$$

where $\vartheta \in \mathbb{W}$, and $\mathbb{W}$ denotes the Banach space of absolutely continuous functions $[-h, 0] \to \mathbb{R}^n$ with square-integrable derivative and with the norm

$$\|\vartheta\|_W^2 = \|\vartheta(0)\|^2 + \int_{-h}^{0} \|\vartheta(\varsigma)\|^2 d\varsigma + \int_{-h}^{0} \|\dot{\vartheta}(\varsigma)\|^2 d\varsigma,$$

where the vector norm $\|\cdot\|$ represents the Euclidean norm.

We now consider the network-based modeling for the UMV by taking intentional packet dropouts into account.

If some control input packets are dropped purposely, suppose that the control inputs which are based on the states of the UMV at the instants $i_k h, \ i_{k+1}h, \ \ldots$ $(k = 0, \ 1, \ 2, \ \ldots)$ are adopted by the steering machine, while the control inputs based on the states of the UMV between the instants $i_k h$ and $i_{k+1}h$ are dropped, where h is the length of the sampling period. Then, for $t \in [i_k h, \ i_{k+1}h)$, the control input used by the UMV in network environments and with single rudder is described as

$$\delta(t) = Le(i_k h), \ k = 0, \ 1, \ 2, \ \ldots, \tag{5.10}$$

where L is a state feedback controller gain. Define ρ as the upper bound of the number of consecutive packet dropouts, and $d(t) = t - i_k h$. It is seen clearly that $d(t) \in [0, \ \eta)$, where $\eta = (\rho + 1)h$. Then, one has

$$\dot{e}(t) = Ae(t) + B_1 Le(t - d(t)) + B_2 \omega(t), \tag{5.11}$$

where $t \in [i_k h, \ i_{k+1}h)$.

The initial condition of $e(t)$ in the network-based system (5.11) is similar to the initial condition presented in (5.9), here it is omitted for brevity.

Remark 5.2 For the control input described by (5.10) and $t \in [i_k h, \ i_{k+1}h)$, the number of dropped control input packets is $i_{k+1} - i_k - 1$. Thus, $i_{k+1} - i_k = 1$ implies that there do not exist packet dropouts. To achieve accurate heading control and rudder oscillation reduction for the UMV in network environments, one should select $i_{k+1} - i_k - 1$ according to the following selection criterion: the selection of $i_{k+1} - i_k - 1$ should ensure that the control input in (5.10) can counteract the wave-induced disturbances.

In order to obtain the accurate heading control and the desired roll damping which are described by $\psi(t) = \psi_d$ and $p_d = \phi_d = 0$, respectively, one can define $y(t) = [\psi(t), \ p(t), \ \phi(t)]^T$, $y_d = [\psi_d, \ 0, \ 0]^T$, where p_d and ϕ_d denote the desired roll velocity and roll angle, respectively. Define $z(t) = y(t) - y_d$, and $C = \begin{bmatrix} 0 & 0 & 1 & 0 & 0 \\ 0 & 0 & 0 & 1 & 0 \\ 0 & 0 & 0 & 0 & 1 \end{bmatrix}$. Then one has $z(t) = Ce(t)$. The requirements for accurate steering and rudder oscillation reduction (which is achieved by accurate steering and roll damping) can be expressed as follows.

(i) The network-based closed-loop system (5.11) with $\omega(t) = 0$ is asymptotically stable;

(ii) The effect of $\omega(t)$ on $z(t)$ is attenuated at a desired level in the H_∞ sense. More specifically, it is required that

$$||z(t)||_2 < \gamma||\omega(t)||_2, \tag{5.12}$$

for all nonzero $\omega(t) \in L_2[t_0, \infty)$ at zero initial condition, where $\gamma > 0$.

Note that $d(t) \in [0, \eta)$. By introducing a scalar $\bar{d} = \eta/2$, it is seen that at any instant t, $d(t) \in [0, \bar{d})$ or $d(t) \in [\bar{d}, \eta)$, where $t \in [i_k h, i_{k+1} h)$. On the other hand, for the specific instant t, $d(t) \in [0, \bar{d})$ and $d(t) \in [\bar{d}, \eta)$ can not occur simultaneously, which phenomenon is named as mutually exclusive distribution in this chapter. To make full use of the mutually exclusive distribution characteristic of the interval time-varying delay $d(t)$, one can define a scalar $\lambda(t)$ as

$$\lambda(t) = \begin{cases} 1, & d(t) \in [0, \bar{d}), \\ 0, & d(t) \in [\bar{d}, \eta). \end{cases} \tag{5.13}$$

The mutually exclusive distribution characteristic of $d(t)$ is taken into account to deal with controller design.

5.3 Controller Design

In this section, we present a new controller design scheme for the network-based closed-loop system (5.11) by considering intentional packet dropouts. In doing so, we construct the following Lyapunov-Krasovskii functional

$$V(t, e_t) = V_1(t, e_t) + V_2(t, e_t) + V_3(t, e_t), \tag{5.14}$$

where

$$\begin{aligned}
V_1(t, e_t) &= e^T(t) P e(t), \\
V_2(t, e_t) &= \int_{t-\bar{d}}^{t} e^T(s) Q_1 e(s) ds + \int_{t-\eta}^{t-\bar{d}} e^T(s) Q_2 e(s) ds, \\
V_3(t, e_t) &= \int_{-\bar{d}}^{0} \int_{t+s}^{t} \dot{e}^T(\theta) R_1 \dot{e}(\theta) d\theta ds + \int_{-\eta}^{-\bar{d}} \int_{t+s}^{t} \dot{e}^T(\theta) R_2 \dot{e}(\theta) d\theta ds,
\end{aligned}$$

and $e_t = e(t+\varsigma)$, $\varsigma \in [t_0 - \eta,\ t_0]$, while P, Q_1, Q_2, R_1, and R_2 are symmetric positive definite matrices. We now state and establish the following result for the network-based system (5.11).

Theorem 5.1 *For given positive scalars h, ρ, and γ, the network-based system (5.11) is asymptotically stable with an H_∞ norm bound γ, if there exist symmetric positive definite matrices W, $\widetilde{Q}_1$, $\widetilde{Q}_2$, $\widetilde{R}_1$, $\widetilde{R}_2$, and a matrix N such that (5.15) holds for every feasible value of $\lambda(t)$ with $\lambda(t) = 1$ or $\lambda(t) = 0$*

$$\begin{bmatrix} \widetilde{\Pi}_{11} & \widetilde{\Pi}_{12} \\ * & \widetilde{\Pi}_{22} \end{bmatrix} < 0, \tag{5.15}$$

where

$$\widetilde{\Pi}_{11} = \begin{bmatrix} \widetilde{\Omega}_{11} & \widetilde{\Omega}_{12} & \widetilde{\Omega}_{13} & 0 & B_2 \\ * & \widetilde{\Omega}_{22} & \widetilde{\Omega}_{23} & \widetilde{\Omega}_{24} & 0 \\ * & * & \widetilde{\Omega}_{33} & \widetilde{\Omega}_{34} & 0 \\ * & * & * & \widetilde{\Omega}_{44} & 0 \\ * & * & * & * & -\gamma I \end{bmatrix},$$

$$\widetilde{\Omega}_{11} = AW + WA^T + \widetilde{Q}_1 - \frac{1}{\bar{d}}\widetilde{R}_1,\ \widetilde{\Omega}_{12} = B_1 N^T + \frac{\lambda(t)}{\bar{d}}\widetilde{R}_1,\ \widetilde{\Omega}_{13} = \frac{1-\lambda(t)}{\bar{d}}\widetilde{R}_1,$$

$$\widetilde{\Omega}_{22} = -\frac{2\lambda(t)}{\bar{d}}\widetilde{R}_1 - \frac{2(1-\lambda(t))}{\eta - \bar{d}}\widetilde{R}_2,\ \widetilde{\Omega}_{23} = \frac{\lambda(t)}{\bar{d}}\widetilde{R}_1 + \frac{1-\lambda(t)}{\eta - \bar{d}}\widetilde{R}_2,$$

$$\widetilde{\Omega}_{24} = \frac{1-\lambda(t)}{\eta - \bar{d}}\widetilde{R}_2,\ \widetilde{\Omega}_{33} = \widetilde{Q}_2 - \widetilde{Q}_1 - \frac{1}{\bar{d}}\widetilde{R}_1 - \frac{1}{\eta - \bar{d}}\widetilde{R}_2,$$

$$\widetilde{\Omega}_{34} = \frac{\lambda(t)}{\eta - \bar{d}}\widetilde{R}_2,\ \widetilde{\Omega}_{44} = -\widetilde{Q}_2 - \frac{1}{\eta - \bar{d}}\widetilde{R}_2,$$

$$\widetilde{\Pi}_{12} = \begin{bmatrix} WA^T & WA^T & WC^T \\ NB_1^T & NB_1^T & 0 \\ 0 & 0 & 0 \\ 0 & 0 & 0 \\ B_2^T & B_2^T & 0 \end{bmatrix},\ \widetilde{\Pi}_{22} = diag\{\Psi_1,\ \Psi_2,\ -\gamma I\},$$

with $\Psi_1 = \bar{d}^{-1}(\widetilde{R}_1 - 2W)$, $\Psi_2 = (\eta - \bar{d})^{-1}(\widetilde{R}_2 - 2W)$.

Moreover, the gain of the controller (5.10) is given by $L = N^T W^{-1}$.

Proof Taking the time derivative of the Lyapunov-Krasovskii functional $V(t, e_t)$ given in (5.14) along the trajectory of the network-based system (5.11), one has

$$\dot{V}_1(t,e_t) = 2e^T(t)P\dot{e}(t), \tag{5.16}$$

$$\dot{V}_2(t,e_t) = e^T(t)Q_1e(t) + e^T(t-\bar{d})(Q_2-Q_1)e(t-\bar{d}) - e^T(t-\eta)Q_2e(t-\eta), \tag{5.17}$$

$$\dot{V}_3(t,e_t) = \dot{e}^T(t)\Theta_1\dot{e}(t) - \int_{t-\bar{d}}^{t}\dot{e}^T(\theta)R_1\dot{e}(\theta)d\theta - \int_{t-\eta}^{t-\bar{d}}\dot{e}^T(\theta)R_2\dot{e}(\theta)d\theta, \tag{5.18}$$

where $\Theta_1 = \bar{d}R_1 + (\eta-\bar{d})R_2$. By using the Jensen integral inequalities [8] and the mutually exclusive distribution characteristic of the interval time-varying delay $d(t)$, one has

$$\begin{aligned}
&-\int_{t-\bar{d}}^{t}\dot{e}^T(\theta)R_1\dot{e}(\theta)d\theta - \int_{t-\eta}^{t-\bar{d}}\dot{e}^T(\theta)R_2\dot{e}(\theta)d\theta\\
=&\lambda(t)[-\int_{t-d(t)}^{t}\dot{e}^T(\theta)R_1\dot{e}(\theta)d\theta - \int_{t-\bar{d}}^{t-d(t)}\dot{e}^T(\theta)R_1\dot{e}(\theta)d\theta\\
&-\int_{t-\eta}^{t-\bar{d}}\dot{e}^T(\theta)R_2\dot{e}(\theta)d\theta] + (1-\lambda(t))[-\int_{t-\bar{d}}^{t}\dot{e}^T(\theta)R_1\dot{e}(\theta)d\theta\\
&-\int_{t-d(t)}^{t-\bar{d}}\dot{e}^T(\theta)R_2\dot{e}(\theta)d\theta - \int_{t-\eta}^{t-d(t)}\dot{e}^T(\theta)R_2\dot{e}(\theta)d\theta]\\
\leq&\lambda(t)(-\frac{1}{\bar{d}}\varphi_1^TR_1\varphi_1 - \frac{1}{\bar{d}}\varphi_2^TR_1\varphi_2 - \frac{1}{\eta-\bar{d}}\varphi_3^TR_2\varphi_3)\\
&+(1-\lambda(t))(-\frac{1}{\bar{d}}\varphi_4^TR_1\varphi_4 - \frac{1}{\eta-\bar{d}}\varphi_5^TR_2\varphi_5 - \frac{1}{\eta-\bar{d}}\varphi_6^TR_2\varphi_6),
\end{aligned} \tag{5.19}$$

where $\varphi_1 = e(t) - e(t-d(t)), \varphi_2 = e(t-d(t)) - e(t-\bar{d}), \varphi_3 = e(t-\bar{d}) - e(t-\eta)$, $\varphi_4 = e(t) - e(t-\bar{d}), \varphi_5 = e(t-\bar{d}) - e(t-d(t)), \varphi_6 = e(t-d(t)) - e(t-\eta)$. Combining (5.16)–(5.19) and the network-based system (5.11) together, one has

$$\dot{V}(t,e_t) + \gamma^{-1}z^T(t)z(t) - \gamma\omega^T(t)\omega(t) \leq \xi^T(t)(\Omega + \Xi)\xi(t), \tag{5.20}$$

where $\xi(t) = [e^T(t)\ \ e^T(t-d(t))\ \ e^T(t-\bar{d})\ \ e^T(t-\eta)\ \ \omega^T(t)]^T$, and

$$\Omega = \begin{bmatrix} \Omega_{11} & \Omega_{12} & \Omega_{13} & 0 & PB_2\\ * & \Omega_{22} & \Omega_{23} & \Omega_{24} & 0\\ * & * & \Omega_{33} & \Omega_{34} & 0\\ * & * & * & \Omega_{44} & 0\\ * & * & * & * & -\gamma I \end{bmatrix},$$

$\Omega_{11} = PA + A^TP + Q_1 - \frac{1}{\bar{d}}R_1$, $\Omega_{12} = PB_1L + \frac{\lambda(t)}{\bar{d}}R_1$, $\Omega_{13} = \frac{1-\lambda(t)}{\bar{d}}R_1$,

$\Omega_{22} = -\frac{2\lambda(t)}{\bar{d}}R_1 - \frac{2(1-\lambda(t))}{\eta-\bar{d}}R_2$, $\Omega_{23} = \frac{\lambda(t)}{\bar{d}}R_1 + \frac{1-\lambda(t)}{\eta-\bar{d}}R_2$,

$$\Omega_{24} = \frac{1-\lambda(t)}{\eta-\bar{d}}R_2,\ \Omega_{33} = Q_2 - Q_1 - \frac{1}{\bar{d}}R_1 - \frac{1}{\eta-\bar{d}}R_2,$$
$$\Omega_{34} = \frac{\lambda(t)}{\eta-\bar{d}}R_2,\ \Omega_{44} = -Q_2 - \frac{1}{\eta-\bar{d}}R_2,$$
$$\Xi = \Lambda_1^T\Theta_1\Lambda_1 + \gamma^{-1}\Lambda_2^T\Lambda_2,\ \Lambda_1 = [A\ \ B_1L\ \ 0\ \ 0\ \ B_2],\ \Lambda_2 = [C\ \ 0\ \ 0\ \ 0\ \ 0].$$

For symmetric positive definite matrices P and R_i $(i = 1, 2)$, the following inequalities hold

$$-R_i^{-1} \le P^{-1}R_iP^{-1} - 2P^{-1}. \tag{5.21}$$

By using Schur complement, the inequalities $-R_i^{-1} \le P^{-1}R_iP^{-1} - 2P^{-1}$, and letting $P^{-1} = W$, $P^{-1}Q_iP^{-1} = \widetilde{Q}_i$ $(i = 1,\ 2)$, $P^{-1}R_iP^{-1} = \widetilde{R}_i$, $P^{-1}L^T = N$, one can see that if (5.15) is satisfied, $\Omega + \Xi < 0$ is also satisfied. Then, for $t \in [i_kh,\ i_{k+1}h)$, if (5.15) is satisfied, one has

$$\dot{V}(t, e_t) + \gamma^{-1}z^T(t)z(t) - \gamma\omega^T(t)\omega(t) < 0. \tag{5.22}$$

Based on the definition of the H_∞ performance, it is easy to prove that if $\omega(t) = 0$, (5.15) can guarantee the asymptotic stability of the system (5.11). If $\omega(t) \neq 0$, one has $||z(t)||_2 < \gamma||\omega(t)||_2$. Then, if (5.15) is satisfied, the network-based system (5.11) is asymptotically stable with an H_∞ norm bound γ. This completes the proof. □

Note that ρ, which denotes the upper bound of the number of consecutive packet dropouts, is employed in the controller design scheme presented in Theorem 5.1. On the other hand, $i_{k+1} - i_k - 1$, which denotes the number of dropped control input packets for $t \in [i_kh,\ i_{k+1}h)$, is implied by the control input presented in (5.10). Combined with the selection criterion of $i_{k+1} - i_k - 1$ in Remark 5.2, the procedure for selecting $i_{k+1} - i_k - 1$ is described as Algorithm 5.1.

Algorithm 5.1 Select $i_{k+1} - i_k - 1$.

Step 1. For the UMV in network environments and given scalars h and γ, solve the controller design criterion presented in Theorem 5.1 to obtain the maximum admissible value of ρ and the controller gain. Let $j = 1$.

Step 2. Set $i_{k+1} - i_k - 1 = j$. Calculate the oscillation ranges of the yaw velocity $r(t)$, the heading angle $\psi(t)$, the rudder angle $\delta(t)$, and the roll angle $\phi(t)$. If the oscillation ranges of the yaw velocity $r(t)$, the heading angle $\psi(t)$, the rudder angle $\delta(t)$, and the roll angle $\phi(t)$ are smaller than the corresponding oscillation ranges of the UMV without dropping control input packets purposely, go to Step 4; otherwise, go to Step 3.

Step 3. Set $j = j + 1$. If $j \le \rho$, go to Step 2; otherwise, stop.

Step 4. Output the number of purposely dropped control input packets, that is $i_{k+1} - i_k - 1$, and stop.

It should be mentioned that different $i_{k+1} - i_k - 1$ leads to different oscillation ranges of the yaw velocity $r(t)$, the heading angle $\psi(t)$, the rudder angle $\delta(t)$, and the roll angle $\phi(t)$. Then, the procedure for selecting $i_{k+1} - i_k - 1$ presented in Algorithm 5.1 is locally optimal.

One can see clearly from Sect. 5.2 that if control input packets are not dropped purposely, the network-based closed-loop system (5.8) is established. When dealing with controller design for the network-based system (5.8), one can take the mutually exclusive distribution characteristic of $\tau(t)$ into account. For this purpose, define $\bar{\tau} = h/2$, and

$$\rho(t) = \begin{cases} 1, & \tau(t) \in [0, \ \bar{\tau}), \\ 0, & \tau(t) \in [\bar{\tau}, \ h). \end{cases} \tag{5.23}$$

Similar to the proof of Theorem 5.1, we state and establish the following result for the network-based system (5.8).

Theorem 5.2 *For given positive scalars h and γ, the network-based system (5.8) is asymptotically stable with an H_∞ norm bound γ, if there exist symmetric positive definite matrices W, $\widetilde{Q}_1$, $\widetilde{Q}_2$, $\widetilde{R}_1$, $\widetilde{R}_2$, and a matrix N of appropriate dimensions such that (5.24) holds for every feasible value of $\rho(t)$ with $\rho(t) = 1$ or $\rho(t) = 0$*

$$\begin{bmatrix} \hat{\Pi}_{11} & \widetilde{\Pi}_{12} \\ * & \hat{\Pi}_{22} \end{bmatrix} < 0, \tag{5.24}$$

where $\hat{\Pi}_{11}$ is derived from $\widetilde{\Pi}_{11}$ in (5.15) by replacing $\lambda(t)$, $\bar{d}$, and η with $\rho(t)$, $\bar{\tau}$, and h, respectively; $\widetilde{\Pi}_{12}$ is the same as the $\widetilde{\Pi}_{12}$ in (5.15); $\hat{\Pi}_{22}$ is derived from $\widetilde{\Pi}_{22}$ in (5.15) by replacing $\bar{d}$ and η with $\bar{\tau}$ and h, respectively.

Moreover, the gain of the controller (5.7) is given by $K = N^T W^{-1}$.

In this chapter, the communication networks are assumed to be reliable, and network-induced packet dropouts, which are denoted as d_k here, do not occur. If there exist network-induced packet dropouts d_k, the procedure for selecting the number of purposely dropped control input packets is described as Algorithm 5.2.

Algorithm 5.2 Select the number of purposely dropped control input packets.

Step 1. Run Algorithm 5.1 to obtain the number of purposely dropped control input packets $i_{k+1} - i_k - 1$ for the UMV without network-induced packet dropouts.

Step 2. For $t \in [i_k h, \ i_{k+1} h)$, if the number of network-induced packet dropouts is equal to zero, that is $d_k = 0$, output $i_{k+1} - i_k - 1$ derived in Step 1 of Algorithm 5.2, and stop; otherwise, go to Step 3.

Step 3. If $d_k > 0$ and $d_k < i_{k+1} - i_k - 1$, choose the number of purposely dropped control input packets as $i_{k+1} - i_k - 1 - d_k$, where $i_{k+1} - i_k - 1$ is derived in Step 1 of Algorithm 5.2, and stop; otherwise, go to Step 4.

Step 4. If $d_k \geq i_{k+1} - i_k - 1$, choose the number of purposely dropped control input packets as zero, and stop.

Remark 5.3 It should be mentioned that if there exist both network-induced packet transmission delays, which are denoted as τ_k here, and network-induced packet dropouts d_k, we can combine τ_k and d_k as one item and utilize an algorithm similar to Algorithm 5.2 to select the number of purposely dropped control input packets.

Remark 5.4 Employing the Lyapunov-Krasovskii functional approach, one can only derive sufficient conditions, which are presented in Theorems 5.1 and 5.2, for asymptotic stability with an H_∞ norm bound γ of the systems (5.11) and (5.8), respectively. If these sufficient conditions are satisfied, one can conclude that the systems (5.11) and (5.8) are asymptotically stable with an H_∞ norm bound γ. Otherwise, no conclusion can be made. Note that the mutually exclusive distribution characteristic of $d(t)$ and $\tau(t)$ is made full use in the proof of Theorems 5.1 and 5.2, respectively, to derive the controller design criteria. In fact, considering such a mutually exclusive distribution characteristic will lead to less conservative controller design criteria, which is analyzed in Sect. 5.4.

5.4 Conservatism Analysis

In the section, we show that considering the mutually exclusive distribution characteristic of $d(t)$ may reduce the conservatism of controller design criteria. For this purpose, choose $Q_1 = Q_2 = Q$, and $R_1 = R_2 = R$ in (5.14). Then the Lyapunov-Krasovskii functional in (5.14) becomes

$$\widetilde{V}(t, e_t) = V_1(t, e_t) + \widetilde{V}_2(t, e_t) + \widetilde{V}_3(t, e_t), \tag{5.25}$$

where

$$\begin{aligned} V_1(t, e_t) &= e^T(t) P e(t), \\ \widetilde{V}_2(t, e_t) &= \int_{t-\eta}^{t} e^T(s) Q e(s) ds, \\ \widetilde{V}_3(t, e_t) &= \int_{-\eta}^{0} \int_{t+s}^{t} \dot{e}^T(\theta) R \dot{e}(\theta) d\theta ds, \end{aligned}$$

with P, Q, and R denoting symmetric positive definite matrices of appropriate dimensions. If the mutually exclusive distribution characteristic of $d(t)$ is utilized to deal with integral inequalities for products of vectors, we state and establish the following result for the network-based system (5.11).

Corollary 5.1 *For given positive scalars h, ρ, and γ, the system (5.11) is asymptotically stable with an H_∞ norm bound γ, if there exist symmetric positive definite* *matrices W, $\widetilde{Q}$, $\widetilde{R}$, and a matrix N of appropriate dimensions such that for every feasible value of $\lambda(t)$ with $\lambda(t)=1$ or $\lambda(t)=0$, the following inequality holds*

$$\begin{bmatrix} \bar{\Pi}_{11} & \bar{\Pi}_{12} \\ * & \bar{\Pi}_{22} \end{bmatrix} < 0, \tag{5.26}$$

where

$$\bar{\Pi}_{11} = \begin{bmatrix} \bar{\Omega}_{11} & \bar{\Omega}_{12} & 0 & B_2 \\ * & \bar{\Omega}_{22} & \bar{\Omega}_{23} & 0 \\ * & * & \bar{\Omega}_{33} & 0 \\ * & * & * & -\gamma I \end{bmatrix}, \ \bar{\Pi}_{12} = \begin{bmatrix} WA^T & WC^T \\ NB_1^T & 0 \\ 0 & 0 \\ B_2^T & 0 \end{bmatrix},$$

$$\begin{aligned}
\bar{\Pi}_{22} =& diag\{\eta^{-1}(\widetilde{R}-2W), \ -\gamma I\}, \\
\bar{\Omega}_{11} =& AW + WA^T + \widetilde{Q} - (\frac{\lambda(t)}{\bar{d}} + \frac{1-\lambda(t)}{\eta})\widetilde{R}, \ \bar{\Omega}_{12} = B_1 N^T + (\frac{\lambda(t)}{\bar{d}} + \frac{1-\lambda(t)}{\eta})\widetilde{R}, \\
\bar{\Omega}_{22} =& -(\frac{\lambda(t)}{\bar{d}} + \frac{1}{\eta} + \frac{1-\lambda(t)}{\eta-\bar{d}})\widetilde{R}, \ \bar{\Omega}_{23} = (\frac{\lambda(t)}{\eta} + \frac{1-\lambda(t)}{\eta-\bar{d}})\widetilde{R}, \\
\bar{\Omega}_{33} =& -\widetilde{Q} - (\frac{\lambda(t)}{\eta} + \frac{1-\lambda(t)}{\eta-\bar{d}})\widetilde{R}.
\end{aligned}$$

Moreover, the gain of the controller (5.10) is given by $L = N^T W^{-1}$.

Proof By considering the mutually exclusive distribution characteristic of $d(t)$, one has

$$\begin{aligned}
\tilde{\dot{V}}_3(t, e_t) =& \eta \dot{e}^T(t) R \dot{e}(t) - \int_{t-\eta}^{t} \dot{e}^T(\theta) R \dot{e}(\theta) d\theta \\
\leq& \eta \dot{e}^T(t) R \dot{e}(t) - (\frac{\lambda(t)}{\bar{d}} + \frac{1-\lambda(t)}{\eta})\varphi_1^T R \varphi_1 \\
& - (\frac{\lambda(t)}{\eta} + \frac{1-\lambda(t)}{\eta-\bar{d}})\varphi_6^T R \varphi_6,
\end{aligned} \tag{5.27}$$

where φ_1 and φ_6 are the same as the corresponding items in (5.19). The rest of the proof is similar to the one in Theorem 5.1 and it is omitted. This completes the proof. □

If the mutually exclusive distribution characteristic of $d(t)$ is not taken into account, the controller design criterion in Corollary 5.2 is followed immediately.

Corollary 5.2 *For given positive scalars h, ρ, and γ, the system (5.11) is asymptotically stable with an H_∞ norm bound γ, if there exist symmetric positive definite matrices W, $\widetilde{Q}$, $\widetilde{R}$, and a matrix N of appropriate dimensions such that the following inequality holds*

$$\begin{bmatrix} \Phi & \bar{\Pi}_{12} \\ * & \bar{\Pi}_{22} \end{bmatrix} < 0, \tag{5.28}$$

where

$$\Phi = \begin{bmatrix} \Phi_{11} & \Phi_{12} & 0 & B_2 \\ * & \Phi_{22} & \Phi_{23} & 0 \\ * & * & \Phi_{33} & 0 \\ * & * & * & -\gamma I \end{bmatrix},$$

$$\Phi_{11} = AW + WA^T + \widetilde{Q} - \frac{1}{\eta}\widetilde{R},\ \Phi_{12} = B_1 N^T + \frac{1}{\eta}\widetilde{R},$$

$$\Phi_{22} = -\frac{2}{\eta}\widetilde{R},\ \Phi_{23} = \frac{1}{\eta}\widetilde{R},\ \Phi_{33} = -\widetilde{Q} - \frac{1}{\eta}\widetilde{R},$$

while $\bar{\Pi}_{12}$ and $\bar{\Pi}_{22}$ are the same as the corresponding items in (5.26). Moreover, the gain of the controller (5.10) is given by $L = N^T W^{-1}$.

Proof If the mutually exclusive distribution characteristic of $d(t)$ is not taken into account, one has

$$\begin{aligned} \tilde{\dot{V}}_3(t, e_t) =& \eta \dot{e}^T(t) R \dot{e}(t) - \int_{t-\eta}^{t} \dot{e}^T(\theta) R \dot{e}(\theta) d\theta \\ \le& \eta \dot{e}^T(t) R \dot{e}(t) - \frac{1}{\eta}\varphi_1^T R \varphi_1 - \frac{1}{\eta}\varphi_6^T R \varphi_6, \end{aligned} \tag{5.29}$$

where φ_1 and φ_6 are the same as the corresponding items in (5.19). The rest of the proof is omitted here for brevity. This completes the proof. □

The following theorem establishes the relationship between the controller design criteria in Corollaries 5.1 and 5.2.

Theorem 5.3 *For the system (5.11), if the inequality (5.28) of Corollary 5.2 is satisfied, then the inequality (5.26) of Corollary 5.1 is also satisfied.*

Proof Note that $-\frac{1}{\eta}\varphi_1^T R \varphi_1 - \frac{1}{\eta}\varphi_6^T R \varphi_6$ in (5.29) can be rewritten as

$$-\frac{1}{\eta}\varphi_1^T R \varphi_1 - \frac{1}{\eta}\varphi_6^T R \varphi_6 = \Gamma + \Delta_1 + \Delta_2,$$

where Γ denotes the right side of the inequality (5.27), $\Delta_1 = \lambda(t)(\frac{1}{d} - \frac{1}{\eta})\varphi_1^T R \varphi_1$, $\Delta_2 = (1 - \lambda(t))(\frac{1}{\eta - d} - \frac{1}{\eta})\varphi_6^T R \varphi_6$.

Noting that $\Delta_1 \ge 0$ and $\Delta_2 \ge 0$, one has that $\Delta_1 + \Delta_2 \ge 0$. From the proof of Theorem 5.1, Corollaries 5.1, and 5.2, one can see that if the inequality (5.28) of Corollary 5.2 is satisfied, then the inequality (5.26) of Corollary 5.1 is also satisfied. This completes the proof. □

Remark 5.5 From Theorem 5.3, one can see that considering the mutually exclusive distribution characteristic of the interval time-varying delay $d(t)$ may reduce the conservatism of controller design criteria. If such a characteristic of $\tau(t)$ is taken into account, a similar conclusion can be drawn. It should be pointed out that based on the Lyapunov-Krasovskii functional (5.14), it is difficult to analyze theoretically the merits of considering the mutually exclusive distribution characteristic of $d(t)$. Instead of the Lyapunov-Krasovskii functional (5.14), we use a simplified Lyapunov-Krasovskii functional (5.25) to study the merits of considering the mutually exclusive distribution characteristic of $d(t)$. Note that the constraints $Q_1 = Q_2 = Q$, and $R_1 = R_2 = R$ are imposed on the Lyapunov-Krasovskii functional (5.25). Thus, one can apply the controller design criteria in Sect. 5.3 based on the Lyapunov-Krasovskii functional (5.14) to achieve better results.

5.5 Performance Analysis and Discussion

We show the merits and effectiveness of the proposed intentional packet dropouts-based controller design for the UMV.

A. Merits and effectiveness of the proposed controller design

In this part, we first show the effectiveness of the proposed intentional packet dropouts-based controller design. Then, the performance of the UMV adopting the intentional packet dropouts-based control inputs is compared with the performance of the UMV without considering intentional packet dropouts. The UMV presented in Fig. 5.1 is investigated. For the system matrices A, B_1, and B_2 in (5.2), choose the parameters the same as the corresponding items in (5.3). Then one has

$$A = \begin{bmatrix} -0.9750 & 0 & 0 & 0 & 0 \\ -0.5980 & -1.3000 & 0 & 0 & 0 \\ 0 & 1.0000 & 0 & 0 & 0 \\ 0.6501 & 0 & 0 & -1.1209 & -0.3969 \\ 0 & 0 & 0 & 1.0000 & 0 \end{bmatrix},$$

$$B_1 = \begin{bmatrix} 0.0760 \\ -6.1114 \\ 0 \\ -0.0338 \\ 0 \end{bmatrix}, \quad B_2 = \begin{bmatrix} 0 & 0 \\ 1.3000 & 0 \\ 0 & 0 \\ 0 & 0.3969 \\ 0 & 0 \end{bmatrix}.$$

For Theorems 5.1 and 5.2 in this chapter, suppose that the length of the sampling period $h = 0.6s$. Combining Theorem 5.1 and Algorithm 5.1 together, one can derive the admissible upper bound of the number of consecutive packet dropouts ρ with $\rho = 4$.

Solving the matrix inequalities in Theorems 5.1 and 5.2, one can derive the controller gains $L = \begin{bmatrix} -0.0036 \ 0.0058 \ 0.0077 \ 0.0000 \ 0.0000 \end{bmatrix}$ and $K = \begin{bmatrix} -0.0654 \ 0.1831 \ 0.2894 \ -0.0010 \ -0.0006 \end{bmatrix}$, respectively.

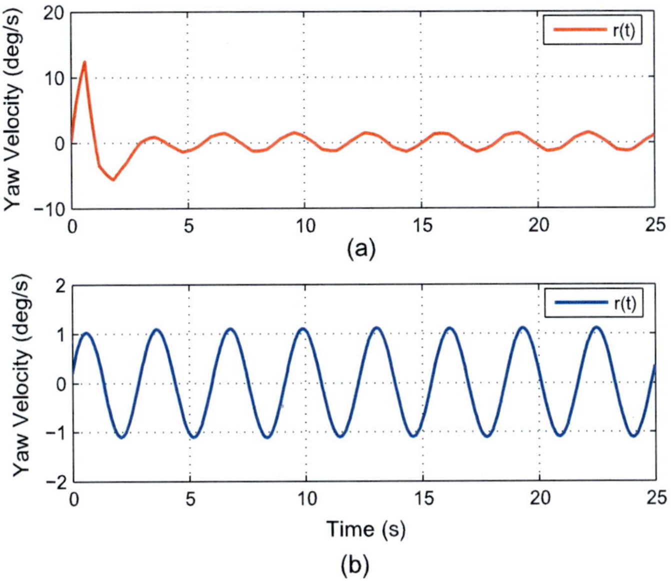

Fig. 5.5 Curves of the yaw velocity for network-based systems (5.8) and (5.11)

Suppose that the initial state of the system (5.2) is $x_0 = [1\ \ 0.2\ \ 15.8\ \ 0.1\ \ 1]^T$, and the desired heading angle $\psi_d = 16$. The external wave-induced disturbances $\omega_\psi(t)$ and $\omega_\phi(t)$ are presented in (5.4).

Applying Algorithm 5.1, one should begin with $i_{k+1} - i_k - 1 = 1$. That is, for $t \in [i_k h,\ i_{k+1} h)$, the number of purposely dropped control input packets for the network-based system (5.11) is chosen to be 1.

Now, we compare the performance of the network-based systems (5.8) and (5.11) to show the effectiveness of the proposed intentional packet dropouts-based method.

The curves of the yaw velocity $r(t)$ for network-based systems (5.8) and (5.11) are plotted in Fig. 5.5a, b, respectively. The curves of the heading angle $\psi(t)$ for systems (5.8) and (5.11) are plotted in Fig. 5.6a, b, respectively. The curves of the rudder angle $\delta(t)$ for systems (5.8) and (5.11) are plotted in Fig. 5.7a, b, respectively. The curves of the roll angle $\phi(t)$ for systems (5.8) and (5.11) are plotted in Fig. 5.8a, b, respectively.

Describe the minimum peak value of $r(t)$, $\psi(t)$, $\delta(t)$, and $\phi(t)$ as $\underline{r}$, $\underline{\psi}$, $\underline{\delta}$, and $\underline{\phi}$, respectively; and describe the maximum peak value of $r(t)$, $\psi(t)$, $\delta(t)$, and $\phi(t)$ as $\bar{r}$, $\bar{\psi}$, $\bar{\delta}$, and $\bar{\phi}$, respectively. Define $\bar{c} - \underline{c}$ as the oscillation range of $c(t)$, where $c(t)$ represents $r(t)$, $\psi(t)$, $\delta(t)$, and $\phi(t)$. Then the peak value and oscillation range of the yaw velocity $r(t)$, the heading angle $\psi(t)$, the rudder angle $\delta(t)$, and the roll angle $\phi(t)$ are given in Tables 5.1, 5.2, 5.3, and 5.4, respectively.

It is clear to see from Figs. 5.5, 5.6, 5.7 and 5.8 and Tables 5.1, 5.2, 5.3 and 5.4 that the oscillation range of the yaw velocity $r(t)$ in Fig. 5.5b is much smaller than the one in Fig. 5.5a; the oscillation range of the heading angle $\psi(t)$ in Fig. 5.6b is

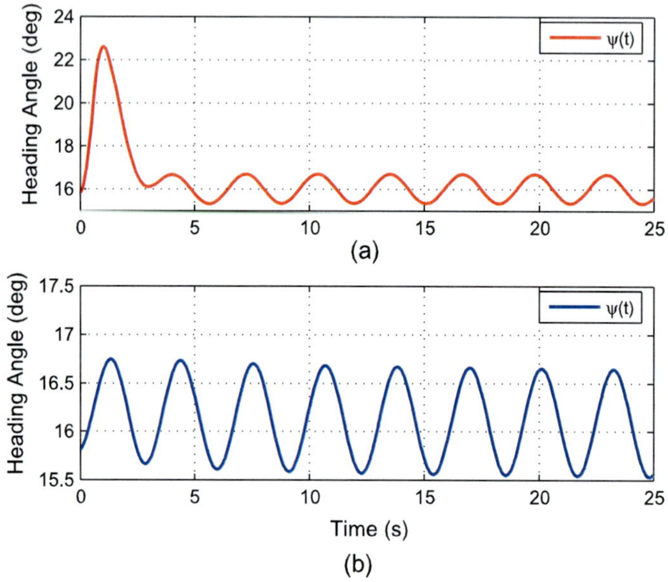

Fig. 5.6 Curves of the heading angle for network-based systems (5.8) and (5.11)

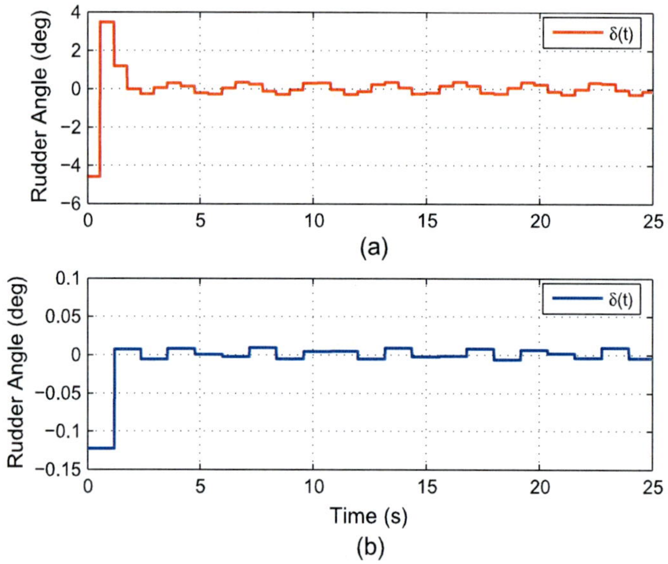

Fig. 5.7 Curves of the rudder angle for network-based systems (5.8) and (5.11)

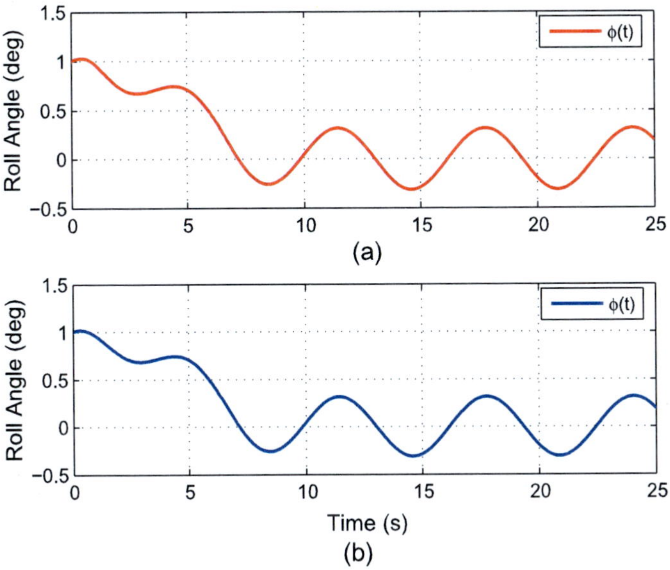

Fig. 5.8 Curves of the roll angle for network-based systems (5.8) and (5.11)

Table 5.1 The peak value and oscillation range of the yaw velocity $r(t)$

	Figure 5.5a	Figure 5.5b
$\underline{r}$	−5.6386	−1.1194
$\bar{r}$	12.4538	1.1074
$\bar{r}$-$\underline{r}$	18.0924	2.2268

Table 5.2 The peak value and oscillation range of the heading angle $\psi(t)$

	Figure 5.6a	Figure 5.6b
$\underline{\psi}$	15.3040	15.5289
$\bar{\psi}$	22.5996	16.7421
$\bar{\psi}$-$\underline{\psi}$	7.2956	1.2132

Table 5.3 The peak value and oscillation range of the rudder angle $\delta(t)$

	Figure 5.7a	Figure 5.7b
$\underline{\delta}$	−4.6304	−0.1232
$\bar{\delta}$	3.4615	0.0086
$\bar{\delta}$-$\underline{\delta}$	8.0919	0.1318

Table 5.4 The peak value and oscillation range of the roll angle $\phi(t)$

	Figure 5.8a	Figure 5.8b
$\underline{\phi}$	−0.3150	−0.3132
$\bar{\phi}$	1.0262	1.0176
$\bar{\phi}$-$\underline{\phi}$	1.3412	1.3308

much smaller than the one in Fig. 5.6a; the rudder angle regulation frequency and the oscillation range of the rudder angle $\delta(t)$ in Fig. 5.7b are much smaller than the rudder angle regulation frequency and the oscillation range of the rudder angle in Fig. 5.7a; the oscillation range of the roll angle $\phi(t)$ in Fig. 5.8b is smaller than the one in Fig. 5.8a. It should be mentioned that the improvement of Fig. 5.8b over Fig. 5.8a is not remarkable, such a phenomenon is induced by the fact that the rudder should bc combined with a fin to reduce the roll angle, which is the subject for future research.

One can also see from Figs. 5.5, 5.6, 5.7 and 5.8 and Tables 5.1, 5.2, 5.3 and 5.4 that the network-based system (5.11) which drops one control input packet purposely (that is, $i_{k+1} - i_k - 1 = 1$) provides better performance than the network-based system (5.8) without dropping control input packets purposely in reducing the heading angle deviation and the oscillation of the rudder angle. Moreover, Algorithm 5.1 for choosing $i_{k+1} - i_k - 1$ is also effective. Note that the transformation value of the rudder angle in Fig. 5.7 is not remarkable. Thus, if the maximum rudder speed is high, the rudder angle transformation delay is negligible.

B. Comparison of the intentional packet dropouts-based and optimal control-based controller design

In what follows, we turn to comparing the performance of the intentional packet dropouts-based networked system (5.11) in this chapter with the system (6.364) in [9]. The heading control performance analysis illustrates the merits of the newly proposed intentional packet dropouts-based controller design. Note that the network-based system (5.11) is derived from (5.2), and the system (5.2) in this chapter and the system (6.364) in [9] are the same. For the system matrices A, B_1, and B_2 in (5.2), choose the parameters as (see also [9])

$$\begin{aligned} U =& 7.8 \text{ (m/s)},\ T_v = 78/U,\ T_r = 13/U,\ K_{dv} = 0.01U, \\ K_{dr} =& -0.0027U,\ K_{dp} = -0.0014U^2,\ K_{vr} = -0.46 \text{ (m/s)}, \\ K_{vp} =& 0.21U,\ \omega_n = 0.63 \text{ (rad/s)},\ \zeta = 0.064 + 0.0038U. \end{aligned} \tag{5.30}$$

Thus, the corresponding system matrices A, B_1, and B_2 for (5.11) in this chapter and (6.364) in [9] are described as

$$A = \begin{bmatrix} -0.1000 & 0 & 0 & 0 & 0 \\ -0.2760 & -0.6000 & 0 & 0 & 0 \\ 0 & 1.0000 & 0 & 0 & 0 \\ 0.6501 & 0 & 0 & -0.1180 & -0.3969 \\ 0 & 0 & 0 & 1.0000 & 0 \end{bmatrix},$$

$$B_1 = b = \begin{bmatrix} 0.0078 \\ -0.0126 \\ 0 \\ -0.0338 \\ 0 \end{bmatrix},\ B_2 = E = \begin{bmatrix} 0 & 0 \\ 0.6000 & 0 \\ 0 & 0 \\ 0 & 0.3969 \\ 0 & 0 \end{bmatrix}.$$

based controller design is effective in reducing the heading angle deviation and the oscillation of the rudder angle.

5.7 Notes

Accurate heading control of UMVs is quite important for marine applications [10]. As one can see in [11, 12], passive packet dropouts are usually considered as the source of system instability and performance degradation. However, for some damping-like systems, purposely introducing a proper time delay may lead to reduced internal oscillation of systems [13–15]. Motivated by this fact, the intentional packet dropouts-based method is proposed in this chapter to reduce the heading angle deviation and the oscillation of the rudder angle of the UMV in network environments. The results in this chapter are based mainly on [16]. For more details about the corresponding analysis and design methods, see also [17–19], etc.

References

1. P.G.M. van der Klugt, *Rudder Roll Stabilization*. Ph.D. thesis, Delft University of Technology, The Netherlands (1987)
2. J. van Amerongen, P.G.M. van der Klugt, H.R. van Nauta Lemke, Rudder roll stabilization for ships. Automatica **26**(4), 679–690 (1990)
3. M. Bibuli, M. Caccia, L. Lapierre, G. Bruzzone, Guidance of unmanned surface vehicles: experiments in vehicle following. IEEE Robot. Autom. Mag. **19**(3), 92–102 (2012)
4. H. Kim, D. Kim, J.-U. Shin, H. Kim, H. Myung, Angular rate-constrained path planning algorithm for unmanned surface vehicles. Ocean Eng. **84**, 37–44 (2014)
5. S.-I. Sohn, J.-H. Oh, Y.-S. Lee, D.-H. Park, I.-K. Oh, Design of a fuel-cell-powered catamaran-type unmanned surface vehicle. IEEE J. Ocean. Eng. **40**(2), 388–396 (2015)
6. Y. Wang, S. Wang, M. Tan, Path generation of autonomous approach to a moving ship for unmanned vehicles. IEEE Trans. Ind. Electron. **62**(9), 5619–5629 (2015)
7. M. Kristan, V.S. Kenk, S. Kovačič, J. Perš, Fast image-based obstacle detection from unmanned surface vehicles. IEEE Trans. Cybern. **46**(3), 641–654 (2016)
8. K. Gu, V.L. Kharitonov, J. Chen, *Stability of Time-Delay Systems* (Birkhäuser, Boston, 2003)
9. T.I. Fossen, *Guidance and Control of Ocean Vehicles* (Wiley, Chichester, U.K., 1994)
10. Z. Li, J. Sun, Disturbance compensating model predictive control with application to ship heading control. IEEE Trans. Control Syst. Technol. **20**(1), 257–265 (2012)
11. X. Ge, F. Yang, Q.-L. Han, Distributed networked control systems: a brief overview. Inf. Sci. **380**, 117–131 (2017)
12. X.-M. Zhang, Q.-L. Han, X. Yu, Survey on recent advances in networked control systems. IEEE Trans. Ind. Inf. **12**(5), 1740–1752 (2016)
13. B.-L. Zhang, Q.-L. Han, Network-based modelling and active control for offshore steel jacket platform with TMD mechanisms. J. Sound Vib. **333**(25), 6796–6814 (2014)
14. B.-L. Zhang, Q.-L. Han, X.-M. Zhang, X. Yu, Sliding mode control with mixed current and delayed states for offshore steel jacket platforms. IEEE Trans. Control Syst. Technol. **22**(5), 1769–1783 (2014)
15. X.-M. Zhang, Q.-L. Han, D. Han, Effects of small time-delays on dynamic output feedback control of offshore steel jacket structures. J. Sound Vib. **330**(16), 3883–3900 (2011)

16. Y.-L. Wang, Q.-L. Han, Network-based heading control and rudder oscillation reduction for unmanned surface vehicles. IEEE Trans. Control Syst. Technol. **25**(5), 1609–1620 (2017)
17. T.I. Fossen, T. Perez, Kalman filtering for positioning and heading control of ships and offshore rigs. IEEE Control Syst. Mag. **29**(6), 32–46 (2009)
18. N.E. Kahveci, P.A. Ioannou, Adaptive steering control for uncertain ship dynamics and stability analysis. Automatica **49**(3), 685–697 (2013)
19. C. Lv, H. Yu, J. Chi, T. Xu, H. Zang, H. Jiang, Z. Zhang, A hybrid coordination controller for speed and heading control of underactuated unmanned surface vehicles system. Ocean Eng. **176**, 222–230 (2019)

Chapter 6
FDF and Controller Coordinated Design for UMVs

This chapter deals with the network-based modeling, and observer-based FDF and controller coordinated design for a UMV in network environments. Network-based models for the UMV subject to actuator faults and wave-induced disturbances are established by introducing an observer-based FDF, and considering network-induced delays and packet dropouts in the sampler-to-control station communication network channel and the control station-to-actuator communication network channel. Based on these models, network-based FDF and controller coordinated design criteria are derived to asymptotically stabilize the residual system. The designed network-based FDF and controller can guarantee the sensitivity of the residual signal to faults and the robustness of the UMV to external disturbances. Fault detection performance analysis verifies the effectiveness of the proposed network-based FDF and controller coordinated design for the networked UMV.

6.1 Network-Based Modeling for a UMV

The motion of a marine vehicle in 6 degrees of freedom includes sway, yaw, roll, surge, heave, and pitch. If the motion of given points on the port side of a marine vehicle and similar points on the starboard side has components in the transverse axis (directed to starboard) direction, it is referred to as an asymmetrical motion, which includes sway, yaw, and roll. If such a motion has no component in the transverse axis direction, it is referred to as a symmetrical motion, which includes surge, heave, and pitch. In this chapter, only the asymmetrical motion is considered. The influences of surge, heave, and pitch is treated as the disturbances.

By writing Newton's laws in a space-fixed coordinate system, one can obtain basic equations for sway, yaw, and roll presented in (5.1). By translating the equations in (5.1) to the coordinate system presented in Fig. 5.1, adopting Taylor expansion and Laplace transformation, and disregarding some hydrodynamic effects, one can

Y.-L. Wang et al., *Network-Based Control of Unmanned Marine Vehicles*,
https://doi.org/10.1007/978-3-031-28605-6_6

achieve the following transfer functions for the motion

$$\begin{cases} v(s) = \dfrac{K_{dv}}{1 + T_v s}\delta(s), \\ \psi(s) = \dfrac{1}{(1 + T_r s)s}[K_{dr}\delta(s) + K_{vr}v(s) + \omega_\psi(s)], \\ \phi(s) = \dfrac{\omega_n^2}{s^2 + 2\zeta\omega_n s + \omega_n^2}[K_{dp}\delta(s) + K_{vp}v(s) + \omega_\phi(s)], \end{cases} \tag{6.1}$$

where $v(s)$, $\psi(s)$, $\phi(s)$, and $\delta(s)$ denote the Laplace transformation of $v(t)$, $\psi(t)$, $\phi(t)$, and $\delta(t)$, respectively; $v(t)$, $\psi(t)$, $\phi(t)$, and $\delta(t)$ denote the sway velocity caused by the rudder motion alone, the heading angle, the roll angle, and the rudder angle, respectively; $\omega_\psi(s)$ and $\omega_\phi(s)$ denote the influence of the wave on $\psi(s)$ and $\phi(s)$, respectively; T_v and T_r denote time constants of transfer functions; K_{vr}, K_{vp}, K_{dv}, K_{dr}, and K_{dp} denote the given gains; ζ and ω_n denote the damping ratio and the natural frequency under no damping, respectively.

Based on (6.1), the state-space model of the sway-yaw and roll subsystems can be described as

$$\begin{cases} \dot{x}(t) = Ax(t) + B\delta(t) + E_1\omega(t), \\ x(t_0) = x_0, \end{cases} \tag{6.2}$$

where $x(t) = [v(t) \quad r(t) \quad \psi(t) \quad p(t) \quad \phi(t)]^T$ with $x(t) \in \mathbb{R}^n$; $v(t)$, $\psi(t)$, and $\phi(t)$ are the same as the corresponding items given above; $r(t)$ and $p(t)$ denote the yaw velocity and the roll velocity, respectively; the rudder angle $\delta(t) \in \mathbb{R}^m$; $\omega(t) = [\omega_\psi(t) \;\; \omega_\phi(t)]^T$ represents the wave-induced disturbances with $\omega(t) \in \mathbb{R}^p$, and $\omega(t)$ is assumed to belong to $L_2[t_0, \infty)$ with t_0 denoting the initial instant; $x_0 \in \mathbb{R}^n$ denotes the initial condition; A, B, and E_1 are the same as A, B_1, and B_2 in (5.2), respectively; T_v, T_r, K_{vr}, K_{vp}, K_{dv}, K_{dr}, K_{dp}, ζ, and ω_n are the same as the items presented above.

Compared with a manned surface vehicle, a UMV in network environments shows some competitive advantages. Throughout this chapter, we consider the case that the marine vehicle is controlled/detected by a remote land-based control station, see Fig. 6.1. The model of the sway-yaw and roll subsystems in (6.2) represents the model for a UMV. The UMV is equipped with the sampler and the actuator. The remote control station and the UMV are linked together through wireless communication networks. The sampled UMV's states are transmitted to the remote control station through the sampler-to-control station wireless communication network channel. Based on the received UMV's states, the remote control station generates the residual signal and control inputs, and transmits control inputs to the steering machine through the control station-to-actuator wireless communication network channel.

Note that the UMV in network environments may encounter faults unavoidably such as saturation, stuck steering machine-type faults, and noise-type faults. It is quite important to study how to detect the occurrence of faults in time. The network-based

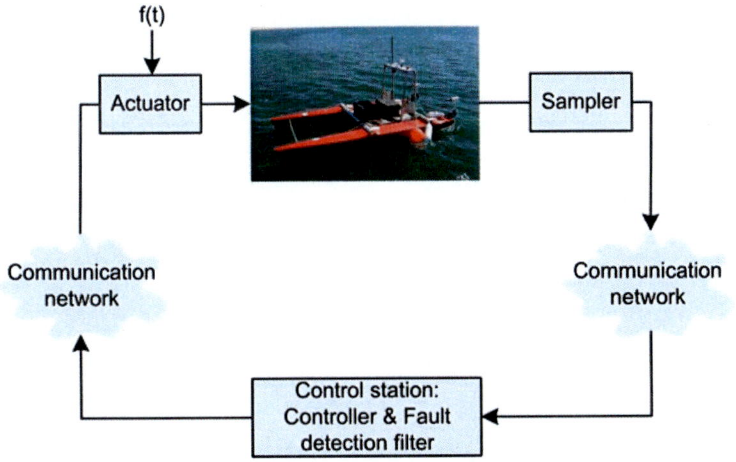

Fig. 6.1 The network-based structure for the UMV subject to faults

structure for the UMV subject to faults is presented in Fig. 6.1. Thus, the model of the sway-yaw and roll subsystems in system (6.2) is converted into

$$\begin{cases} \dot{x}(t) = Ax(t) + B\delta(t) + E_1\omega(t) + E_2 f(t), \\ y(t) = Cx(t), \\ x(t_0) = x_0, \end{cases} \tag{6.3}$$

where $y(t) \in \mathbb{R}^l$ and $f(t) \in \mathbb{R}^q$ denote the measurement output and the actuator fault signal, respectively; E_2 and C are known constant matrices of appropriate dimensions; C^T is assumed to be full column rank; (A, C) is detectable.

For the UMV in Fig. 6.1, the control station consists of the controller and the FDF. In this chapter, we assume that the sampler and the control station are time-driven whose sampling period is denoted as h, while the actuator is event-driven; the sampler and the control station are clock-synchronized with the same sampling instants and sampling period; packet disordering is not considered.

Firstly, we consider the case that the sampler is connected to the control station directly. That is, the UMV is controlled through a one-channel network. Network-induced delays and packet dropouts in the control station-to-actuator communication network channel are taken into account.

Let $t_k,\ t_k + h,\ t_k + 2h,\ t_k + \cdots,\ t_{k+1},\ t_{k+1} + h,\ t_{k+1} + 2h,\ t_{k+1} + \cdots\ (k = 0, 1, 2, \ldots)$ denote the sampling instants of the sampler. Suppose that the control inputs generated at the instants $t_k, t_{k+1}, \ldots$ are transmitted to the UMV successfully, while the control inputs generated between the instants t_k and t_{k+1} $(k = 0, 1, 2, \ldots)$ are dropped. ρ_{ca} denotes the upper bound of the control station-to-actuator consecutive packet dropouts. Then, the signal transmission for the UMV considering control station-to-actuator network-induced delays and packet dropouts is presented in Fig. 6.2, where the dashed lines denote that the corresponding data are dropped.

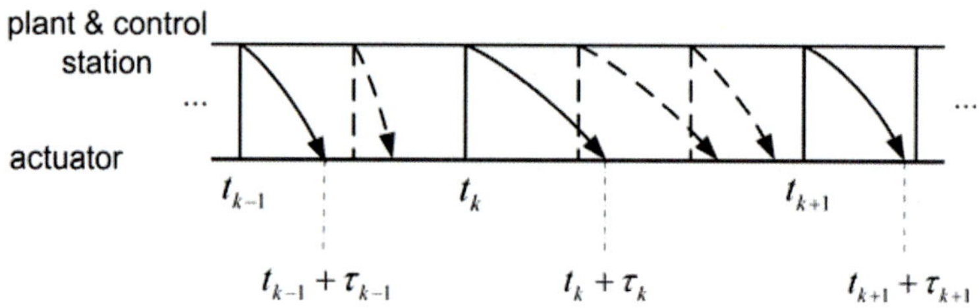

Fig. 6.2 The signal transmission for the UMV controlled through a one-channel network

The following observer-based FDF and controller are adopted to generate the residual signal and control inputs

$$\begin{cases} \dot{\hat{x}}(t) = A\hat{x}(t) + B\hat{u}(t) + L(y(t) - \hat{y}(t)), \\ \hat{y}(t) = C\hat{x}(t), \\ r(t) = V(y(t) - \hat{y}(t)), \\ \hat{u}(t) = K\hat{x}(t), \end{cases} \tag{6.4}$$

where $\hat{x}(t) \in \mathbb{R}^n$, $\hat{y}(t) \in \mathbb{R}^l$, $\hat{u}(t) \in \mathbb{R}^m$, and $r(t) \in \mathbb{R}^q$ are the state, the output, the control input, and the residual signal of the FDF, respectively; L, V, and K are to be designed.

Then, for $t \in [t_k + \tau_k, t_{k+1} + \tau_{k+1})$, the control input utilized by the UMV is given by

$$\delta(t) = \hat{u}(t_k) = K\hat{x}(t_k), \tag{6.5}$$

where the controller gain matrix K in (6.5) is the same as the K in (6.4); τ_k denotes the control station-to-actuator network-induced delays. Suppose that $0 < \tau_m \le \tau_k < \tau_M \le \bar{\tau}h$, where τ_m and τ_M are given scalars satisfying $\tau_m > 0$, $\tau_M > 0$, and $\bar{\tau}$ is a positive integer.

For $t \in [t_k + \tau_k, t_{k+1} + \tau_{k+1})$, the state equation of the UMV presented in (6.3) is converted into

$$\dot{x}(t) = Ax(t) + BK\hat{x}(t_k) + E_1\omega(t) + E_2 f(t). \tag{6.6}$$

Define $e(t) = x(t) - \hat{x}(t)$, $\tau(t) = t - t_k$. For $t \in [t_k + \tau_k, t_{k+1} + \tau_{k+1})$, one has $\tau(t) \in [\tau_k, t_{k+1} - t_k + \tau_{k+1})$. Considering that $\tau_m \le \tau_k < \tau_M$ and $t_{k+1} - t_k \le (\rho_{ca} + 1)h$, one has $\tau(t) \in [\tau_m, \eta)$, where $\eta = (\rho_{ca} + 1)h + \tau_M$. Combining (6.3), (6.4), and (6.6) together, one has

$$\begin{cases} \dot{e}(t) = (A - LC + BK)e(t) - BKx(t) - BKe(t - \tau(t)), \\ \qquad + BKx(t - \tau(t)) + E_1\omega(t) + E_2 f(t), \\ r(t) = VCe(t). \end{cases} \tag{6.7}$$

Then, by defining $\xi(t) = [e^T(t)\ x^T(t)]^T$, $\nu(t) = [\omega^T(t)\ f^T(t)]^T$, $r_e(t) = r(t) - f(t)$, and combining (6.4), (6.6), and (6.7) together, one can establish the following network-based closed-loop system

$$\begin{cases} \dot{\xi}(t) = \widetilde{A}\xi(t) + \widetilde{B}\xi(t-\tau(t)) + \widetilde{E}\nu(t), \\ r_e(t) = \widetilde{C}\xi(t) + \widetilde{F}\nu(t), \end{cases} \tag{6.8}$$

where

$$\widetilde{A} = \begin{bmatrix} A - LC + BK & -BK \\ 0 & A \end{bmatrix}, \ \widetilde{B} = \begin{bmatrix} -BK & BK \\ -BK & BK \end{bmatrix},$$
$$\widetilde{E} = \begin{bmatrix} E_1 & E_2 \\ E_1 & E_2 \end{bmatrix}, \ \widetilde{C} = \begin{bmatrix} VC & 0 \end{bmatrix}, \ \widetilde{F} = \begin{bmatrix} 0 & -I \end{bmatrix}.$$

Note that only the control station-to-actuator network-induced delays and packet dropouts are considered in the networked system (6.8).

In what follows, we turn to network-based modeling for the UMV controlled through two-channel communication networks. More precisely, the sampler-to-control station communication network channel and the control station-to-actuator communication network channel. Network-induced delays and packet dropouts in the aforementioned two communication network channels are considered simultaneously. In this situation, the signal transmission in Fig. 6.2 is converted into that in Fig. 6.3.

Suppose that the definitions for t_k, t_{k+1}, ... are the same as the ones presented above; ρ_{sc} and ρ_{ca} denote the upper bounds of the sampler-to-control station communication network channel and the control station-to-actuator communication network channel consecutive packet dropouts, respectively; d_k and τ_k denote the length of the sampler-to-control station network-induced and the control station-to-actuator network-induced delays, respectively; the lower bound and the upper bound for τ_k are the same as the ones presented above; $0 < d_m \le d_k < d_M \le \bar{d}h$, where d_m and d_M are given scalars satisfying $d_m > 0$, $d_M > 0$, and $\bar{d}$ is a positive integer.

Then, for $t \in [t_k + \tau_k,\ t_{k+1} + \tau_{k+1})$, the control input utilized by the UMV under two-channel networks is the same as the control input in (6.5). The state equation

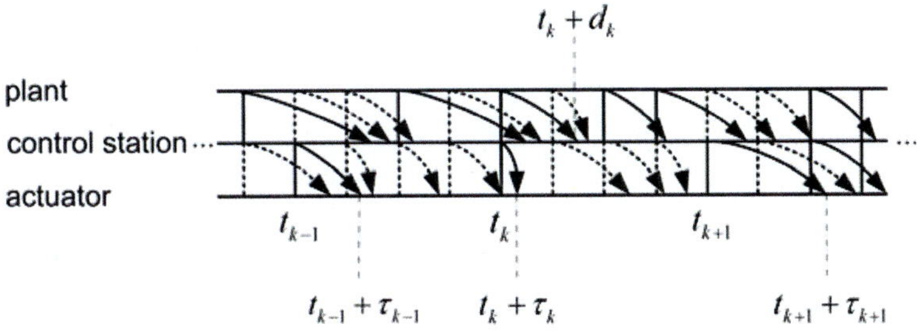

Fig. 6.3 The signal transmission for the UMV controlled through two-channel networks

presented in (6.3) for the UMV is converted into the one in (6.6). However, as observed from Fig. 6.3 and for $t \in [t_k + \tau_k,\ t_{k+1} + \tau_{k+1})$, the measured output $y(t)$ available to the FDF is variable, which induces some difficulty for describing $y(t)$ in a uniform form and constructing the corresponding closed-loop system. Then, one should propose an appropriate modeling approach for the UMV controlled through two-channel networks.

Assume that the latest available measured output $y(\tilde{t}_k)$ at the instant $t_k + \tau_k$ is utilized by the FDF for $t \in [t_k + \tau_k,\ t_{k+1} + \tau_{k+1})$, where $\tilde{t}_k$ denotes the sampling instant of the sampler. Note that both the sampler-to-control station network-induced and the control station-to-actuator network-induced packet dropouts are considered, $0 < \tau_k < \bar{\tau}h$, $0 < d_k < \bar{d}h$, $\tilde{t}_k + d_k \le t_k + \tau_k$. One can see that $t_k - (\bar{d} + \rho_{sc})h \le \tilde{t}_k \le t_k + (\bar{\tau} - 1)h$. More precisely, the feasible earliest and latest available measured outputs for the FDF are $y(t_k - (\bar{d} + \rho_{sc})h)$ and $y(t_k + (\bar{\tau} - 1)h)$, respectively.

Remark 6.1 When the actuator receives the control input $\hat{u}(t_k)$ at the instant $t_k + \tau_k$, the actuator sends an acknowledgement signal to the control station, where the acknowledgement signal is assigned the highest transmission priority and its transmission delay is negligible. Once the control station receives the acknowledgement signal, the latest available measured output at the instant $t_k + \tau_k$ is utilized by the FDF for $t \in [t_k + \tau_k,\ t_{k+1} + \tau_{k+1})$.

For the purpose of description, if $\tilde{t}_k = t_k,\ t_{k+1},\ \ldots$, or $t_k + (\bar{\tau} - 1)h$, define $\tilde{t}_k$ as $t_{k,lat}$; if $\tilde{t}_k = t_k - (\bar{d} + \rho_{sc})h,\ t_k - (\bar{d} + \rho_{sc} - 1)h,\ \ldots$, or $t_k - h$, define $\tilde{t}_k$ as $t_{k,pre}$. Then, for $t \in [t_k + \tau_k,\ t_{k+1} + \tau_{k+1})$, one can see that $\tilde{t}_k = t_{k,lat}$ or $\tilde{t}_k = t_{k,pre}$. On the other hand, for $t \in [t_k + \tau_k,\ t_{k+1} + \tau_{k+1})$, $\tilde{t}_k = t_{k,lat}$ and $\tilde{t}_k = t_{k,pre}$ can not occur simultaneously. We refer to such a phenomenon as mutually exclusive occurrence. In this chapter, we take the mutually exclusive occurrence characteristic of $\tilde{t}_k$ into full consideration to establish a model for the UMV controlled through two-channel networks. For this purpose, define a scalar $\alpha(t)$ as follows

$$\alpha(t) = \begin{cases} 1, & \text{if } \tilde{t}_k = t_{k,lat}, \\ 0, & \text{if } \tilde{t}_k = t_{k,pre}. \end{cases} \tag{6.9}$$

Define $\tau(t) = t - t_k$, $d_1(t) = t - t_{k,pre}$, $d_2(t) = t - t_{k,lat}$. For $t \in [t_k + \tau_k,\ t_{k+1} + \tau_{k+1})$, one has $\tau(t) \in [\tau_k,\ t_{k+1} - t_k + \tau_{k+1})$, $d_1(t) \in [t_k + \tau_k - t_{k,pre},\ t_{k+1} + \tau_{k+1} - t_{k,pre})$, $d_2(t) \in [t_k + \tau_k - t_{k,lat},\ t_{k+1} + \tau_{k+1} - t_{k,lat})$. From the definitions for ρ_{sc} and ρ_{ca}, and $\tau_m \le \tau_k < \tau_M \le \bar{\tau}h$, $d_m \le d_k < d_M \le \bar{d}h$, one has $\tau(t) \in [\tau_m,\ \eta)$, $d_1(t) \in [d_{1,min},\ d_{1,max})$, $d_2(t) \in [d_{2,min},\ \eta)$, where $\eta = (\rho_{ca} + 1)h + \tau_M$, $d_{1,min} = h + \tau_m$, $d_{1,max} = (\rho_{sc} + \rho_{ca} + \bar{d} + 1)h + \tau_M$, $d_{2,min} = \min\{d_m,\ \tau_m\}$. Then, the FDF in (6.4) is converted into

$$\begin{cases} \dot{\hat{x}}(t) = (A + BK - LC)\hat{x}(t) + (1 - \alpha(t))LCx(t - d_1(t)) \\ \qquad + \alpha(t)LCx(t - d_2(t)), \\ r(t) = (1 - \alpha(t))VCx(t - d_1(t)) + \alpha(t)VCx(t - d_2(t)) \\ \qquad - VC\hat{x}(t). \end{cases} \tag{6.10}$$

Define $e(t) = x(t) - \hat{x}(t)$. Combining (6.3), (6.6), and (6.10) together, one has

$$\begin{cases} \dot{e}(t) = Ae(t) - (BK - LC)x(t) - BKe(t - \tau(t)) \\ \qquad + BKx(t - \tau(t)) - (1 - \alpha(t))LCx(t - d_1(t)) \\ \qquad - \alpha(t)LCx(t - d_2(t)) + E_1\omega(t) + E_2 f(t), \\ r(t) = VCe(t) - VCx(t) + (1 - \alpha(t))VCx(t - d_1(t)) \\ \qquad + \alpha(t)VCx(t - d_2(t)). \end{cases} \tag{6.11}$$

By defining $\xi(t) = [e^T(t)\ x^T(t)]^T$, $\nu(t) = [\omega^T(t)\ f^T(t)]^T$, $r_e(t) = r(t) - f(t)$, and combining (6.6), (6.10), and (6.11) together, one can establish the following network-based closed-loop system

$$\begin{cases} \dot{\xi}(t) = \bar{A}\xi(t) + \bar{B}_1\xi(t - \tau(t)) + \bar{B}_2\xi(t - d_1(t)) \\ \qquad + \bar{B}_3\xi(t - d_2(t)) + \widetilde{E}\nu(t), \\ r_e(t) = \bar{C}\xi(t) + \bar{D}_1\xi(t - d_1(t)) + \bar{D}_2\xi(t - d_2(t)) + \widetilde{F}\nu(t), \end{cases} \tag{6.12}$$

where

$$\bar{A} = \begin{bmatrix} A & LC - BK \\ 0 & A \end{bmatrix},\ \bar{B}_1 = \begin{bmatrix} -BK & BK \\ -BK & BK \end{bmatrix},$$
$$\bar{B}_2 = \begin{bmatrix} 0 & -(1 - \alpha(t))LC \\ 0 & 0 \end{bmatrix},\ \bar{B}_3 = \begin{bmatrix} 0 & -\alpha(t)LC \\ 0 & 0 \end{bmatrix},$$
$$\tilde{E} = \begin{bmatrix} E_1 & E_2 \\ E_1 & E_2 \end{bmatrix},\ \bar{C} = \begin{bmatrix} VC & -VC \end{bmatrix},\ \tilde{F} = \begin{bmatrix} 0 & -I \end{bmatrix},$$
$$\bar{D}_1 = \begin{bmatrix} 0 & (1 - \alpha(t))VC \end{bmatrix},\ \bar{D}_2 = \begin{bmatrix} 0 & \alpha(t)VC \end{bmatrix}.$$

Remark 6.2 For the UMV equipped with an observer-based FDF, how to take into consideration the network-induced delays and packet dropouts in the sampler-to-control station communication network channel and the control station-to-actuator communication network channel, and establish the network-based models are significant and unresolved in the literature. Based on the mutually exclusive occurrence characteristic of $\tilde{t}_k$, a network-based model for the UMV controlled through two-channel networks is established in (6.12).

Remark 6.3 It should be pointed out that the observer-based FDF and controller adopted in (6.4) are motivated by the integrated fault detection and robust control scheme in [1]. However, they are different from the scheme in [1]. Moreover, network-induced delays and packet dropouts in the sampler-to-control station communication network channel and the control station-to-actuator communication network channel are taken into account in this chapter. Then, the network-based closed-loop systems (6.8) and (6.12) in this chapter are different from the ones in [1–3].

To detect the occurrence of faults as they occur, one should construct a suitable residual evaluation function $J(t)$, a residual evaluation function threshold J_{th}, and a fault detection logic. Since they are the same as the corresponding items in (3.8), (3.9), and (3.10), the detailed definitions are omitted here for brevity.

Based on the alarm of faults, some measures can be taken to guarantee the sailing safety of the UMV in network environments.

It should be mentioned that $\tau(t)$, which is an artificial interval time-varying delay, is different from the network-induced delays τ_k. Note that the interval time-varying delay $\tau(t) \in [\tau_m, \ \eta)$. Introducing a scalar $\tau_{mid} = (\tau_m + \eta)/2$, one can see that at any instant t, $\tau(t) \in [\tau_m, \ \tau_{mid})$ or $\tau(t) \in [\tau_{mid}, \ \eta)$, where $t \in [t_k + \tau_k, \ t_{k+1} + \tau_{k+1})$. On the other hand, for the specific instant t, $\tau(t) \in [\tau_m, \ \tau_{mid})$ and $\tau(t) \in [\tau_{mid}, \ \eta)$ do not occur simultaneously, which phenomenon is named as mutually exclusive distribution. The convex analysis method was proposed in [4] to deal with integral inequalities for products of vectors. In this chapter, we propose the combined convex analysis and mutually exclusive distribution approach to deal with integral inequalities for products of vectors. For this purpose, define a scalar $\lambda(t)$, where

$$\lambda(t) = \begin{cases} 1, & \tau(t) \in [\tau_m, \ \tau_{mid}), \\ 0, & \tau(t) \in [\tau_{mid}, \ \eta). \end{cases} \tag{6.13}$$

Based on the networked system (6.8) and the fault detection logic (3.10), this chapter investigates the problem of network-based FDF and controller coordinated design for the motion process of the UMV. The proposed design method can be extended to investigate the system (6.12), and the corresponding results are omitted for brevity.

6.2 Network-Based FDF and Controller Coordinated Design

In this section, we first present a network-based FDF and controller coordinated design for the network-based closed-loop system (6.8). Then, the merits of the combined convex analysis and mutually exclusive distribution approach are analyzed.

6.2.1 *FDF and Controller Coordinated Design*

We state and establish the following result dealing with the network-based FDF and controller coordinated design for the networked system (6.8).

Theorem 6.1 *For given scalars τ_m, τ_M, ρ_{ca}, h, ε_1, and ε_2, the residual system (6.8) is asymptotically stable with an H_∞ norm bound γ and the FDF gains $L = \bar{L}^T N^{-T}$ and $V = \bar{V}^T N^{-T}$, and the controller gain $K = \bar{K}^T W^{-1}$, if there exist symmetric positive definite matrices $\widetilde{Q}_1$, $\widetilde{Q}_2$, $\widetilde{R}_1$, $\widetilde{R}_2$, W, and matrices $\bar{K}$, $\bar{V}$, $\bar{L}$, N, such that (6.14) and (6.15) hold for each feasible value of $\lambda(t)$*

$$\begin{bmatrix} \widetilde{\Pi}_{11,i} & \widetilde{\Pi}_{12} \\ * & \widetilde{\Pi}_{22} \end{bmatrix} < 0, \tag{6.14}$$

$$WC^T = C^T N, \tag{6.15}$$

where $i = 1,\ 2,$ *and*

$$\widetilde{\Pi}_{11,1} = \widetilde{\Pi} - \tilde{\theta}_1 - \tilde{\theta}_3,\ \widetilde{\Pi}_{11,2} = \widetilde{\Pi} - \tilde{\theta}_2 - \tilde{\theta}_4,$$

$$\widetilde{\Pi} = \begin{bmatrix} \widetilde{\Omega}_{11} & \widetilde{R}_1 & H_2^T & 0 & \widetilde{E} \\ * & \widetilde{\Omega}_{22} & \widetilde{R}_2 & 0 & 0 \\ * & * & \widetilde{\Omega}_{33} & \widetilde{R}_2 & 0 \\ * & * & * & \widetilde{\Omega}_{44} & 0 \\ * & * & * & * & -\gamma I \end{bmatrix},$$

$$\tilde{\theta}_1 = \psi_1^T \widetilde{R}_2 \psi_1,\ \tilde{\theta}_3 = (\eta - \tau_m)(\frac{1-\lambda(t)}{\eta - \tau_{mid}} - \frac{1-\lambda(t)}{\eta - \tau_m})\tilde{\theta}_1,$$

$$\tilde{\theta}_2 = \psi_2^T \widetilde{R}_2 \psi_2,\ \tilde{\theta}_4 = (\eta - \tau_m)(\frac{\lambda(t)}{\tau_{mid} - \tau_m} - \frac{\lambda(t)}{\eta - \tau_m})\tilde{\theta}_2,$$

$$\widetilde{\Pi}_{12} = \begin{bmatrix} \widetilde{\Phi}_1 & \widetilde{\Phi}_1 & \widetilde{\Phi}_2 \end{bmatrix},\ \widetilde{\Pi}_{22} = diag\{\mathscr{X}_1,\ \mathscr{X}_2,\ -\gamma I\},$$

$$\widetilde{\Omega}_{11} = H_1 + H_1^T + \widetilde{Q}_1 + \widetilde{Q}_2 - \widetilde{R}_1,\ \widetilde{\Omega}_{22} = -\widetilde{Q}_1 - \widetilde{R}_1 - \widetilde{R}_2,$$

$$\widetilde{\Omega}_{33} = -2\widetilde{R}_2,\ \widetilde{\Omega}_{44} = -\widetilde{Q}_2 - \widetilde{R}_2,$$

$$\psi_1 = \begin{bmatrix} 0 & 0 & I & -I & 0 \end{bmatrix},\ \psi_2 = \begin{bmatrix} 0 & I & -I & 0 & 0 \end{bmatrix},$$

$$\widetilde{\Phi}_1 = \begin{bmatrix} H_1^T & 0 & H_2^T & 0 & \widetilde{E} \end{bmatrix}^T,\ \widetilde{\Phi}_2 = \begin{bmatrix} H_3^T & 0 & 0 & 0 & \widetilde{F} \end{bmatrix}^T,$$

$$\mathscr{X}_1 = \tau_m^{-2}(\widetilde{R}_1 - 2\Upsilon),\ \mathscr{X}_2 = (\eta - \tau_m)^{-2}(\widetilde{R}_2 - 2\Upsilon),$$

$$H_1 = \begin{bmatrix} \varepsilon_1 W A^T - \varepsilon_1 C^T \bar{L} + \varepsilon_1 \bar{K} B^T & 0 \\ -\varepsilon_2 \bar{K} B^T & \varepsilon_2 W A^T \end{bmatrix},$$

$$H_2 = \begin{bmatrix} -\varepsilon_1 \bar{K} B^T & -\varepsilon_1 \bar{K} B^T \\ \varepsilon_2 \bar{K} B^T & \varepsilon_2 \bar{K} B^T \end{bmatrix},\ H_3 = \begin{bmatrix} \varepsilon_1 C^T \bar{V} \\ 0 \end{bmatrix},\ \Upsilon = \begin{bmatrix} \varepsilon_1 W & 0 \\ 0 & \varepsilon_2 W \end{bmatrix}.$$

Proof Construct the following Lyapunov-Krasovskii functional

$$V(t, \xi_t) = \sum_{j=1}^{3} V_j(t, \xi_t), \tag{6.16}$$

where

$$V_1(t,\xi_t)=\xi(t)^T P\xi(t),$$
$$V_2(t,\xi_t)=\int_{t-\tau_m}^{t}\xi^T(s)Q_1\xi(s)ds+\int_{t-\eta}^{t}\xi^T(s)Q_2\xi(s)ds,$$
$$V_3(t,\xi_t)=\tau_m\int_{-\tau_m}^{0}\int_{t+s}^{t}\dot{\xi}^T(\theta)R_1\dot{\xi}(\theta)d\theta ds + (\eta-\tau_m)\int_{-\eta}^{-\tau_m}\int_{t+s}^{t}\dot{\xi}^T(\theta)R_2\dot{\xi}(\theta)d\theta ds,$$

with $\xi_t=\xi(t+\varsigma), \varsigma\in[t_0-\eta,\ t_0]$, P, Q_1, Q_2, R_1, and R_2 being symmetric positive definite matrices with appropriate dimensions. Taking the time derivative of the Lyapunov functional $V(t,\xi_t)$ given in (6.16) along the trajectory of the system (6.8), one has

$$\dot{V}_1(t,\xi_t)=2\xi^T(t)P\dot{\xi}(t), \tag{6.17}$$
$$\dot{V}_2(t,\xi_t)=\xi^T(t)(Q_1+Q_2)\xi(t)-\xi^T(t-\tau_m)Q_1\xi(t-\tau_m) - \xi^T(t-\eta)Q_2\xi(t-\eta), \tag{6.18}$$
$$\dot{V}_3(t,\xi_t)=\dot{\xi}^T(t)\Theta\dot{\xi}(t)-\tau_m\int_{t-\tau_m}^{t}\dot{\xi}^T(\theta)R_1\dot{\xi}(\theta)d\theta - (\eta-\tau_m)\int_{t-\eta}^{t-\tau_m}\dot{\xi}^T(\theta)R_2\dot{\xi}(\theta)d\theta, \tag{6.19}$$

where $\Theta=\tau_m^2R_1+(\eta-\tau_m)^2R_2$.

Note that

$$\begin{aligned}&-(\eta-\tau_m)\int_{t-\eta}^{t-\tau_m}\dot{\xi}^T(\theta)R_2\dot{\xi}(\theta)d\theta\\ &=-(\eta-\tau(t))\int_{t-\tau(t)}^{t-\tau_m}\dot{\xi}^T(\theta)R_2\dot{\xi}(\theta)d\theta-(\tau(t)-\tau_m)\int_{t-\tau(t)}^{t-\tau_m}\dot{\xi}^T(\theta)R_2\dot{\xi}(\theta)d\theta\\ &\quad-(\eta-\tau(t))\int_{t-\eta}^{t-\tau(t)}\dot{\xi}^T(\theta)R_2\dot{\xi}(\theta)d\theta-(\tau(t)-\tau_m)\int_{t-\eta}^{t-\tau(t)}\dot{\xi}^T(\theta)R_2\dot{\xi}(\theta)d\theta.\end{aligned}$$

Based on the Jensen integral inequality in [5], one has

$$-(\eta-\tau(t))\int_{t-\eta}^{t-\tau(t)}\dot{\xi}^T(\theta)R_2\dot{\xi}(\theta)d\theta\le-\varphi_1^TR_2\varphi_1, \tag{6.20}$$
$$-(\tau(t)-\tau_m)\int_{t-\tau(t)}^{t-\tau_m}\dot{\xi}^T(\theta)R_2\dot{\xi}(\theta)d\theta\le-\varphi_2^TR_2\varphi_2, \tag{6.21}$$
$$-\tau_m\int_{t-\tau_m}^{t}\dot{\xi}^T(\theta)R_1\dot{\xi}(\theta)d\theta\le-\varphi_3^TR_1\varphi_3, \tag{6.22}$$

where $\varphi_1 = [\xi(t - \tau(t)) - \xi(t - \eta)], \varphi_2 = [\xi(t - \tau_m) - \xi(t - \tau(t))], \varphi_3 = [\xi(t) - \xi(t - \tau_m)]$.

Define $\rho(t) = \frac{\tau(t)-\tau_m}{\eta-\tau_m}$. Taking the mutually exclusive distribution characteristic of $\tau(t)$ into account, one has

$$\begin{aligned} & -(\tau(t) - \tau_m) \int_{t-\eta}^{t-\tau(t)} \dot{\xi}^T(\theta) R_2 \dot{\xi}(\theta) d\theta \\ & \leq -\rho(t)\varphi_1(t)^T R_2 \varphi_1(t) - \rho(t)(\eta - \tau_m)(\frac{1-\lambda(t)}{\eta - \tau_{mid}} - \frac{1-\lambda(t)}{\eta - \tau_m})\varphi_1(t)^T R_2 \varphi_1(t) \end{aligned} \tag{6.23}$$

and

$$\begin{aligned} & -(\eta - \tau(t)) \int_{t-\tau(t)}^{t-\tau_m} \dot{\xi}^T(\theta) R_2 \dot{\xi}(\theta) d\theta \\ & \leq -(1 - \rho(t))\varphi_2(t)^T R_2 \varphi_2(t) - (1 - \rho(t))(\eta - \tau_m) \\ & \quad \times (\frac{\lambda(t)}{\tau_{mid} - \tau_m} - \frac{\lambda(t)}{\eta - \tau_m})\varphi_2(t)^T R_2 \varphi_2(t). \end{aligned} \tag{6.24}$$

By assuming that $P^{-1} = \begin{bmatrix} \varepsilon_1 W & 0 \\ 0 & \varepsilon_2 W \end{bmatrix}$ with ε_1 and ε_2 denoting given scalars, adopting the convex analysis method and the Schur complement, and using the definition for H_∞ performance, one can achieve the result in Theorem 6.1. This completes the proof. □

Remark 6.4 From the system (6.8), one can see that the network-based closed-loop system is established as a time-delay system. Note that the Lyapunov-Krasovskii functional approach [5] is widely used to prove the stability of a time-delay system. In this chapter, we employ the Lyapunov-Krasovskii functional approach to derive the FDF and controller coordinated design criterion, i.e. Theorem 6.1.

Note that the equality constraint in (6.15) induces some difficulty for numerical calculation. We consider a method for eliminating the equality constraint in (6.15). For the matrix C^T of full column rank, there always exist two orthogonal matrices $X \in \mathbb{R}^{n\times n}$ and $Y \in \mathbb{R}^{l\times l}$ such that

$$XC^TY = \begin{bmatrix} X_1 \\ X_2 \end{bmatrix} C^TY = \begin{bmatrix} \Phi \\ 0 \end{bmatrix}, \tag{6.25}$$

where $X_1 \in \mathbb{R}^{l\times n}$, $X_2 \in \mathbb{R}^{(n-l)\times n}$, $\Phi = \text{diag}\{\sigma_1, \ \ldots, \ \sigma_l\}$, and $\sigma_1, \ \ldots, \ \sigma_l$ are nonzero singular values of C^T. Then, based on Lemma 2 in [6], suppose that the matrix C^T is full column rank, if the matrix W can be written as

$$W = X^T \begin{bmatrix} W_{11} & 0 \\ 0 & W_{22} \end{bmatrix} X = X_1^T W_{11} X_1 + X_2^T W_{22} X_2, \tag{6.26}$$

where W_{11} and W_{22} are symmetric positive definite matrices with appropriate dimensions, X_1 and X_2 are defined in (6.25), then there exists a nonsingular matrix N such that $WC^T = C^T N$.

Based on Theorem 6.1 and the statement presented above, we can state and establish the following FDF and controller coordinated design criterion.

Theorem 6.2 *For given scalars τ_m, τ_M, ρ_{ca}, h, ε_1, and ε_2, the residual system (6.8) is asymptotically stable with an H_∞ norm bound γ and the FDF gains $L = \bar{L}^T Y \Phi W_{11}^{-1} \Phi^{-1} Y^T$ and $V = \bar{V}^T Y \Phi W_{11}^{-1} \Phi^{-1} Y^T$, and the controller gain $K = \bar{K}^T (X_1^T W_{11} X_1 + X_2^T W_{22} X_2)^{-1}$, if there exist symmetric positive definite matrices $\widetilde{Q}_1$, $\widetilde{Q}_2$, $\widetilde{R}_1$, $\widetilde{R}_2$, W_{11}, W_{22}, and matrices $\bar{K}$, $\bar{V}$, $\bar{L}$, such that the inequalities in (6.27) hold for each feasible value of $\lambda(t)$*

$$\begin{bmatrix} \hat{\Pi}_{11,i} & \hat{\Pi}_{12} \\ * & \hat{\Pi}_{22} \end{bmatrix} < 0, \tag{6.27}$$

where $\hat{\Pi}_{11,i}$, $\hat{\Pi}_{12}$, and $\hat{\Pi}_{22}$ are derived from $\widetilde{\Pi}_{11,i}$, $\widetilde{\Pi}_{12}$, and $\widetilde{\Pi}_{22}$ in (6.14), respectively, by substituting W with $X_1^T W_{11} X_1 + X_2^T W_{22} X_2$.

It should be mentioned that the FDF and controller coordinated design criteria in Theorems 6.1 and 6.2 should hold for each feasible value of $\lambda(t)$ with $\lambda(t) = 1$ or $\lambda(t) = 0$. On the other hand, the FDF and controller coordinated design for the networked system (6.12) is an extension of the result in Theorem 6.2, here it is omitted for brevity.

Remark 6.5 For a specific UMV, the feasibility of the matrix inequalities in (6.27) is simultaneously affected by scalars τ_m, τ_M, ρ_{ca}, h, ε_1, and ε_2. If the scalars τ_m, ε_1, and ε_2 are predefined, one can see that the smaller the scalars ρ_{ca}, h, and τ_M, the higher the probability for the feasibility of the matrix inequalities in (6.27).

6.2.2 Merits of the Approach Dealing with Integral Inequalities

The combined convex analysis and mutually exclusive distribution approach is proposed to deal with integral inequalities for products of vectors. Then, a natural question is whether the proposed approach can introduce better results than the existing convex analysis method.

If the existing convex analysis method is adopted to deal with integral inequalities, the FDF and controller coordinated design criterion in Theorem 6.2 can be stated as that for given scalars τ_m, τ_M, ρ_{ca}, h, ε_1, and ε_2, the residual system (6.8) is asymptotically stable with an H_∞ norm bound γ and the FDF gains $L = \bar{L}^T Y \Phi W_{11}^{-1} \Phi^{-1} Y^T$ and $V = \bar{V}^T Y \Phi W_{11}^{-1} \Phi^{-1} Y^T$, and the controller gain $K = \bar{K}^T (X_1^T W_{11} X_1 + X_2^T W_{22} X_2)^{-1}$, if there exist symmetric positive definite matrices

$\widetilde{Q}_1$, $\widetilde{Q}_2$, $\widetilde{R}_1$, $\widetilde{R}_2$, W_{11}, W_{22}, and matrices $\bar{K}$, $\bar{V}$, $\bar{L}$, such that the inequalities in (6.28) hold

$$\begin{bmatrix} \bar{\Pi}_{11,i} & \hat{\Pi}_{12} \\ * & \hat{\Pi}_{22} \end{bmatrix} < 0, \tag{6.28}$$

where $i = 1,\ 2$, while $\bar{\Pi}_{11,1}$ and $\bar{\Pi}_{11,2}$ are derived from the matrices $\hat{\Pi}_{11,1}$ and $\hat{\Pi}_{11,2}$ in (6.27) by deleting $\tilde{\theta}_3$ and $\tilde{\theta}_4$, respectively; $\hat{\Pi}_{12}$ and $\hat{\Pi}_{22}$ are the same as the corresponding items in (6.27).

The following theorem establishes the relationship between the design criterion presented above and the one in Theorem 6.2.

Theorem 6.3 *Consider the residual system (6.8). For given scalars τ_m, τ_M, ρ_{ca}, h, ε_1, and ε_2, if there exist symmetric positive definite matrices $\widetilde{Q}_1$, $\widetilde{Q}_2$, $\widetilde{R}_1$, $\widetilde{R}_2$, W_{11}, W_{22}, and matrices $\bar{K}$, $\bar{V}$, $\bar{L}$, such that the inequalities in (6.28) are satisfied, then the inequalities in (6.27) are also satisfied for each feasible value of $\lambda(t)$.*

Proof Note that $\hat{\Pi}_{11,1}$ and $\hat{\Pi}_{11,2}$ in (6.27) can be rewritten as $\hat{\Pi}_{11,1} = \bar{\Pi}_{11,1} - \tilde{\theta}_3$ and $\hat{\Pi}_{11,2} = \bar{\Pi}_{11,2} - \tilde{\theta}_4$, respectively, where $\bar{\Pi}_{11,1}$ and $\bar{\Pi}_{11,2}$ are the same as the items in (6.28), $\tilde{\theta}_3$ and $\tilde{\theta}_4$ are the same as the items in (6.14). Considering that $\tilde{\theta}_3 \geq 0$, $\tilde{\theta}_4 \geq 0$, while $\tilde{\theta}_3$ and $\tilde{\theta}_4$ are not equal to zero simultaneously, one can conclude that if the inequalities in (6.28) are satisfied, then the inequalities in (6.27) are also satisfied for each feasible value of $\lambda(t)$. This completes the proof. □

Remark 6.6 It is proved in Theorem 6.3 that if the inequalities in (6.28) are satisfied, then the inequalities in (6.27) are also satisfied for each feasible value of $\lambda(t)$. This implies that the combined convex analysis and mutually exclusive distribution approach can provide more relaxed FDF and controller coordinated design criteria than the existing convex analysis method. If the combined convex analysis and mutually exclusive distribution approach is adopted to study the problems in [4], better results are expected to be obtained. The corresponding results are omitted here for brevity.

6.3 Performance Analysis and Discussion

As one can see in Chap. 2 of [7], the forward speed is an important parameter for the motion process of a marine vehicle. In this section, we present the fault detection performance analysis for the UMV with different forward speeds.

6.3.1 Performance Analysis for the Low-Forward-Speed UMV

Consider the sway-yaw and roll subsystems for the UMV. For the system matrices A, B, and E_1 in (6.2), choose the parameters as

$$\begin{aligned} U &= 3.8 \text{ (m/s)},\ T_v = 2/U,\ T_r = 1.6/U,\ K_{dr} = -0.0027U, \\ K_{dp} &= -0.0014U^2,\ K_{dv} = 0.01U,\ K_{vp} = 0.21U, \\ \omega_n &= 1.63 \text{ (rad/s)},\ \zeta = 0.64 + 0.38U,\ K_{vr} = -0.46 \text{ (m/s)}, \end{aligned} \tag{6.29}$$

where U is the forward speed of the UMV. Then, we have

$$A = \begin{bmatrix} -1.9000 & 0 & 0 & 0 & 0 \\ -1.0925 & -2.3750 & 0 & 0 & 0 \\ 0 & 1.0000 & 0 & 0 & 0 \\ 2.1202 & 0 & 0 & -6.7938 & -2.6569 \\ 0 & 0 & 0 & 1.0000 & 0 \end{bmatrix},$$

$$B = \begin{bmatrix} 0.0722 \\ -0.0244 \\ 0 \\ -0.0537 \\ 0 \end{bmatrix},\quad E_1 = \begin{bmatrix} 0 & 0 \\ 2.3750 & 0 \\ 0 & 0 \\ 0 & 2.6569 \\ 0 & 0 \end{bmatrix}.$$

If there exist faults, suppose that the matrices E_2 and C in (6.3) are $E_2 = \begin{bmatrix} 0.6 & -1 & 2 & 0.8 & 1 \end{bmatrix}^T, C = \begin{bmatrix} 1 & 0.8 & 1 & -1 & 0.6 \end{bmatrix}$.

By using matrix singular value decomposition and from (6.25), one gets

$$\begin{aligned} X_1 &= \begin{bmatrix} -0.5000 & -0.4000 & -0.5000 & 0.5000 & -0.3000 \end{bmatrix}, \\ X_2 &= \begin{bmatrix} -0.4000 & 0.8933 & -0.1333 & 0.1333 & -0.0800 \\ -0.5000 & -0.1333 & 0.8333 & 0.1667 & -0.1000 \\ 0.5000 & 0.1333 & 0.1667 & 0.8333 & 0.1000 \\ -0.3000 & -0.0800 & -0.1000 & 0.1000 & 0.9400 \end{bmatrix}, \\ Y &= -1,\ \Phi = 2, \end{aligned}$$

where the definitions for X_1, X_2, Y and Φ are presented in (6.25).

For Theorem 6.2, suppose that $\tau_m = 0.05$s, $\tau_M = 0.1$s, $\rho_{ca} = 3$, $h = 0.05$s. To avoid that some elements of the obtained matrix V are close to zero, we assume that $W_{11} < 100I$. Similar to Algorithm 2 in [8], we assume that the initial values $\varepsilon_{1,0} = 0.6$, $\varepsilon_{2,0} = 1.8$, the final values $\varepsilon_{1,ult} = 0.4$, $\varepsilon_{2,ult} = 1.6$, and the step lengths $\varepsilon_{1,dec} = \varepsilon_{2,dec} = 0.1$ for ε_1 and ε_2 which are introduced in the proof of Theorem 6.1. Then, one can get the locally optimal parameters $\varepsilon_{1,opt} = 0.6$ and $\varepsilon_{2,opt} = 1.8$. By choosing $\varepsilon_1 = 0.6$, $\varepsilon_2 = 1.8$, and solving the FDF and controller coordinated design

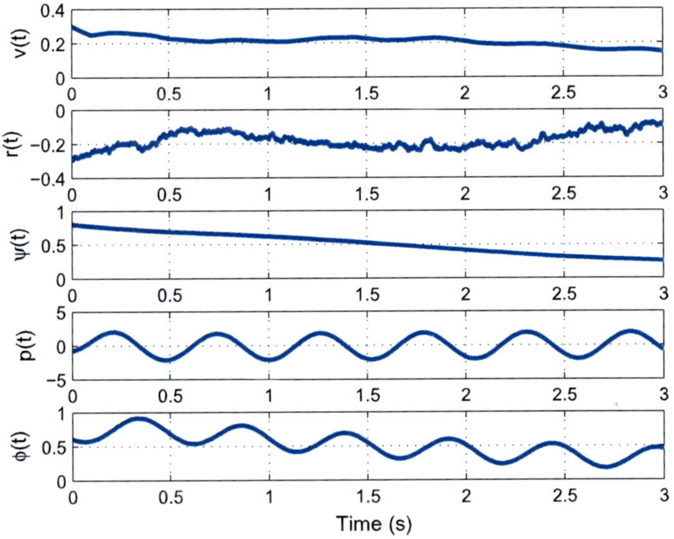

Fig. 6.4 The UMV state response $x(t)$ with no fault

criterion in Theorem 6.2 for each feasible value of $\lambda(t)$ with $\lambda(t) = 1$ or $\lambda(t) = 0$, one obtains

$$\begin{aligned} L &= \begin{bmatrix}-0.0417 & -2.0804 & 2.2427 & 0.5323 & 0.8374\end{bmatrix}^T, \\ V &= 0.0050, \\ K &= \begin{bmatrix}1.1118 & 3.4650 & 6.6327 & -1.5691 & 3.1662\end{bmatrix}. \end{aligned} \tag{6.30}$$

Suppose that the initial state of the augmented system (6.8) is $\xi_0 = [0.2 \;\; 0 \;\; -0.2 \;\; 0.1 \;\; -0.1 \;\; 0.3 \;\; -0.3 \;\; 0.8 \;\; -0.8 \;\; 0.6]^T$. The disturbance input $\omega_\psi(t)$ is a normally (Gaussian) distributed random signal whose mean and variance are 0 and 1, respectively, and $\omega_\phi(t) = 10\sin(t)$. The UMV state response $x(t)$ with no fault is presented in Fig. 6.4. From Fig. 6.4, one can see that the sway velocity $v(t)$, the yaw velocity $r(t)$, the roll velocity $p(t)$, and the roll angle $\phi(t)$ are affected by the disturbance input $\omega(t)$ with $\omega(t) = [\omega_\psi(t) \;\; \omega_\phi(t)]^T$, while the heading angle $\psi(t)$ is almost not affected by $\omega(t)$. It should be mentioned that although $v(t)$, $r(t)$, $p(t)$, and $\phi(t)$ are affected by $\omega(t)$, the prescribed H_∞ performance requirement is still guaranteed, which illustrates the effectiveness of the proposed controller design scheme.

Suppose that the stuck-type actuator fault $f(t)$ is described as

$$f(t) = \begin{cases} 0.6, & t \in [0.5\text{s}, \;1.5\text{s}], \\ 0, & \text{otherwise}. \end{cases} \tag{6.31}$$

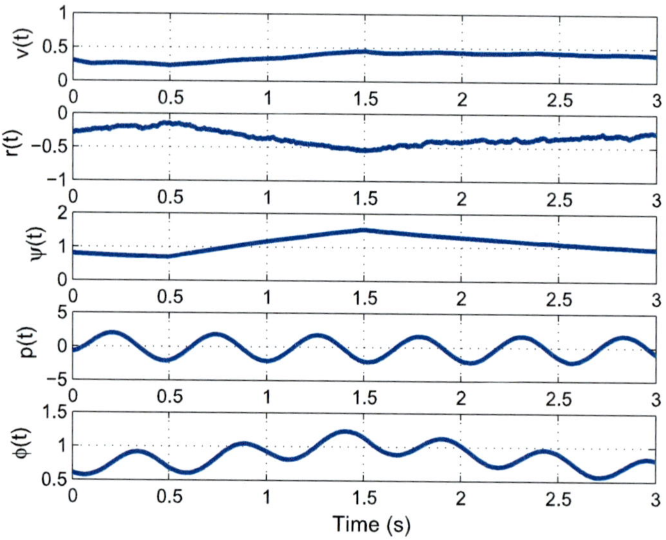

Fig. 6.5 The UMV state response $x(t)$ with $f(t)$ in (6.31)

Then, Figs. 6.5, 6.6, and 6.7 show the UMV state response $x(t)$, the residual response $r_e(t)$, and the residual evaluation function response $J(t)$, respectively. Figure 6.5 demonstrates that the sway velocity $v(t)$, the yaw velocity $r(t)$, and the heading angle $\psi(t)$ are affected by the occurrence of the fault in (6.31), while the roll velocity $p(t)$ and the roll angle $\phi(t)$ are almost not affected by the fault in (6.31). Figure 6.6 shows that the occurrence of the stuck-type actuator fault $f(t)$ imposes a timely influence on the residual response $r_e(t)$, which helps to reduce the needed time for fault detection. Although the occurrence of the fault in (6.31) can affect the sway velocity $v(t)$, the yaw velocity $r(t)$, and the heading angle $\psi(t)$, Fig. 6.7 illustrates that the newly proposed FDF and controller coordinated design scheme can not only reflect the occurrence of the fault in time, but also recognize the fault without confusing it with the disturbance $\omega(t)$.

In the following, we consider a noise-type fault which is given in (6.32),

$$f(t) = \begin{cases} 0.3\omega_\psi(t)\sin(t), & t \in [0.5\text{s},\ 1.5\text{s}], \\ 0, & \text{otherwise}. \end{cases} \tag{6.32}$$

The disturbance input $\omega(t) = [\omega_\psi(t) \ \ \omega_\phi(t)]^T$ with $\omega_\psi(t)$ and $\omega_\phi(t)$ being the same as the corresponding items presented above.

Then, Figs. 6.8, 6.9, and 6.10 show the UMV state response $x(t)$, the residual response $r_e(t)$, and the residual evaluation function response $J(t)$, respectively. Compared with Fig. 6.4, one can see that the sway velocity $v(t)$, the yaw velocity $r(t)$, the heading angle $\psi(t)$, the roll velocity $p(t)$, and the roll angle $\phi(t)$ in Fig. 6.8 are robust to the occurrence of the noise-type fault presented in (6.32). This demon-

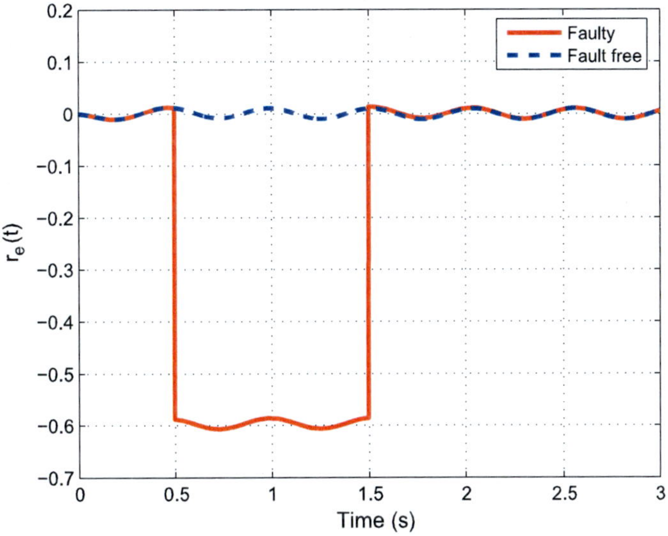

Fig. 6.6 The residual response $r_e(t)$ with $f(t)$ in (6.31)

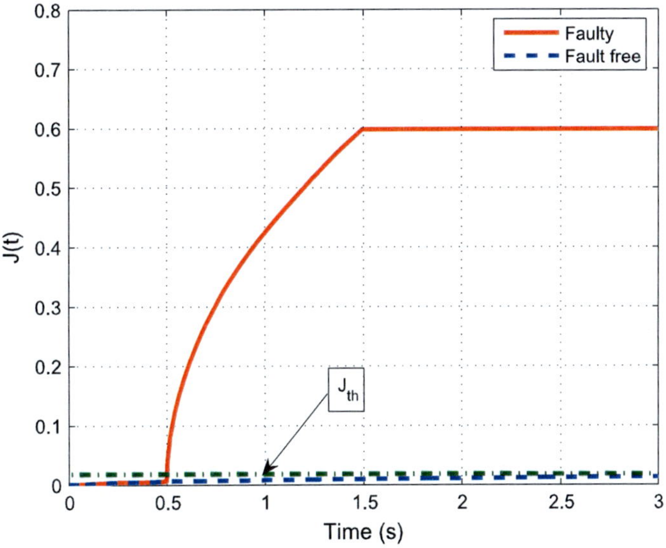

Fig. 6.7 The residual evaluation function response $J(t)$ with $f(t)$ in (6.31)

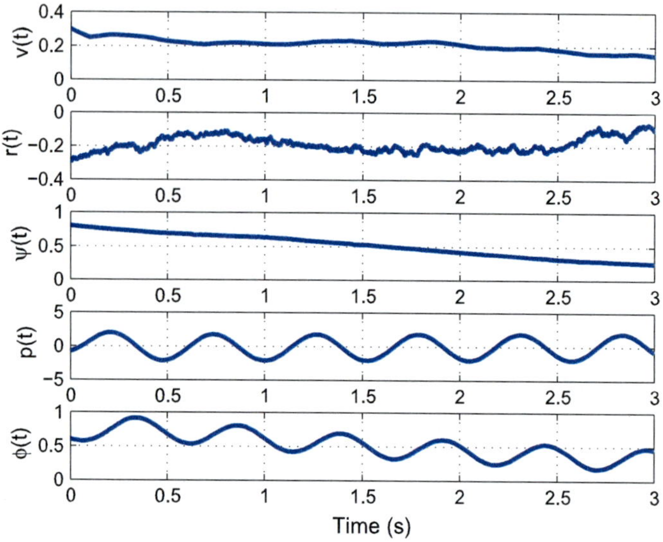

Fig. 6.8 The UMV state response $x(t)$ with $f(t)$ in (6.32)

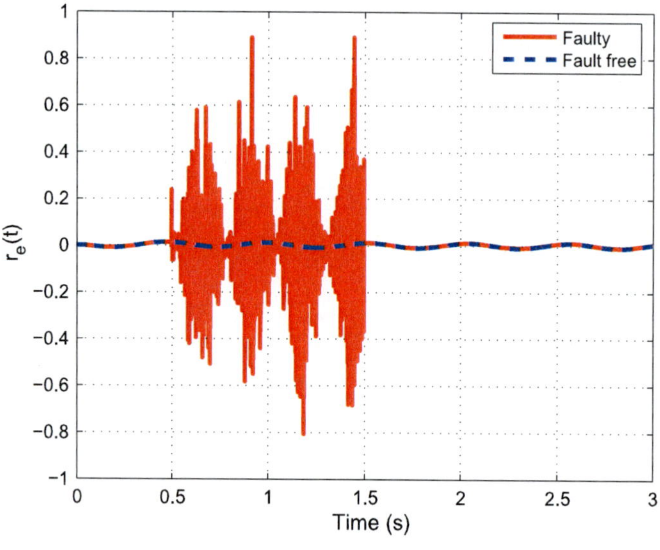

Fig. 6.9 The residual response $r_e(t)$ with $f(t)$ in (6.32)

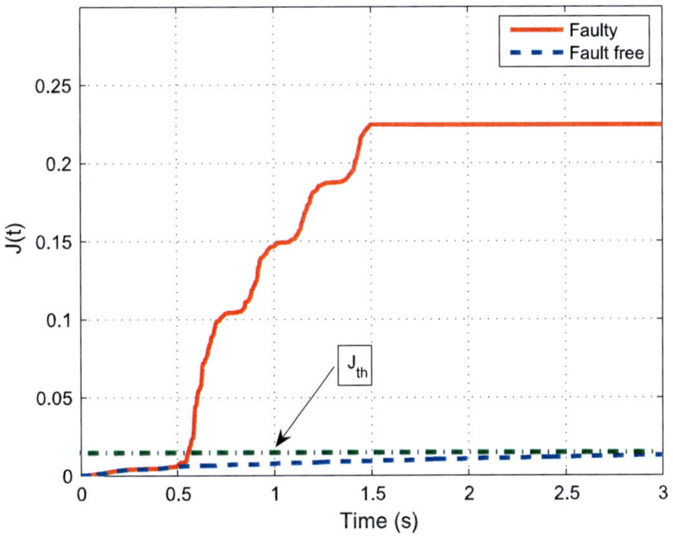

Fig. 6.10 The residual evaluation function response $J(t)$ with $f(t)$ in (6.32)

strates the effectiveness of the proposed FDF and controller coordinated design. Figure 6.9 illustrates that the residual response $r_e(t)$ can reflect the occurrence of the fault given in (6.32) in a timely manner, which helps to reduce the needed time for fault detection. Figure 6.10 verifies that the proposed design scheme can both reflect the occurrence of the fault in time and recognize the fault without confusing it with the disturbance $\omega(t)$.

From Figs. 6.7 and 6.10, and the fault detection logic presented in (3.10), one can see that the faults in (6.31) and (6.32) are detected at the instants 0.5005s and 0.5552s, respectively. That is, the fault detection time in Fig. 6.7 and 6.10 is 0.0005s and 0.0552s, respectively. Note that Fig. 6.7 provides a shorter fault detection time than Fig. 6.10, which phenomenon is induced by the fact that a more complex fault is considered in (6.32). Once the occurrence of faults is detected, an alarm of faults is generated and some measures can be taken to guarantee the sailing reliability of the networked UMV.

6.3.2 Performance Analysis for the High-Forward-Speed UMV

The results in Sect. 6.3.1 demonstrate the effectiveness of the proposed FDF and controller coordinated design scheme for the UMV with a low forward speed. In this part, we show that for the UMV with a high forward speed, the proposed FDF and controller coordinated design is still effective, which is verified by the following results.

For the system matrices A, B, and E_1 in (6.2), choose the parameters as

$$\begin{aligned} &U = 7.8\ (\text{m/s}),\ T_v = 1.8/U,\ T_r = 2/U,\ K_{dr} = -0.0036U, \\ &K_{dp} = -0.0022U^2,\ K_{dv} = 0.06U,\ K_{vp} = 0.16U, \\ &\omega_n = 2.2\ (\text{rad/s}),\ \zeta = 0.58 + 0.67U,\ K_{vr} = -0.58\ (\text{m/s}), \end{aligned} \tag{6.33}$$

where U is the forward speed of the UMV. Then, we have

$$A = \begin{bmatrix} -4.3333 & 0 & 0 & 0 & 0 \\ -2.2620 & -3.9000 & 0 & 0 & 0 \\ 0 & 1.0000 & 0 & 0 & 0 \\ 6.0403 & 0 & 0 & -25.5464 & -4.8400 \\ 0 & 0 & 0 & 1.0000 & 0 \end{bmatrix},$$

$$B = \begin{bmatrix} 2.0280 \\ -0.1095 \\ 0 \\ -0.6478 \\ 0 \end{bmatrix},\ E_1 = \begin{bmatrix} 0 & 0 \\ 3.9000 & 0 \\ 0 & 0 \\ 0 & 4.8400 \\ 0 & 0 \end{bmatrix}.$$

Suppose that E_2, C, τ_m, τ_M, ρ_{ca}, h, W_{11}, ε_1, ε_2, and ξ_0 are the same as the corresponding items in Sect. 6.3.1. Since the matrix C is unchanged, one can conclude that X_1, X_2, Y and Φ are also the same as the corresponding items in Sect. 6.3.1.

By solving the FDF and controller coordinated design criterion in Theorem 6.2 for each feasible value of $\lambda(t)$ with $\lambda(t) = 1$ or $\lambda(t) = 0$, we have

$$\begin{aligned} L &= \begin{bmatrix} -0.5059 & -1.3356 & 2.0722 & 1.1118 & 0.8629 \end{bmatrix}^T, \\ V &= 0.0055, \\ K &= \begin{bmatrix} 0.0763 & 0.2385 & 0.5374 & -0.0198 & 0.3963 \end{bmatrix}. \end{aligned} \tag{6.34}$$

We take into account the disturbances of the yaw and roll motion which are similar to (13) and (14) in [9]. The disturbance of the yaw motion $\omega_\psi(t) = h(s)\omega_1(t)$, and the disturbance of roll motion $\omega_\phi(t) = h(s)\omega_2(t)$, where $\omega_1(t)$ is a normally (Gaussian) distributed random signal whose mean and variance are 0 and 1, respectively; $\omega_2(t)$ is a normally (Gaussian) distributed random signal whose mean and variance are 0.2 and 0.7, respectively; the shaping filter $h(s) = \frac{K_\omega s}{s^2 + 2\mu_0\omega_0 s + \omega_0^2}$ with K_ω, μ_0, and ω_0 denoting the dominate wave strength coefficient, the damping coefficient, and the encountering wave frequency, respectively, and $K_\omega = 0.7$, $\mu_0 = 0.8$, $\omega_0 = 1$. The fault is described as

$$f(t) = \begin{cases} -1.2\omega_\psi(t)\cos(2t), & t \in [0.5\text{s},\ 1.5\text{s}], \\ 0, & \text{otherwise}, \end{cases} \tag{6.35}$$

where $\omega_\psi(t) = h(s)\omega_1(t)$.

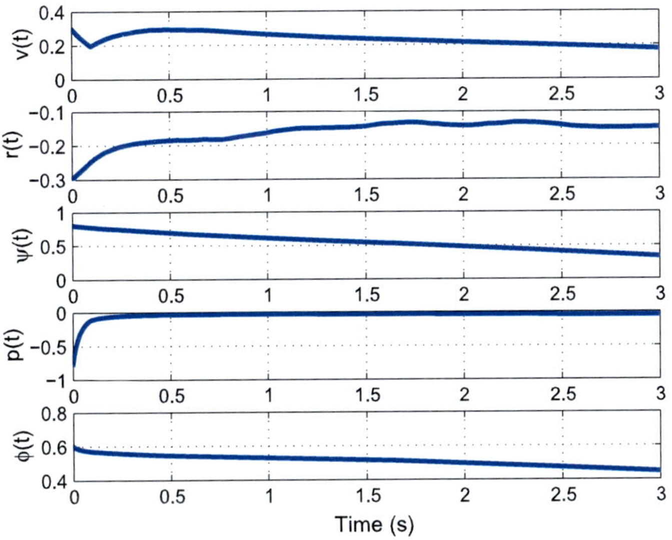

Fig. 6.11 The UMV state response $x(t)$ with $f(t)$ in (6.35)

Then, Figs. 6.11, 6.12, and 6.13 show the UMV state response $x(t)$, the residual response $r_e(t)$, and the residual evaluation function response $J(t)$, respectively. One can see that the sway velocity $v(t)$, the yaw velocity $r(t)$, the heading angle $\psi(t)$, the roll velocity $p(t)$, and the roll angle $\phi(t)$ in Fig. 6.11 are robust to wave-induced disturbances and almost not affected by the fault in (6.35). Figures 6.12 and 6.13 demonstrate that the residual response $r_e(t)$ and the residual evaluation function response $J(t)$ can reflect the occurrence of the fault given in (6.35) in time.

We now take different disturbances of the yaw and roll motion into consideration. The disturbance of the yaw motion $\omega_\psi(t) = \tilde{h}(s)\omega_1(t)$, and the disturbance of roll motion $\omega_\phi(t) = \tilde{h}(s)\omega_2(t)$, where $\omega_1(t)$ and $\omega_2(t)$ are the same as the corresponding items presented in this part; the shaping filter $\tilde{h}(s) = \frac{\tilde{K}_\omega s}{s^2+2\tilde{\mu}_0\tilde{\omega}_0 s+\tilde{\omega}_0^2}$ with the dominate wave strength coefficient $\tilde{K}_\omega = 1.1$, the damping coefficient $\tilde{\mu}_0 = 0.5$, and the encountering wave frequency $\tilde{\omega}_0 = 0.2$. The following fault is taken into account

$$f(t) = \begin{cases} 0.1\omega_\phi(t)\sin(t)\cos(5t), & t \in [0.5\text{s},\ 1.5\text{s}], \\ 0, & \text{otherwise}, \end{cases} \tag{6.36}$$

where $\omega_\phi(t) = \tilde{h}(s)\omega_2(t)$.

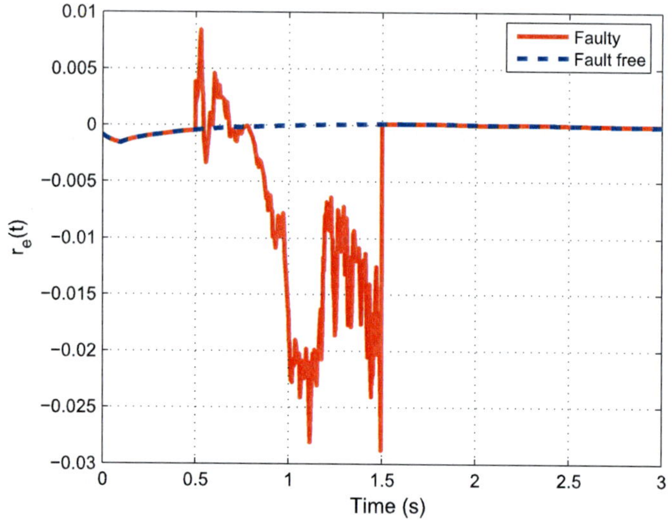

Fig. 6.12 The residual response $r_e(t)$ with $f(t)$ in (6.35)

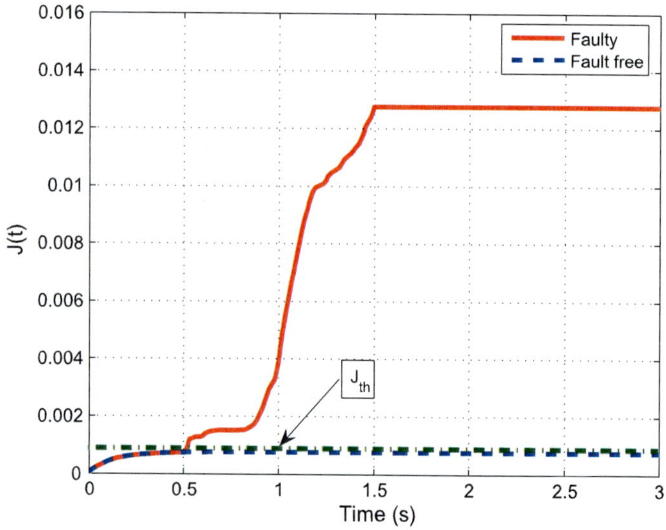

Fig. 6.13 The residual evaluation function response $J(t)$ with $f(t)$ in (6.35)

Then, Figs. 6.14, 6.15, and 6.16 show the UMV state response $x(t)$, the residual response $r_e(t)$, and the residual evaluation function response $J(t)$, respectively. As observed from Fig. 6.14, the sway velocity $v(t)$, the yaw velocity $r(t)$, the heading angle $\psi(t)$, the roll velocity $p(t)$, and the roll angle $\phi(t)$ are robust to wave-induced disturbances and the fault presented in (6.36). Figure 6.15 shows that the residual

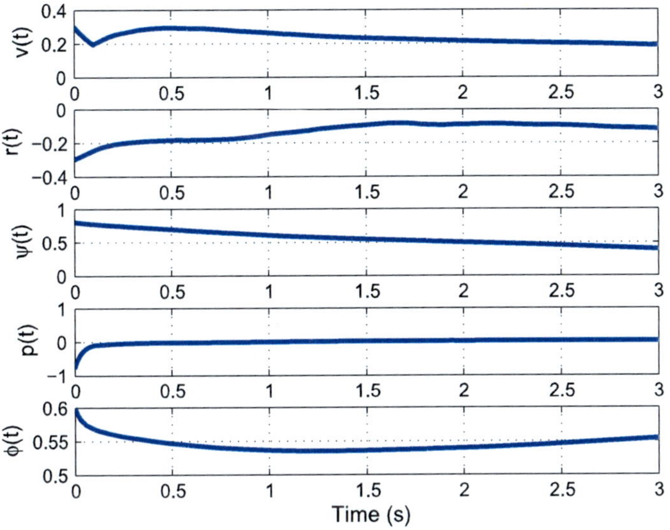

Fig. 6.14 The UMV state response $x(t)$ with $f(t)$ in (6.36)

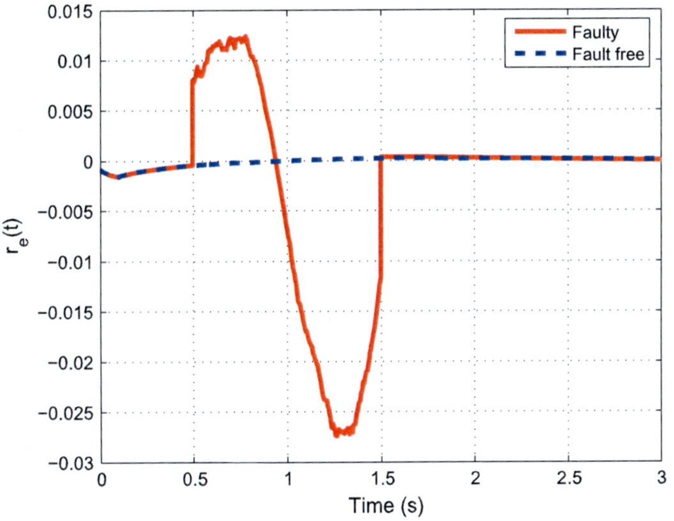

Fig. 6.15 The residual response $r_e(t)$ with $f(t)$ in (6.36)

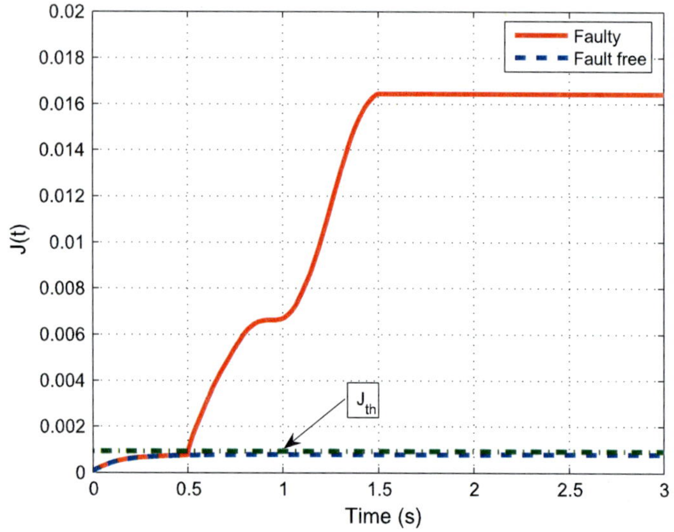

Fig. 6.16 The residual evaluation function response $J(t)$ with $f(t)$ in (6.36)

response $r_e(t)$ can reflect the occurrence of the fault presented in (6.36) in a timely manner. Figure 6.16 illustrates that the proposed FDF and controller coordinated design scheme can guarantee a short fault detection time.

From Figs. 6.13 and 6.16, and the fault detection logic presented in (3.10), one can see that the faults in (6.35) and (6.36) are detected at the instants 0.5085s and 0.5021s, respectively. That is, the fault detection time in Figs. 6.13 and 6.16 is 0.0085s and 0.0021s, respectively. When the occurrence of faults is detected, an alarm of fault occurrence is generated to remind the operating engineers of the UMV.

6.4 Conclusions

The network-based FDF and controller coordinated design for the UMV subject to actuator faults, and network-induced delays and packet dropouts in the sampler-to-control station communication network channel and the control station-to-actuator communication network channel has been studied. The mutually exclusive occurrence characteristic of the sampling instant $\tilde{t}_k$ has been taken into account to establish a networked model. Based on the established models and the combined convex analysis and mutually exclusive distribution approach, FDF and controller coordinated design criteria have been obtained to asymptotically stabilize the residual system. The effectiveness of the proposed FDF and controller coordinated design scheme has been verified by the fault detection performance analysis.

6.5 Notes

For a UMV in network environments, the occurrence of faults is usually unavoidable. Then it is significant to detect the occurrence of faults in time. However, for the UMV subject to faults, and the sampler-to-control station and the control station-to-actuator network-induced characteristics, how to establish a network-based model and coordinately design network-based FDF and controller are still unresolved. Moreover, when investigating network-based FDF and controller coordinated design for the UMV, if the mutually exclusive distribution characteristic of the interval time-varying delay $\tau(t)$ is considered, relaxed design criteria are expected to be derived. Motivated by these facts, the network-based FDF and controller coordinated design schemes for the UMV are proposed. The results in this chapter are based mainly on [10]. For more details about the corresponding analysis and design methods, see also [2, 3, 11, 12], etc.

References

1. G.-H. Yang, H. Wang, L. Xie, Fault detection for output feedback control systems with actuator stuck faults: a steady-state-based approach. Int. J. Robust Nonlinear Control **20**(15), 1739–1757 (2010)
2. X.-J. Li, G.-H. Yang, Dynamic observer-based robust control and fault detection for linear systems. IET Control Theory Appl. **6**(17), 2657–2666 (2012)
3. G.-X. Zhong, G.-H. Yang, Robust control and fault detection for continuous-time switched systems subject to a dwell time constraint. Int. J. Robust Nonlinear Control **25**(18), 3799–3817 (2015)
4. P. Park, J.W. Ko, Stability and robust stability for systems with a time-varying delay. Automatica **43**(10), 1855–1858 (2007)
5. K. Gu, V.L. Kharitonov, J. Chen, *Stability of Time-Delay Systems* (Birkhäuser, Boston, 2003)
6. F. Yang, Z. Wang, Y.S. Hung, M. Gani, H_∞ control for networked systems with random communication delays. IEEE Trans. Autom. Control **51**(3), 511–518 (2006)
7. P.G.M. van der Klugt, *Rudder Roll Stabilization*. Ph.D. thesis, Delft University of Technology, The Netherlands (1987)
8. Y.-L. Wang, Q.-L. Han, Modelling and observer-based H_∞ controller design for networked control systems. IET Control Theory Appl. **8**(15), 1478–1486 (2014)
9. R.-Y. Ren, Z.-J. Zou, X.-G. Wang, A two-time scale control law based on singular perturbations used in rudder roll stabilization of ships. Ocean Eng. **88**, 488–498 (2014)
10. Y.-L. Wang, Q.-L. Han, Network-based fault detection filter and controller coordinated design for unmanned surface vehicles in network environments. IEEE Trans. Ind. Inf. **12**(5), 1753–1765 (2016)
11. X. He, Z. Wang, Y. Liu, D.H. Zhou, Least-squares fault detection and diagnosis for networked sensing systems using a direct state estimation approach. IEEE Trans. Ind. Inf. **9**(3), 1670–1679 (2013)
12. J. You, S. Yin, H. Gao, Fault detection for discrete systems with network-induced nonlinearities. IEEE Trans. Ind. Inf. **10**(4), 2216–2223 (2014)

Chapter 7
T-S Fuzzy Dynamic Positioning Controller Design for UMVs

This chapter deals with T-S fuzzy dynamic positioning controller design for a UMV in network environments. Network-based T-S fuzzy DPS models for the UMV are first established. Then, by taking into consideration an asynchronous difference between the normalized membership function of the T-S fuzzy DPS and that of the controller, stability and stabilization criteria are derived. The proposed stabilization criteria can stabilize states of the UMV. The dynamic positioning performance analysis verifies the effectiveness of the networked modeling and the dynamic positioning controller design.

7.1 Network-Based T-S Fuzzy Modeling

This section aims to establish network-based T-S fuzzy dynamic positioning models for the UMV equipped with thrusters. For the normalized model of horizontal motion in a DPS, motion components such as surge, sway and yaw were investigated in [1]. Consider the body-fixed and earth-fixed reference frames presented in Fig. 7.1, where x, y, and z denote the longitudinal axis, transverse axis, and normal axis, respectively; X, Y, and Z denote earth-fixed reference frames. The origin of the coordinates is chosen to be at the center line of the marine vehicle.

The body-fixed equations of motion in surge, sway, and yaw are described as

$$M\dot{\nu}(t) + N\nu(t) + G\varphi(t) = u(t) + \omega(t), \tag{7.1}$$

where $\nu(t) = [\rho(t)\ \upsilon(t)\ r(t)]^T$ is the body-fixed linear and angular velocity vector with $\rho(t)$, $\upsilon(t)$, and $r(t)$ denoting the surge velocity, sway velocity, and yaw velocity, respectively; $\varphi(t) = [x(t)\ y(t)\ \psi(t)]^T$ is the earth-fixed orientation vector with $x(t)$ and $y(t)$ denoting positions and $\psi(t)$ denoting the yaw angle. The control input vector $u(t) = [u_1(t)\ u_2(t)\ u_3(t)]^T$ with $u_1(t)$ and $u_2(t)$ denoting the forces provided

Y.-L. Wang et al., *Network-Based Control of Unmanned Marine Vehicles*,
https://doi.org/10.1007/978-3-031-28605-6_7

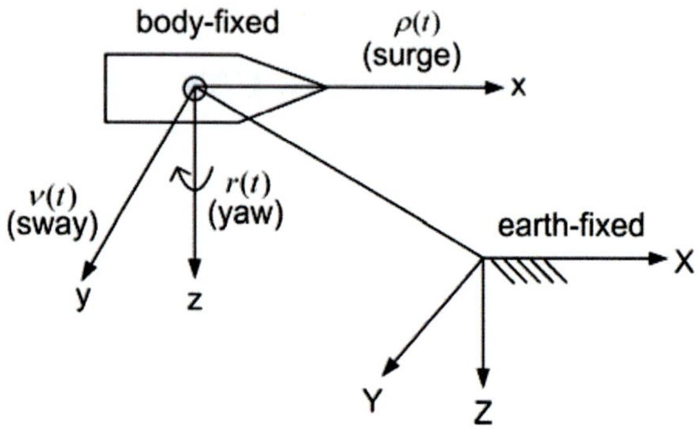

Fig. 7.1 Body-fixed and earth-fixed reference frames

by main propellers aft of the marine vehicle and by tunnel thrusters, respectively, and $u_3(t)$ denoting the moment in yaw provided by azimuth thrusters; $\omega(t)$ is the wave-induced disturbance; M denotes the matrix of inertia which is invertible with $M = M^T > 0$; N introduces damping; the matrix $G = \text{diag}\{g_{11}, \ g_{22}, \ g_{33}\}$ represents mooring forces; and

$$\dot{\varphi}(t) = \Omega(\psi(t))\nu(t), \tag{7.2}$$

where

$$\Omega(\psi(t)) = \begin{bmatrix} \cos(\psi(t)) & -\sin(\psi(t)) & 0 \\ \sin(\psi(t)) & \cos(\psi(t)) & 0 \\ 0 & 0 & 1 \end{bmatrix}.$$

The starboard-port symmetry of the marine vehicle implies that M and N own the following structure

$$M = \begin{bmatrix} m_{11} & 0 & 0 \\ 0 & m_{22} & m_{23} \\ 0 & m_{32} & m_{33} \end{bmatrix}, \ N = \begin{bmatrix} n_{11} & 0 & 0 \\ 0 & n_{22} & n_{23} \\ 0 & n_{32} & n_{33} \end{bmatrix}.$$

Let

$$A = -M^{-1}G = \begin{bmatrix} a_{11} & a_{12} & a_{13} \\ a_{21} & a_{22} & a_{23} \\ a_{31} & a_{32} & a_{33} \end{bmatrix}, \ B = -M^{-1}N = \begin{bmatrix} b_{11} & b_{12} & b_{13} \\ b_{21} & b_{22} & b_{23} \\ b_{31} & b_{32} & b_{33} \end{bmatrix},$$

$$D = M^{-1} = \begin{bmatrix} d_{11} & d_{12} & d_{13} \\ d_{21} & d_{22} & d_{23} \\ d_{31} & d_{32} & d_{33} \end{bmatrix}.$$

Then the system (7.1) can be expressed as

$$\dot{\nu}(t) = A\varphi(t) + B\nu(t) + Du(t) + D\omega(t). \tag{7.3}$$

Define $\xi(t) = [\xi_1(t)\ \ \xi_2(t)\ \ \xi_3(t)\ \ \xi_4(t)\ \ \xi_5(t)\ \ \xi_6(t)]^T = [x(t)\ \ y(t)\ \ \psi(t)\ \ \rho(t)\ \ \upsilon(t)\ r(t)]^T$, where $\xi_1(t)$ and $\xi_4(t)$ denote the earth-fixed position on the X-axis and the body-fixed velocity on the x-axis, respectively; $\xi_2(t)$ and $\xi_5(t)$ denote the earth-fixed position on the Y-axis and the body-fixed velocity on the y-axis, respectively; $\xi_3(t)$ and $\xi_6(t)$ denote the yaw angle and yaw angular velocity, respectively. Combining (7.2) and (7.3) together, one can obtain state equations described as follows

$$\begin{aligned}
\dot{\xi}_1(t) &= \cos(\xi_3(t))\xi_4(t) - \sin(\xi_3(t))\xi_5(t),\\
\dot{\xi}_2(t) &= \sin(\xi_3(t))\xi_4(t) + \cos(\xi_3(t))\xi_5(t),\\
\dot{\xi}_3(t) &= \xi_6(t),\\
\dot{\xi}_4(t) &= a_{11}\xi_1(t) + a_{12}\xi_2(t) + a_{13}\xi_3(t) + b_{11}\xi_4(t)\\
&\quad + b_{12}\xi_5(t) + b_{13}\xi_6(t) + d_{11}u_1(t) + d_{12}u_2(t)\\
&\quad + d_{13}u_3(t) + d_{11}\omega_1(t) + d_{12}\omega_2(t) + d_{13}\omega_3(t),\\
\dot{\xi}_5(t) &= a_{21}\xi_1(t) + a_{22}\xi_2(t) + a_{23}\xi_3(t) + b_{21}\xi_4(t)\\
&\quad + b_{22}\xi_5(t) + b_{23}\xi_6(t) + d_{21}u_1(t) + d_{22}u_2(t)\\
&\quad + d_{23}u_3(t) + d_{21}\omega_1(t) + d_{22}\omega_2(t) + d_{23}\omega_3(t),\\
\dot{\xi}_6(t) &= a_{31}\xi_1(t) + a_{32}\xi_2(t) + a_{33}\xi_3(t) + b_{31}\xi_4(t)\\
&\quad + b_{32}\xi_5(t) + b_{33}\xi_6(t) + d_{31}u_1(t) + d_{32}u_2(t)\\
&\quad + d_{33}u_3(t) + d_{31}\omega_1(t) + d_{32}\omega_2(t) + d_{33}\omega_3(t).
\end{aligned} \tag{7.4}$$

Without loss of generality, suppose that the yaw angle $\psi(t)$, which is also known as $\xi_3(t)$, varies between $-\frac{\pi}{6}$ and $\frac{\pi}{6}$, and let $\theta_1(t) = \sin(\xi_3(t))$, $\theta_2(t) = \cos(\xi_3(t))$. Then $\theta_1(t) \in [-\frac{1}{2},\ \frac{1}{2}]$, $\theta_2(t) \in [\frac{\sqrt{3}}{2},\ 1]$. The T-S fuzzy DPS can be obtained by introducing the following rules

Plant Rule i: IF $\theta_1(t)$ is W_{i1} and $\theta_2(t)$ is W_{i2}
THEN

$$\begin{cases}
\dot{\xi}(t) = \widetilde{A}_i\xi(t) + \widetilde{D}_i u(t) + \widetilde{D}_i\omega(t),\\
z(t) = C_{2i}\xi(t) + F_i\omega(t),\\
y(t) = C_1\xi(t),
\end{cases} \tag{7.5}$$

where $i = 1,\ 2,\ 3,$ and 4, $\theta_1(t) = \sin(\xi_3(t))$ and $\theta_2(t) = \cos(\xi_3(t))$ are premise variables, W_{i1} and W_{i2} are fuzzy sets, $z(t)$ and $y(t)$ denote the controlled output and the measured output, respectively, C_{2i}, F_i, and C_1 are known matrices with appropriate dimensions, while

$$\widetilde{A}_1 = \begin{bmatrix}
0 & 0 & 0 & 1 & -\frac{1}{2} & 0\\
0 & 0 & 0 & \frac{1}{2} & 1 & 0\\
0 & 0 & 0 & 0 & 0 & 1\\
a_{11} & a_{12} & a_{13} & b_{11} & b_{12} & b_{13}\\
a_{21} & a_{22} & a_{23} & b_{21} & b_{22} & b_{23}\\
a_{31} & a_{32} & a_{33} & b_{31} & b_{32} & b_{33}
\end{bmatrix},\quad
\widetilde{A}_2 = \begin{bmatrix}
0 & 0 & 0 & \frac{\sqrt{3}}{2} & -\frac{1}{2} & 0\\
0 & 0 & 0 & \frac{1}{2} & \frac{\sqrt{3}}{2} & 0\\
0 & 0 & 0 & 0 & 0 & 1\\
a_{11} & a_{12} & a_{13} & b_{11} & b_{12} & b_{13}\\
a_{21} & a_{22} & a_{23} & b_{21} & b_{22} & b_{23}\\
a_{31} & a_{32} & a_{33} & b_{31} & b_{32} & b_{33}
\end{bmatrix},$$

$$\widetilde{A}_3 = \begin{bmatrix} 0 & 0 & 0 & 1 & \frac{1}{2} & 0 \\ 0 & 0 & 0 & -\frac{1}{2} & 1 & 0 \\ 0 & 0 & 0 & 0 & 0 & 1 \\ a_{11} & a_{12} & a_{13} & b_{11} & b_{12} & b_{13} \\ a_{21} & a_{22} & a_{23} & b_{21} & b_{22} & b_{23} \\ a_{31} & a_{32} & a_{33} & b_{31} & b_{32} & b_{33} \end{bmatrix}, \ \widetilde{A}_4 = \begin{bmatrix} 0 & 0 & 0 & \frac{\sqrt{3}}{2} & \frac{1}{2} & 0 \\ 0 & 0 & 0 & -\frac{1}{2} & \frac{\sqrt{3}}{2} & 0 \\ 0 & 0 & 0 & 0 & 0 & 1 \\ a_{11} & a_{12} & a_{13} & b_{11} & b_{12} & b_{13} \\ a_{21} & a_{22} & a_{23} & b_{21} & b_{22} & b_{23} \\ a_{31} & a_{32} & a_{33} & b_{31} & b_{32} & b_{33} \end{bmatrix},$$

$$\widetilde{D}_1 = \widetilde{D}_2 = \widetilde{D}_3 = \widetilde{D}_4 = \widetilde{D} = \begin{bmatrix} 0 & 0 & 0 \\ 0 & 0 & 0 \\ 0 & 0 & 0 \\ d_{11} & d_{12} & d_{13} \\ d_{21} & d_{22} & d_{23} \\ d_{31} & d_{32} & d_{33} \end{bmatrix}.$$

Remark 7.1 Note that the measured output $y(t)$ is described as $y(t) = C_1\xi(t)$ instead of $y(t) = C_{1i}\xi(t)$. In fact, when dealing with observer-based controller design, an equality constraint $JC_{1i}^T = C_{1i}^T\bar{J}$ is usually introduced with J and $\bar{J}$ denoting unknown matrices. This equality constraint leads to much difficulty for designing the observer-based controller. Thus, this chapter chooses $C_{1i} = C_1$.

From the definition of $\widetilde{A}_i$, one can conclude that $W_{11}(\theta_1(t)) = W_{21}(\theta_1(t))$, $W_{31}(\theta_1(t)) = W_{41}(\theta_1(t))$, $W_{12}(\theta_2(t)) = W_{32}(\theta_2(t))$, $W_{22}(\theta_2(t)) = W_{42}(\theta_2(t))$. Note that

$$\begin{aligned} &W_{11}(\theta_1(t)) + W_{31}(\theta_1(t)) = 1, \\ &\tfrac{1}{2}W_{11}(\theta_1(t)) - \tfrac{1}{2}W_{31}(\theta_1(t)) = \theta_1(t). \end{aligned} \tag{7.6}$$

Then one has

$$\begin{aligned} &W_{11}(\theta_1(t)) = W_{21}(\theta_1(t)) = \tfrac{1}{2} + \theta_1(t), \\ &W_{31}(\theta_1(t)) = W_{41}(\theta_1(t)) = \tfrac{1}{2} - \theta_1(t). \end{aligned} \tag{7.7}$$

Similarly,

$$\begin{aligned} &W_{12}(\theta_2(t)) = W_{32}(\theta_2(t)) = -3 - 2\sqrt{3} + (4 + 2\sqrt{3})\theta_2(t), \\ &W_{22}(\theta_2(t)) = W_{42}(\theta_2(t)) = 4 + 2\sqrt{3} - (4 + 2\sqrt{3})\theta_2(t). \end{aligned} \tag{7.8}$$

From (7.5), one can derive the following T-S fuzzy DPS

$$\begin{cases} \dot{\xi}(t) = \sum\limits_{i=1}^{4} h_i(\theta(t))[\widetilde{A}_i\xi(t) + \widetilde{D}_i u(t) + \widetilde{D}_i\omega(t)], \\ z(t) = \sum\limits_{i=1}^{4} h_i(\theta(t))[C_{2i}\xi(t) + F_i\omega(t)], \\ y(t) = C_1\xi(t), \end{cases} \tag{7.9}$$

where

$$h_i(\theta(t)) = \frac{\vartheta_i(\theta(t))}{\sum\limits_{j=1}^{4}\vartheta_j(\theta(t))}, \ \vartheta_i(\theta(t)) = W_{i1}(\theta_1(t))W_{i2}(\theta_2(t)),$$

$$h_i(\theta(t)) \geq 0, \ \sum_{i=1}^{4} h_i(\theta(t)) = 1.$$

Remark 7.2 As observed from (7.5), a different variation scope of the yaw angle $\psi(t)$ will lead to different $\widetilde{A}_i$. This chapter assumes that the yaw angle $\psi(t)$ varies between $-\frac{\pi}{6}$ and $\frac{\pi}{6}$. If a different variation scope for the yaw angle $\psi(t)$ is chosen, the corresponding T-S fuzzy DPS is also different from (7.9).

Since communication networks are introduced between the UMV and the observer-based controller, the premise membership function structure of the observer may be different from that of the fuzzy system (7.9). Thus, it is quite important to construct an observer-based controller under imperfect premise matching. The rule of the observer is given as follows

Observer Rule j: IF $\hat{\theta}_1(t)$ is $\hat{W}_{j1}$ and $\hat{\theta}_2(t)$ is $\hat{W}_{j2}$
THEN

$$\begin{cases} \dot{\hat{\xi}}(t) = \widetilde{A}_j\hat{\xi}(t) + \widetilde{D}_j\hat{u}(t) + L_j(\bar{y}(t) - \hat{y}(t)), \\ \hat{y}(t) = C_1\hat{\xi}(t), \end{cases} \tag{7.10}$$

where $\hat{\theta}_1(t) = \sin(\hat{\xi}_3(t))$ and $\hat{\theta}_2(t) = \cos(\hat{\xi}_3(t))$ are premise variables, $\hat{W}_{j1}$ and $\hat{W}_{j2}$ are fuzzy sets, $j = 1,\ 2,\ 3,$ and 4; $\hat{\xi}(t)$ is the estimated observer state, $\bar{y}(t)$ is the measured output received by the observer, L_j is the observer gain to be designed. The global dynamics of the observer can be described as

$$\begin{cases} \dot{\hat{\xi}}(t) = \sum\limits_{j=1}^{4}\phi_j(\hat{\theta}(t))[\widetilde{A}_j\hat{\xi}(t) + \widetilde{D}_j\hat{u}(t) + L_j(\bar{y}(t) - \hat{y}(t))], \\ \hat{y}(t) = C_1\hat{\xi}(t), \end{cases} \tag{7.11}$$

where

$$\phi_j(\hat{\theta}(t)) = \frac{\hat{\vartheta}_j(\hat{\theta}(t))}{\sum\limits_{s=1}^{4}\hat{\vartheta}_s(\hat{\theta}(t))}, \ \hat{\vartheta}_j(\hat{\theta}(t)) = \hat{W}_{j1}(\hat{\theta}_1(t))\hat{W}_{j2}(\hat{\theta}_2(t)),$$

$$\phi_j(\hat{\theta}(t)) \geq 0, \ \sum_{j=1}^{4}\phi_j(\hat{\theta}(t)) = 1.$$

Note that no communication network is introduced between the observer and the controller. It is reasonable to assume that premise variables of the fuzzy observer and the controller are the same. Then the observer-based fuzzy control law can be represented as

Controller Rule l: IF $\hat{\theta}_1(t)$ is $\hat{W}_{l1}$ and $\hat{\theta}_2(t)$ is $\hat{W}_{l2}$
THEN

$$\hat{u}(t) = K_l\hat{\xi}(t), \tag{7.12}$$

where $\hat{\theta}_1(t) = \sin(\hat{\xi}_3(t))$ and $\hat{\theta}_2(t) = \cos(\hat{\xi}_3(t))$ are premise variables, $\hat{W}_{l1}$ and $\hat{W}_{l2}$ are fuzzy sets, $l = 1, 2, 3,$ and 4; K_l is the controller gain to be determined. Then the fuzzy controller is

$$\hat{u}(t) = \sum_{l=1}^{4} \phi_l(\hat{\theta}(t)) K_l \hat{\xi}(t), \tag{7.13}$$

where the definition for $\phi_l(\hat{\theta}(t))$ is similar to that of $\phi_j(\hat{\theta}(t))$.

In what follows, we aims to establish network-based models for the UMV.

Throughout this chapter, we consider the case that the UMV and the remote control station are connected through communication networks; there exist packet dropouts, network-induced delays, and packet disordering; if packet disordering occurs, the latest available data packets will be utilized by the observer or the actuator, and disordered packets will be dropped; the sampler is time-driven, while the observer-based controller and the actuator are event-driven; the actuator is chosen as the zero order holder which is connected to the propeller and thruster system; the sampler, the observer-based controller, and the actuator are assumed to be clock synchronized.

Figure 7.2 depicts the signal transmission for the UMV subject to sampler-to-controller and controller-to-actuator packet dropouts, network-induced delays, and packet disordering, where the solid lines denote successful data packet transmission, while the dashed lines and the dotted lines denote packet dropouts and packet disordering, respectively.

As observed from Fig. 7.2, sampled data based on the measured outputs at the instants $t_k h, t_{k+1}h, \ldots (k = 0, 1, 2, \ldots)$ are transmitted to the receivers successfully, while the data sampled at the instants $\tilde{t}_{k1}h$ and $\tilde{t}_{k2}h$ are dropped due to communication network unreliability and packet disordering, respectively; h denotes the length of the sampling period; $\varpi - 1$ denotes the upper bound of consecutive packet dropouts; τ_k^{sc} and τ_k denote the length of sampler-to-controller and sampler-to-actuator network-induced delays, respectively; τ_k^{ca} is defined as $\tau_k - \tau_k^{sc}$; $\tau_m^{sc} \leq \tau_k^{sc} \leq \tau_M^{sc}$, $\tau_m^{ca} \leq \tau_k^{ca} \leq \tau_M^{ca}$, $\tau_m \leq \tau_k \leq \tau_M$, where $\tau_m^{sc} \geq 0$, $\tau_m^{ca} \geq 0$, $\tau_m \geq 0$, and $\tau_m = \tau_m^{sc} + \tau_m^{ca}$.

Thus, for $t \in [t_k h + \tau_k^{sc}, t_{k+1}h + \tau_{k+1}^{sc})$, the measurement output utilized by the observer is described as

$$\bar{y}(t) = y(t_k h). \tag{7.14}$$

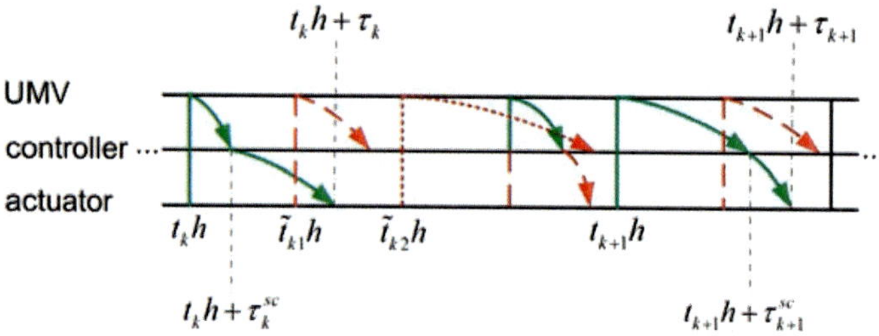

Fig. 7.2 Signal transmission for the UMV

For $t \in [t_k h + \tau_k,\ t_{k+1}h + \tau_{k+1})$, the control input available to the actuator is

$$u(t) = \hat{u}(t_k h + \tau_k^{sc}). \tag{7.15}$$

Remark 7.3 Note that for $t \in [t_k h + \tau_k,\ t_{k+1}h + \tau_{k+1})$, the measurement outputs available to the observer are variable. Then one should choose an appropriate measurement output for the observer. At the instant $t_k h + \tau_k$, the measurement outputs utilized by the observer may be sampled at instants $t_k h,\ t_k h + h,\ \ldots,\ t_k h + \lfloor \frac{\tau_k}{h} \rfloor h$, where $\lfloor \frac{\tau_k}{h} \rfloor$ is the largest integer smaller than or equal to $\frac{\tau_k}{h}$. A similar conclusion can be drawn for the instant $t_{k+1}h + \tau_{k+1}$. When the control input $\hat{u}(t_k h + \tau_k^{sc})$ is received by the actuator at the instant $t_k + \tau_k$, the actuator sends an acknowledgement signal to the observer-based controller. The acknowledgement signal is assigned the highest transmission priority and its network-induced delays are negligible. Thus, for $t \in [t_k h + \tau_k,\ t_{k+1}h + \tau_{k+1})$, the observer can choose to use the measurement output $y(t_k h)$ as its input.

Define $d(t) = t - t_k h$ and $\tau(t) = t - t_k h - \tau_k^{sc}$. Then one has $d(t) \in [d_{low},\ \bar{d})$ and $\tau(t) \in [\tau_{low},\ \bar{\tau})$ with $d_{low} = \tau_m$, $\bar{d} = \varpi h + \tau_M$, $\tau_{low} = \tau_m^{ca}$, and $\bar{\tau} = \varpi h + \tau_M - \tau_m^{sc}$. Define $\tilde{\xi}(t) = \xi(t) - \hat{\xi}(t)$, and $\bar{\xi}(t) = [\hat{\xi}^T(t)\ \ \tilde{\xi}^T(t)]^T$. The following networked T-S fuzzy DPS can be established readily

$$\begin{cases} \dot{\bar{\xi}}(t) = \sum\limits_{i=1}^{4}\sum\limits_{j=1}^{4}\sum\limits_{l=1}^{4}\sum\limits_{s=1}^{4} h_i(\theta(t))\phi_j(\hat{\theta}(t))\phi_l(\hat{\theta}(t)) \\ \qquad\qquad \phi_s(\hat{\theta}(t-\tau(t)))[\Pi_{1ijl}\bar{\xi}(t) + \Pi_{2s}\bar{\xi}(t-\tau(t)) \\ \qquad\qquad +\Pi_{3j}\bar{\xi}(t-d(t)) + \bar{D}\omega(t)], \\ z(t) = \sum\limits_{i=1}^{4} h_i(\theta(t))[C_{2i}\widetilde{I}\bar{\xi}(t) + F_i\omega(t)], \end{cases} \tag{7.16}$$

where

$$\Pi_{1ijl} = \begin{bmatrix} \widetilde{A}_j + \widetilde{D}K_l - L_jC_1 & 0 \\ \widetilde{A}_i - \widetilde{A}_j - \widetilde{D}K_l + L_jC_1 & \widetilde{A}_i \end{bmatrix},\ \Pi_{2s} = \begin{bmatrix} 0 & 0 \\ \widetilde{D}K_s & 0 \end{bmatrix},$$
$$\Pi_{3j} = \begin{bmatrix} L_jC_1 & L_jC_1 \\ -L_jC_1 & -L_jC_1 \end{bmatrix},\ \bar{D} = \begin{bmatrix} 0 \\ \widetilde{D} \end{bmatrix},\ \widetilde{I} = [I\ I].$$

Remark 7.4 Note that $\widetilde{D}_1 = \widetilde{D}_2 = \widetilde{D}_3 = \widetilde{D}_4 = \widetilde{D}$. Then $\widetilde{D}_i$ $(i = 1,\ 2,\ 3,\ 4)$ is written as $\widetilde{D}$ in (7.16) and the followed expressions for brevity.

If communication networks are introduced between the UMV and the observer-based controller, it is natural to take into account the imperfect premise matching. However, this will lead to increased computational complexity inevitably, which statement is verified by the stability criterion in Theorem 7.1. Without loss of generality, we turn to consider the case of premise matching, and the rule of the observer is described as

Observer Rule i: IF $\theta_1(t)$ is W_{i1} and $\theta_2(t)$ is W_{i2}

THEN

$$\begin{cases} \dot{\hat{\xi}}(t) = \widetilde{A}_i\hat{\xi}(t) + \widetilde{D}\hat{u}(t) + L_i(\bar{y}(t) - \hat{y}(t)), \\ \hat{y}(t) = C_1\hat{\xi}(t), \end{cases} \tag{7.17}$$

where $\theta_1(t)$, $\theta_2(t)$, W_{i1}, and W_{i2} are the same as the corresponding items in Plant Rule i; $\hat{\xi}(t)$, $\bar{y}(t)$, and L_i are the same as the corresponding items in (7.10). The global dynamics of the observer is described as

$$\begin{cases} \dot{\hat{\xi}}(t) = \sum_{j=1}^{4} h_j(\theta(t))[\widetilde{A}_j\hat{\xi}(t) + \widetilde{D}\hat{u}(t) + L_j(\bar{y}(t) - \hat{y}(t))], \\ \hat{y}(t) = C_1\hat{\xi}(t), \\ \hat{u}(t) = K\hat{\xi}(t), \end{cases} \tag{7.18}$$

where the definition of $h_j(\theta(t))$ is similar to $h_i(\theta(t))$ in (7.9).

Motivated by the networked T-S fuzzy DPS in (7.16), one can establish the fuzzy DPS under premise matching as follows

$$\begin{cases} \dot{\bar{\xi}}(t) = \sum_{i=1}^{4}\sum_{j=1}^{4} h_i(\theta(t))h_j(\theta(t))[\Pi_{1ij}\bar{\xi}(t) \\ \qquad +\Pi_2\bar{\xi}(t-\tau(t)) + \Pi_{3j}\bar{\xi}(t-d(t)) + \bar{D}\omega(t)], \\ z(t) = \sum_{i=1}^{4} h_i(\theta(t))[C_{2i}\widetilde{I}\bar{\xi}(t) + F_i\omega(t)], \end{cases} \tag{7.19}$$

where

$$\Pi_{1ij} = \begin{bmatrix} \widetilde{A}_j + \widetilde{D}K - L_jC_1 & 0 \\ \widetilde{A}_i - \widetilde{A}_j - \widetilde{D}K + L_jC_1 & \widetilde{A}_i \end{bmatrix}, \Pi_2 = \begin{bmatrix} 0 & 0 \\ \widetilde{D}K & 0 \end{bmatrix},$$
$$\Pi_{3j} = \begin{bmatrix} L_jC_1 & L_jC_1 \\ -L_jC_1 & -L_jC_1 \end{bmatrix}, \bar{D} = \begin{bmatrix} 0 \\ \widetilde{D} \end{bmatrix}, \widetilde{I} = \begin{bmatrix} I & I \end{bmatrix}.$$

Note that sampler-to-controller and controller-to-actuator network-induced characteristics are taken into full consideration in closed-loop systems (7.16) and (7.19). If one considers only controller-to-actuator network-induced characteristics, the global dynamics of the observer in (7.18) reduces to

$$\begin{cases} \dot{\hat{\xi}}(t) = \sum_{j=1}^{4} h_j(\theta(t))[\widetilde{A}_j\hat{\xi}(t) + \widetilde{D}\hat{u}(t) + L_j(y(t) - \hat{y}(t))], \\ \hat{y}(t) = C_1\hat{\xi}(t), \\ \hat{u}(t) = K\hat{\xi}(t), \end{cases} \tag{7.20}$$

where $h_j(\theta(t))$ is similar to $h_i(\theta(t))$ in (7.9). Then the networked T-S fuzzy DPS in (7.19) is converted to

$$\begin{cases} \dot{\bar{\xi}}(t) = \sum_{i=1}^{4}\sum_{j=1}^{4} h_i(\theta(t))h_j(\theta(t))[\bar{\Pi}_{1ij}\bar{\xi}(t) + \Pi_2\bar{\xi}(t-\tau(t)) + \bar{D}\omega(t)], \\ z(t) = \sum_{i=1}^{4} h_i(\theta(t))[C_{2i}\widetilde{I}\bar{\xi}(t) + F_i\omega(t)], \end{cases} \tag{7.21}$$

where

$$\widetilde{\Pi}_{1ij} = \begin{bmatrix} \widetilde{A}_j + \widetilde{D}K & L_j C_1 \\ \widetilde{A}_i - \widetilde{A}_j - \widetilde{D}K & \widetilde{A}_i - L_j C_1 \end{bmatrix}, \; \Pi_2 = \begin{bmatrix} 0 & 0 \\ \widetilde{D}K & 0 \end{bmatrix}, \; \bar{D} = \begin{bmatrix} 0 \\ \widetilde{D} \end{bmatrix}, \; \widetilde{I} = \begin{bmatrix} I & I \end{bmatrix},$$

and $\tau(t) \in [\tau_l, \ \tau^u)$, where $\tau_l = \tau_m^{ca}$, $\tau^u = \varpi h + \tau_M^{ca}$.

Remark 7.5 Note that the T-S fuzzy-model-based dynamic positioning of a marine vehicle is investigated in [2] by using a state feedback fuzzy controller. In this chapter, we consider an observer-based output feedback fuzzy control for dynamic positioning of a marine vehicle in network environments. As a consequence, network-induced characteristics are taken into full consideration to establish network-based models and then investigate the stability analysis and controller design for networked T-S fuzzy DPSs.

7.2 Stability Analysis for Networked T-S Fuzzy DPSs

In this section, we analyse stability of network-based T-S fuzzy DPSs (7.16) and (7.19) for the UMV. In doing so, we construct the Lyapunov-Krasovskii functional as

$$V(t, \bar{\xi}_t) = \sum_{i=1}^{4} V_i(t, \bar{\xi}_t), \tag{7.22}$$

where

$$\begin{aligned}
V_1(t, \bar{\xi}_t) &= \bar{\xi}^T(t) P \bar{\xi}(t), \\
V_2(t, \bar{\xi}_t) &= (\bar{\tau} - \tau(t)) \int_{t-\tau(t)}^{t} \dot{\bar{\xi}}^T(s) Q_1 \dot{\bar{\xi}}(s) ds + (\bar{d} - d(t)) \int_{t-d(t)}^{t} \dot{\bar{\xi}}^T(s) Q_2 \dot{\bar{\xi}}(s) ds, \\
V_3(t, \bar{\xi}_t) &= \int_{t-\tau_{low}}^{t} \bar{\xi}^T(s) R_1 \bar{\xi}(s) ds + \int_{t-\bar{\tau}}^{t-\tau_{low}} \bar{\xi}^T(s) R_2 \bar{\xi}(s) ds + \int_{t-d_{low}}^{t} \bar{\xi}^T(s) R_3 \bar{\xi}(s) ds \\
&\quad + \int_{t-\bar{d}}^{t-d_{low}} \bar{\xi}^T(s) R_4 \bar{\xi}(s) ds, \\
V_4(t, \bar{\xi}_t) &= \int_{-\bar{\tau}}^{-\tau_{low}} \int_{t+s}^{t} \dot{\bar{\xi}}^T(\theta) S_1 \dot{\bar{\xi}}(\theta) d\theta ds + \int_{-\bar{d}}^{-d_{low}} \int_{t+s}^{t} \dot{\bar{\xi}}^T(\theta) S_2 \dot{\bar{\xi}}(\theta) d\theta ds,
\end{aligned}$$

and $\bar{\xi}_t = \bar{\xi}(t + \varsigma)$ with $\varsigma \in [t_0 h + \tau_0 - max\{\bar{\tau}, \ \bar{d}\}, \ t_0 h + \tau_0]$, while P, Q_1, Q_2, R_1, R_2, R_3, R_4, S_1, and S_2 are symmetric positive definite matrices of appropriate dimensions. We now derive the following stability criterion for the network-based system (7.16).

Theorem 7.1 *For given scalars* $\tau_m^{ca} \geq 0$, $\tau_m^{sc} \geq 0$, $\tau_M > 0$, $h > 0$, $\varpi > 0$, $\gamma > 0$, $\sigma_m \in [0, \ 1]$, *membership functions* $\phi_m(\hat{\theta}(t - \tau(t)))$ *and* $\phi_m(\hat{\theta}(t))$ *satisfying*

$|\phi_m(\hat{\theta}(t-\tau(t))) - \phi_m(\hat{\theta}(t))| \le \sigma_m$, *and matrices* L_j *and* K *of appropriate dimensions, the network-based T-S fuzzy DPS (7.16) is asymptotically stable with an* H_∞ *norm bound* γ, *if there exist symmetric positive definite matrices* P, Q_1, Q_2, R_1, R_2, R_3, R_4, S_1, S_2, *and* Z_{ijl} *such that*

$$\Gamma_{ijlm} + \Xi_{ijlm} + Z_{ijl} > 0, \tag{7.23}$$

$$\Gamma_{ijjs} + \Xi_{ijjs} + \sum_{m=1}^{4} \sigma_m(\Gamma_{ijjm} + \Xi_{ijjm} + Z_{ijj}) < 0, \tag{7.24}$$

$$\Gamma_{ijls} + \Gamma_{iljs} + \Xi_{ijls} + \Xi_{iljs} + \sum_{m=1}^{4} \sigma_m(\Gamma_{ijlm} + \Gamma_{iljm} + \Xi_{ijlm} + \Xi_{iljm} + Z_{ijl} + Z_{ilj}) < 0, \tag{7.25}$$

where $i,\ j,\ l,\ s,\ m = 1,\ 2,\ 3,$ *and* $4\ (j < l)$, *and*

$$\Gamma_{ijls} = \begin{bmatrix} \Gamma_{1,ijls} & \Gamma_2 \\ * & \Gamma_3 \end{bmatrix},\ \Xi_{ijls} = U_{1ijls}^T \Theta U_{1ijls} + \gamma^{-1} U_{2i}^T U_{2i},$$

$$\Gamma_{1,ijls} = \begin{bmatrix} \Gamma_{ijl}^{11} & 0 & 0 & 0 & 0 & \Gamma_s^{16} & \Gamma_j^{17} \\ * & \Gamma^{22} & 0 & 0 & 0 & \Gamma^{26} & 0 \\ * & * & \Gamma^{33} & 0 & 0 & \Gamma^{36} & 0 \\ * & * & * & \Gamma^{44} & 0 & 0 & \Gamma^{47} \\ * & * & * & * & \Gamma^{55} & 0 & \Gamma^{57} \\ * & * & * & * & * & \Gamma^{66} & 0 \\ * & * & * & * & * & 0 & \Gamma^{77} \end{bmatrix},\ \Gamma_2 = \begin{bmatrix} \frac{6Q_1}{\bar{\tau}} & \frac{6Q_2}{\bar{d}} & 0 & 0 & 0 & 0 & P\bar{D} \\ 0 & 0 & \hat{S}_1 & 0 & 0 & 0 & 0 \\ 0 & 0 & 0 & \hat{S}_1 & 0 & 0 & 0 \\ 0 & 0 & 0 & 0 & \hat{S}_2 & 0 & 0 \\ 0 & 0 & 0 & 0 & 0 & \hat{S}_2 & 0 \\ \frac{6Q_1}{\bar{\tau}} & 0 & \hat{S}_1 & \hat{S}_1 & 0 & 0 & 0 \\ 0 & \frac{6Q_2}{\bar{d}} & 0 & 0 & \hat{S}_2 & \hat{S}_2 & 0 \end{bmatrix},$$

$$\begin{aligned}
\Gamma_3 =& \text{diag}\{-\frac{12Q_1}{\bar{\tau}},\ -\frac{12Q_2}{\bar{d}},\ -\frac{12S_1}{\bar{\tau}-\tau_{low}},\ -\frac{12S_1}{\bar{\tau}-\tau_{low}},\ -\frac{12S_2}{\bar{d}-d_{low}}, \\
& -\frac{12S_2}{\bar{d}-d_{low}},\ -\gamma I\}, \\
\Gamma_{ijl}^{11} =& P\Pi_{1ijl} + \Pi_{1ijl}^T P - \frac{4Q_1}{\bar{\tau}} - \frac{4Q_2}{\bar{d}} + R_1 + R_3,\ \Gamma_s^{16} = P\Pi_{2s} - \frac{2Q_1}{\bar{\tau}}, \\
\Gamma_j^{17} =& P\Pi_{3j} - \frac{2Q_2}{\bar{d}},\ \Gamma^{22} = R_2 - R_1 - \frac{4S_1}{\bar{\tau}-\tau_{low}},\ \Gamma^{26} = -\frac{2S_1}{\bar{\tau}-\tau_{low}}, \\
\Gamma^{33} =& -R_2 - \frac{4S_1}{\bar{\tau}-\tau_{low}},\ \Gamma^{36} = -\frac{2S_1}{\bar{\tau}-\tau_{low}},\ \Gamma^{44} = R_4 - R_3 - \frac{4S_2}{\bar{d}-d_{low}}, \\
\Gamma^{47} =& -\frac{2S_2}{\bar{d}-d_{low}},\ \Gamma^{55} = -R_4 - \frac{4S_2}{\bar{d}-d_{low}},\ \Gamma^{57} = -\frac{2S_2}{\bar{d}-d_{low}}, \\
\Gamma^{66} =& -\frac{4Q_1}{\bar{\tau}} - \frac{8S_1}{\bar{\tau}-\tau_{low}},\ \Gamma^{77} = -\frac{4Q_2}{\bar{d}} - \frac{8S_2}{\bar{d}-d_{low}},\ \hat{S}_1 = \frac{6S_1}{\bar{\tau}-\tau_{low}},\ \hat{S}_2 = \frac{6S_2}{\bar{d}-d_{low}}, \\
U_{1ijls} =& \Big[\Pi_{1ijl}\ \underbrace{0,\ \ldots,\ 0}_{4}\ \Pi_{2s}\ \Pi_{3j}\ \underbrace{0,\ \ldots,\ 0}_{6}\ \bar{D}\Big],\ U_{2i} = \Big[C_{2i}\tilde{I}\ \underbrace{0,\ \ldots,\ 0}_{12}\ F_i\Big], \\
\tilde{I} =& [I\ I],\ \Theta = (\bar{\tau}-\tau_{low})(Q_1+S_1) + (\bar{d}-d_{low})(Q_2+S_2),
\end{aligned} \tag{7.26}$$

while Γ_{ijlm} and Ξ_{ijlm} are derived from Γ_{ijls} and Ξ_{ijls}, respectively, by substituting the subscript s with m; Γ_{ijjs}, Ξ_{ijjs}, Γ_{ijjm}, Ξ_{ijjm}, and Z_{ijj} are derived from Γ_{ijls}, Ξ_{ijls}, Γ_{ijlm}, Ξ_{ijlm}, and Z_{ijl}, respectively, by substituting the subscript l with j; Γ_{iljs}, Ξ_{iljs}, Γ_{iljm}, Ξ_{iljm}, and Z_{ilj} are derived from Γ_{ijls}, Ξ_{ijls}, Γ_{ijlm}, Ξ_{ijlm}, and Z_{ijl}, respectively, by interchanging the subscripts j and l.

Proof Taking the time derivative of $V(t, \bar{\xi}_t)$ given in (7.22), one has

$$\dot{V}_1(t, \bar{\xi}_t) = 2\bar{\xi}^T(t)P\dot{\bar{\xi}}(t). \tag{7.27}$$

Applying the Wirtinger-based integral inequality [3, 4], we have

$$\begin{aligned}\dot{V}_2(t, \bar{\xi}_t) &= (\bar{\tau} - \tau(t))\dot{\bar{\xi}}^T(t)Q_1\dot{\bar{\xi}}(t) + (\bar{d} - d(t))\dot{\bar{\xi}}^T(t)Q_2\dot{\bar{\xi}}(t)\\ &\quad - \int_{t-\tau(t)}^{t} \dot{\bar{\xi}}^T(s)Q_1\dot{\bar{\xi}}(s) - \int_{t-d(t)}^{t} \dot{\bar{\xi}}^T(s)Q_2\dot{\bar{\xi}}(s)\\ &\leq (\bar{\tau} - \tau_{low})\dot{\bar{\xi}}^T(t)Q_1\dot{\bar{\xi}}(t) + (\bar{d} - d_{low})\dot{\bar{\xi}}^T(t)Q_2\dot{\bar{\xi}}(t)\\ &\quad - \frac{1}{\bar{\tau}}\zeta_1^T Q_1\zeta_1 - \frac{3}{\bar{\tau}}\zeta_2^T Q_1\zeta_2 - \frac{1}{\bar{d}}\zeta_3^T Q_2\zeta_3 - \frac{3}{\bar{d}}\zeta_4^T Q_2\zeta_4,\end{aligned} \tag{7.28}$$

$$\begin{aligned}\dot{V}_3(t, \bar{\xi}_t) &= \bar{\xi}^T(t)R_1\bar{\xi}(t) + \bar{\xi}^T(t - \tau_{low})(R_2 - R_1)\bar{\xi}(t - \tau_{low}) - \bar{\xi}^T(t - \bar{\tau})R_2\bar{\xi}(t - \bar{\tau})\\ &\quad + \bar{\xi}^T(t)R_3\bar{\xi}(t) + \bar{\xi}^T(t - d_{low})(R_4 - R_3)\bar{\xi}(t - d_{low}) - \bar{\xi}^T(t - \bar{d})R_4\bar{\xi}(t - \bar{d}).\end{aligned} \tag{7.29}$$

Similar to the inequality in (7.28), we have

$$\begin{aligned}\dot{V}_4(t, \bar{\xi}_t) &\leq (\bar{\tau} - \tau_{low})\dot{\bar{\xi}}^T(t)S_1\dot{\bar{\xi}}(t) + (\bar{d} - d_{low})\dot{\bar{\xi}}^T(t)S_2\dot{\bar{\xi}}(t) - \frac{1}{\bar{\tau} - \tau_{low}}\zeta_5^T S_1\zeta_5\\ &\quad - \frac{3}{\bar{\tau} - \tau_{low}}\zeta_6^T S_1\zeta_6 - \frac{1}{\bar{\tau} - \tau_{low}}\zeta_7^T S_1\zeta_7 - \frac{3}{\bar{\tau} - \tau_{low}}\zeta_8^T S_1\zeta_8 - \frac{1}{\bar{d} - d_{low}}\zeta_9^T S_2\zeta_9\\ &\quad - \frac{3}{\bar{d} - d_{low}}\zeta_{10}^T S_2\zeta_{10} - \frac{1}{\bar{d} - d_{low}}\zeta_{11}^T S_2\zeta_{11} - \frac{3}{\bar{d} - d_{low}}\zeta_{12}^T S_2\zeta_{12},\end{aligned} \tag{7.30}$$

where

$$\begin{aligned}\zeta_1 &= \bar{\xi}(t) - \bar{\xi}(t - \tau(t)),\ \zeta_2 = \bar{\xi}(t) + \bar{\xi}(t - \tau(t)) - \frac{2}{\tau(t)}\int_{t-\tau(t)}^{t} \bar{\xi}(s)ds,\\ \zeta_3 &= \bar{\xi}(t) - \bar{\xi}(t - d(t)),\ \zeta_4 = \bar{\xi}(t) + \bar{\xi}(t - d(t)) - \frac{2}{d(t)}\int_{t-d(t)}^{t} \bar{\xi}(s)ds,\\ \zeta_5 &= \bar{\xi}(t - \tau_{low}) - \bar{\xi}(t - \tau(t)),\\ \zeta_6 &= \bar{\xi}(t - \tau_{low}) + \bar{\xi}(t - \tau(t)) - \frac{2}{\tau(t) - \tau_{low}}\int_{t-\tau(t)}^{t-\tau_{low}} \bar{\xi}(s)ds,\end{aligned}$$

$$\begin{aligned}
\zeta_7 &= \bar{\xi}(t-\tau(t)) - \bar{\xi}(t-\bar{\tau}),\\
\zeta_8 &= \bar{\xi}(t-\tau(t)) + \bar{\xi}(t-\bar{\tau}) - \frac{2}{\bar{\tau}-\tau(t)}\int_{t-\bar{\tau}}^{t-\tau(t)} \bar{\xi}(s)ds,\\
\zeta_9 &= \bar{\xi}(t-d_{low}) - \bar{\xi}(t-d(t)),\\
\zeta_{10} &= \bar{\xi}(t-d_{low}) + \bar{\xi}(t-d(t)) - \frac{2}{d(t)-d_{low}}\int_{t-d(t)}^{t-d_{low}} \bar{\xi}(s)ds,
\end{aligned}$$

$$\begin{aligned}
\zeta_{11} &= \bar{\xi}(t-d(t)) - \bar{\xi}(t-\bar{d}),\\
\zeta_{12} &= \bar{\xi}(t-d(t)) + \bar{\xi}(t-\bar{d}) - \frac{2}{\bar{d}-d(t)}\int_{t-\bar{d}}^{t-d(t)} \bar{\xi}(s)ds.
\end{aligned} \tag{7.31}$$

Define

$$\begin{aligned}
\eta(t) = [&\bar{\xi}^T(t),\ \bar{\xi}^T(t-\tau_{low}),\ \bar{\xi}^T(t-\bar{\tau}),\ \bar{\xi}^T(t-d_{low}),\ \bar{\xi}^T(t-\bar{d}),\ \bar{\xi}^T(t-\tau(t)),\\
&\bar{\xi}^T(t-d(t)),\ \frac{1}{\tau(t)}\int_{t-\tau(t)}^{t} \bar{\xi}^T(s)ds,\ \frac{1}{d(t)}\int_{t-d(t)}^{t} \bar{\xi}^T(s)ds,\\
&\frac{1}{\tau(t)-\tau_{low}}\int_{t-\tau(t)}^{t-\tau_{low}} \bar{\xi}^T(s)ds,\ \frac{1}{\bar{\tau}-\tau(t)}\int_{t-\bar{\tau}}^{t-\tau(t)} \bar{\xi}^T(s)ds,\\
&\frac{1}{d(t)-d_{low}}\int_{t-d(t)}^{t-d_{low}} \bar{\xi}^T(s)ds,\ \frac{1}{\bar{d}-d(t)}\int_{t-\bar{d}}^{t-d(t)} \bar{\xi}^T(s)ds,\ \omega^T(t)]^T.
\end{aligned}$$

Then we can obtain the following inequality

$$\begin{aligned}
&\dot{V}(t,\bar{\xi}_t) + \gamma^{-1}z^T(t)z(t) - \gamma\omega^T(t)\omega(t)\\
&\le \sum_{i=1}^{4}\sum_{j=1}^{4}\sum_{l=1}^{4}\sum_{s=1}^{4} h_i(\theta(t))\phi_j(\hat{\theta}(t))\phi_l(\hat{\theta}(t))\phi_s(\hat{\theta}(t-\tau(t)))\eta^T(t)[\Gamma_{ijls}+\Xi_{ijls}]\eta(t),
\end{aligned} \tag{7.32}$$

where Γ_{ijls} and Ξ_{ijls} are the same as the corresponding items in (7.26).

Note that the right side of the inequality (7.32) can be rewritten as

$$\begin{aligned}
\Delta = &\sum_{i=1}^{4}\sum_{j=1}^{4}\sum_{l=1}^{4}\sum_{s=1}^{4} h_i(\theta(t))\phi_j(\hat{\theta}(t))\phi_l(\hat{\theta}(t))\phi_s(\hat{\theta}(t))\eta^T(t)[\Gamma_{ijls}+\Xi_{ijls}]\eta(t)\\
&+\sum_{i=1}^{4}\sum_{j=1}^{4}\sum_{l=1}^{4}\sum_{s=1}^{4} h_i(\theta(t))\phi_j(\hat{\theta}(t))\phi_l(\hat{\theta}(t))(\phi_s(\hat{\theta}(t-\tau(t))) - \phi_s(\hat{\theta}(t)))\\
&\times \eta^T(t)[\Gamma_{ijls}+\Xi_{ijls}+Z_{ijl}]\eta(t).
\end{aligned} \tag{7.33}$$

By assuming that $|\phi_m(\hat{\theta}(t-\tau(t))) - \phi_m(\hat{\theta}(t))| \le \sigma_m$, one can see that

$$\begin{aligned}
&\sum_{i=1}^{4}\sum_{j=1}^{4}\sum_{l=1}^{4}\sum_{m=1}^{4} h_i(\theta(t))\phi_j(\hat{\theta}(t))\phi_l(\hat{\theta}(t))(\phi_m(\hat{\theta}(t-\tau(t))) \\
&\qquad\qquad - \phi_m(\hat{\theta}(t)))\eta^T(t)[\Gamma_{ijlm} + \Xi_{ijlm} + Z_{ijl}]\eta(t) \\
&\leq \sum_{i=1}^{4}\sum_{j=1}^{4}\sum_{l=1}^{4}\sum_{s=1}^{4} h_i(\theta(t))\phi_j(\hat{\theta}(t))\phi_l(\hat{\theta}(t))\phi_s(\hat{\theta}(t)) \\
&\quad \times \eta^T(t)[\sum_{m=1}^{4}\sigma_m(\Gamma_{ijlm} + \Xi_{ijlm} + Z_{ijl})]\eta(t).
\end{aligned} \tag{7.34}$$

Thus, the Δ in (7.33) satisfies

$$\begin{aligned}
\Delta \leq &\sum_{i=1}^{4}\sum_{j=1}^{4}\sum_{l=1}^{4}\sum_{s=1}^{4} h_i(\theta(t))\phi_j(\hat{\theta}(t))\phi_l(\hat{\theta}(t))\phi_s(\hat{\theta}(t)) \\
&\times \eta^T(t)[\Gamma_{ijls} + \Xi_{ijls} + \sum_{m=1}^{4}\sigma_m(\Gamma_{ijlm} + \Xi_{ijlm} + Z_{ijl})]\eta(t) \\
= &\sum_{i=1}^{4}\sum_{j=1}^{4}\sum_{s=1}^{4} h_i(\theta(t))\phi_j(\hat{\theta}(t))\phi_j(\hat{\theta}(t))\phi_s(\hat{\theta}(t)) \\
&\times \eta^T(t)[\Gamma_{ijjs} + \Xi_{ijjs} + \sum_{m=1}^{4}\sigma_m(\Gamma_{ijjm} + \Xi_{ijjm} + Z_{ijj})]\eta(t) \\
&+ \sum_{i=1}^{4}\sum_{j=1}^{3}\sum_{s=1}^{4}\sum_{j<l} h_i(\theta(t))\phi_j(\hat{\theta}(t))\phi_l(\hat{\theta}(t))\phi_s(\hat{\theta}(t)) \\
&\times \eta^T(t)[\Gamma_{ijls} + \Gamma_{iljs} + \Xi_{ijls} + \Xi_{iljs} + \sum_{m=1}^{4}\sigma_m(\Gamma_{ijlm} \\
&+ \Gamma_{iljm} + \Xi_{ijlm} + \Xi_{iljm} + Z_{ijl} + Z_{ilj})]\eta(t).
\end{aligned} \tag{7.35}$$

From the inequalities presented above, one can conclude that if the inequalities in (7.23)–(7.25) are satisfied, then $\dot{V}(t, \xi_t) + \gamma^{-1}z^T(t)z(t) - \gamma\omega^T(t)\omega(t) < 0$. Based on the definition for H_∞ performance, one can get the stability criterion in Theorem 7.1. This completes the proof. □

Remark 7.6 Note that the assumption $|\phi_m(\hat{\theta}(t-\tau(t))) - \phi_m(\hat{\theta}(t))| \leq \sigma_m$ is introduced in the proof of Theorem 7.1. As discussed in [5], considering the asynchronous difference between the normalized membership function of the controlled plant and that of the controller will lead to less conservatism. Thus, the stability criterion in Theorem 7.1 is expected to provide good dynamic positioning performance.

If the networked T-S fuzzy DPS (7.19) under the premise matching is considered, the following stability criterion is followed immediately.

Theorem 7.2 *For given scalars $\tau_m^{ca} \geq 0$, $\tau_m^{sc} \geq 0$, $\tau_M > 0$, $h > 0$, $\varpi > 0$, $\gamma > 0$, and matrices L_j and K of appropriate dimensions, the network-based T-S fuzzy DPS (7.19) is asymptotically stable with an H_∞ norm bound γ, if there exist symmetric positive definite matrices P, Q_1, Q_2, R_1, R_2, R_3, R_4, S_1, and S_2 such that the inequalities (7.36) and (7.37) hold for $1 \leq i \neq j \leq 4$,*

$$\begin{bmatrix} \Gamma_{ii} & \Gamma_{4,ii} \\ * & \Gamma_5 \end{bmatrix} < 0, \tag{7.36}$$

$$\begin{bmatrix} \bar{\Gamma}_{ij} & \Gamma_{4,ii} & \Gamma_{4,ij} & \Gamma_{4,ji} \\ * & 3\Gamma_5 & 0 & 0 \\ * & * & 2\Gamma_5 & 0 \\ * & * & * & 2\Gamma_5 \end{bmatrix} < 0, \tag{7.37}$$

where

$$\begin{aligned}
\bar{\Gamma}_{ij} &= \frac{1}{3}\Gamma_{ii} + \frac{1}{2}\Gamma_{ij} + \frac{1}{2}\Gamma_{ji}, \ \Gamma_{4,ij} = \begin{bmatrix} U_{3ij} & U_{3ij} & U_{3ij} & U_{3ij} & U_{2i} \end{bmatrix}, \\
\Gamma_5 &= \text{diag}\{-(\bar{\tau} - \tau_{low})^{-1} Q_1^{-1}, \ -(\bar{\tau} - \tau_{low})^{-1} S_1^{-1}, \\
&\quad -(\bar{d} - d_{low})^{-1} Q_2^{-1}, \ -(\bar{d} - d_{low})^{-1} S_2^{-1}, \ -\gamma I\}, \\
U_{3ij} &= \begin{bmatrix} \Pi_{1ij} & \underbrace{0, \ \ldots, \ 0}_{4} & \Pi_2 & \Pi_{3j} & \underbrace{0, \ \ldots, \ 0}_{6} & \bar{D} \end{bmatrix}, \\
U_{2i} &= \begin{bmatrix} C_{2i}\widetilde{I} & \underbrace{0, \ \ldots, \ 0}_{12} & F_i \end{bmatrix}, \ \widetilde{I} = \begin{bmatrix} I & I \end{bmatrix},
\end{aligned}$$

while Γ_{ij} is derived from Γ_{ijls} in (7.26) by substituting Π_{1ijl} and Π_{2s} with Π_{1ij} and Π_2 in (7.19), respectively; Γ_{ii} and $\Gamma_{4,ii}$ are derived from Γ_{ij} and $\Gamma_{4,ij}$, respectively, by substituting the subscript j with i; Γ_{ji} and $\Gamma_{4,ji}$ are derived from Γ_{ij} and $\Gamma_{4,ij}$, respectively, by interchanging the subscripts i and j.

Proof By combining Theorem 2.2 in [6] and the proof of Theorem 7.1 in this chapter, one can derive the stability criterion presented above. The detailed proof is omitted here for brevity. This completes the proof. □

7.3 Controller Design for Networked T-S Fuzzy DPSs

We state and establish the following controller design criterion for the network-based T-S fuzzy DPS (7.19) of the UMV.

Theorem 7.3 *For given scalars $\tau_m^{ca} \geq 0$, $\tau_m^{sc} \geq 0$, $\tau_M > 0$, $h > 0$, $\varpi > 0$, $\gamma > 0$, the network-based T-S fuzzy DPS (7.19) is asymptotically stable with an H_∞ norm bound γ and the controller gain $K = V_1^T J^{-1}$, and the observer gain $L_j = V_{2j}^T \bar{J}^{-T}$, if there exist symmetric positive definite matrices J, $\widetilde{Q}_1$, $\widetilde{Q}_2$, $\widetilde{R}_1$, $\widetilde{R}_2$, $\widetilde{R}_3$, $\widetilde{R}_4$, $\widetilde{S}_1$, and $\widetilde{S}_2$, matrices V_1, V_{2j}, and $\bar{J}$ such that*

$$\begin{bmatrix} \widetilde{\Gamma}_{ii} & \widetilde{\Gamma}_{4,ii} \\ * & \widetilde{\Gamma}_5 \end{bmatrix} < 0, \tag{7.38}$$

$$\begin{bmatrix} \widetilde{\Gamma}_{ij}^{com} & \widetilde{\Gamma}_{4,ii} & \widetilde{\Gamma}_{4,ij} & \widetilde{\Gamma}_{4,ji} \\ * & 3\widetilde{\Gamma}_5 & 0 & 0 \\ * & * & 2\widetilde{\Gamma}_5 & 0 \\ * & * & * & 2\widetilde{\Gamma}_5 \end{bmatrix} < 0, \tag{7.39}$$

$$JC_1^T = C_1^T \bar{J}, \tag{7.40}$$

where $\widetilde{\Gamma}_{ij} = \begin{bmatrix} \widetilde{\Gamma}_{1,ij} & \widetilde{\Gamma}_2 \\ * & \widetilde{\Gamma}_3 \end{bmatrix}$, *and*

$$\widetilde{\Gamma}_{1,ij} = \begin{bmatrix} \widetilde{\Gamma}_{ij}^{11} & 0 & 0 & 0 & 0 & \widetilde{\Gamma}^{16} & \widetilde{\Gamma}_j^{17} \\ * & \widetilde{\Gamma}^{22} & 0 & 0 & 0 & \widetilde{\Gamma}^{26} & 0 \\ * & * & \widetilde{\Gamma}^{33} & 0 & 0 & \widetilde{\Gamma}^{36} & 0 \\ * & * & * & \widetilde{\Gamma}^{44} & 0 & 0 & \widetilde{\Gamma}^{47} \\ * & * & * & * & \widetilde{\Gamma}^{55} & 0 & \widetilde{\Gamma}^{57} \\ * & * & * & * & * & \widetilde{\Gamma}^{66} & 0 \\ * & * & * & * & * & * & \widetilde{\Gamma}^{77} \end{bmatrix}, \quad \widetilde{\Gamma}_2 = \begin{bmatrix} \frac{6\widetilde{Q}_1}{\bar{\tau}} & \frac{6\widetilde{Q}_2}{\bar{d}} & 0 & 0 & 0 & 0 & \bar{D} \\ 0 & 0 & \bar{S}_1 & 0 & 0 & 0 & 0 \\ 0 & 0 & 0 & \bar{S}_1 & 0 & 0 & 0 \\ 0 & 0 & 0 & 0 & \bar{S}_2 & 0 & 0 \\ 0 & 0 & 0 & 0 & 0 & \bar{S}_2 & 0 \\ \frac{6\widetilde{Q}_1}{\bar{\tau}} & 0 & \bar{S}_1 & \bar{S}_1 & 0 & 0 & 0 \\ 0 & \frac{6\widetilde{Q}_2}{\bar{d}} & 0 & 0 & \bar{S}_2 & \bar{S}_2 & 0 \end{bmatrix},$$

$$\widetilde{\Gamma}_3 = \text{diag}\{-\frac{12\widetilde{Q}_1}{\bar{\tau}}, -\frac{12\widetilde{Q}_2}{\bar{d}}, -\frac{12\widetilde{S}_1}{\bar{\tau}-\tau_{low}}, -\frac{12\widetilde{S}_1}{\bar{\tau}-\tau_{low}}, -\frac{12\widetilde{S}_2}{\bar{d}-d_{low}}, -\frac{12\widetilde{S}_2}{\bar{d}-d_{low}}, -\gamma I\},$$

$$\widetilde{\Gamma}_{ij}^{11} = H_{1ij} + H_{1ij}^T - \frac{4\widetilde{Q}_1}{\bar{\tau}} - \frac{4\widetilde{Q}_2}{\bar{d}} + \widetilde{R}_1 + \widetilde{R}_3, \ \widetilde{\Gamma}^{16} = H_2^T - \frac{2\widetilde{Q}_1}{\bar{\tau}},$$

$$\widetilde{\Gamma}_j^{17} = H_{3j}^T - \frac{2\widetilde{Q}_2}{\bar{d}}, \ \widetilde{\Gamma}^{22} = \widetilde{R}_2 - \widetilde{R}_1 - \frac{4\widetilde{S}_1}{\bar{\tau}-\tau_{low}}, \ \widetilde{\Gamma}^{26} = -\frac{2\widetilde{S}_1}{\bar{\tau}-\tau_{low}},$$

$$\widetilde{\Gamma}^{33} = -\widetilde{R}_2 - \frac{4\widetilde{S}_1}{\bar{\tau}-\tau_{low}}, \ \widetilde{\Gamma}^{36} = -\frac{2\widetilde{S}_1}{\bar{\tau}-\tau_{low}}, \ \widetilde{\Gamma}^{44} = \widetilde{R}_4 - \widetilde{R}_3 - \frac{4\widetilde{S}_2}{\bar{d}-d_{low}},$$

$$\widetilde{\Gamma}^{47} = -\frac{2\widetilde{S}_2}{\bar{d}-d_{low}}, \ \widetilde{\Gamma}^{55} = -\widetilde{R}_4 - \frac{4\widetilde{S}_2}{\bar{d}-d_{low}}, \ \widetilde{\Gamma}^{57} = -\frac{2\widetilde{S}_2}{\bar{d}-d_{low}},$$

$$\widetilde{\Gamma}^{66} = -\frac{4\widetilde{Q}_1}{\bar{\tau}} - \frac{8\widetilde{S}_1}{\bar{\tau}-\tau_{low}}, \ \widetilde{\Gamma}^{77} = -\frac{4\widetilde{Q}_2}{\bar{d}} - \frac{8\widetilde{S}_2}{\bar{d}-d_{low}}, \ \bar{S}_1 = \frac{6\widetilde{S}_1}{\bar{\tau}-\tau_{low}}, \ \bar{S}_2 = \frac{6\widetilde{S}_2}{\bar{d}-d_{low}},$$

$$\widetilde{\Gamma}_{4,ij} = \begin{bmatrix} \widetilde{U}_{3ij} & \widetilde{U}_{3ij} & \widetilde{U}_{3ij} & \widetilde{U}_{3ij} & \widetilde{U}_{2i} \end{bmatrix},$$

$$\widetilde{\Gamma}_5 = \text{diag}\{(\bar{\tau}-\tau_{low})^{-1}(\widetilde{Q}_1 - 2\Upsilon), \ (\bar{\tau}-\tau_{low})^{-1}(\widetilde{S}_1 - 2\Upsilon), \ (\bar{d}-d_{low})^{-1}(\widetilde{Q}_2 - 2\Upsilon), \ (\bar{d}-d_{low})^{-1}(\widetilde{S}_2 - 2\Upsilon), \ -\gamma I\},$$

$$\widetilde{U}_{3ij} = \begin{bmatrix} H_{1ij}^T & \underbrace{0, \ldots, 0}_{4} & H_2^T & H_{3j}^T & \underbrace{0, \ldots, 0}_{6} & \bar{D} \end{bmatrix}^T, \ \widetilde{U}_{2i} = \begin{bmatrix} H_{4i}^T & \underbrace{0, \ldots, 0}_{12} & F_i \end{bmatrix}^T,$$

$$\widetilde{\Gamma}_{ij}^{com} = \frac{1}{3}\widetilde{\Gamma}_{ii} + \frac{1}{2}\widetilde{\Gamma}_{ij} + \frac{1}{2}\widetilde{\Gamma}_{ji}, \ \Upsilon = \text{diag}\{J, \ J\},$$

$$H_{1ij} = \begin{bmatrix} \Lambda_1 & \Lambda_2 \\ 0 & J\widetilde{A}_i^T \end{bmatrix}, \ H_2 = \begin{bmatrix} 0 & V_1\widetilde{D}^T \\ 0 & 0 \end{bmatrix}, \ H_{3j} = \begin{bmatrix} C_1^T V_{2j} & -C_1^T V_{2j} \\ C_1^T V_{2j} & -C_1^T V_{2j} \end{bmatrix}, \ H_{4i} = \begin{bmatrix} JC_{2i}^T \\ JC_{2i}^T \end{bmatrix},$$

$$\Lambda_1 = J\widetilde{A}_j^T + V_1\widetilde{D}^T - C_1^T V_{2j}, \ \Lambda_2 = J\widetilde{A}_i^T - J\widetilde{A}_j^T - V_1\widetilde{D}^T + C_1^T V_{2j},$$

while $\widetilde{\Gamma}_{ii}$ *and* $\widetilde{\Gamma}_{4,ii}$ *are derived from* $\widetilde{\Gamma}_{ij}$ *and* $\widetilde{\Gamma}_{4,ij}$*, respectively, by substituting the subscript* j *with* i*;* $\widetilde{\Gamma}_{ji}$ *and* $\widetilde{\Gamma}_{4,ji}$ *are derived from* $\widetilde{\Gamma}_{ij}$ *and* $\widetilde{\Gamma}_{4,ij}$*, respectively, by interchanging the subscripts* i *and* j*.*

Proof Let $P = \text{diag}\{\hat{P},\ \hat{P}\}$, and $\hat{P}^{-1} = J$. Pre- and post-multiplying both sides of (7.36) with $\text{diag}\{\underbrace{P^{-1},\ \ldots,\ P^{-1}}_{13},\ \underbrace{I,\ \ldots,\ I}_{6}\}$ and its transpose, and pre- and post-multiplying both sides of (7.37) with $\text{diag}\{\underbrace{P^{-1},\ \ldots,\ P^{-1}}_{13},\ \underbrace{I,\ \ldots,\ I}_{16}\}$ and its transpose, supposing that there exists a matrix $\bar{J}$ such that $JC_1^T = C_1^T\bar{J}$, defining $P^{-1}Q_1P^{-1} = \widetilde{Q}_1$, $P^{-1}Q_2P^{-1} = \widetilde{Q}_2$, $P^{-1}R_1P^{-1} = \widetilde{R}_1$, $P^{-1}R_2P^{-1} = \widetilde{R}_2$, $P^{-1}R_3P^{-1} = \widetilde{R}_3$, $P^{-1}R_4P^{-1} = \widetilde{R}_4$, $P^{-1}S_1P^{-1} = \widetilde{S}_1$, $P^{-1}S_2P^{-1} = \widetilde{S}_2$, $JK^T = V_1$, $\bar{J}L_j^T = V_{2j}$, and considering that $-Q_1^{-1} \leq \widetilde{Q}_1 - 2\Upsilon$, $-Q_2^{-1} \leq \widetilde{Q}_2 - 2\Upsilon$, $-S_1^{-1} \leq \widetilde{S}_1 - 2\Upsilon$, $-S_2^{-1} \leq \widetilde{S}_2 - 2\Upsilon$, one can conclude that if (7.38)–(7.40) are satisfied, the stability criterion in Theorem 7.2 is also satisfied. This completes the proof. □

It should be mentioned that the equality constraint in (7.40) induces some difficulty for numerical calculation. We turn to eliminate such an equality constraint. For the matrix C_1^T of full column rank, there always exist two orthogonal matrices $X \in \mathbb{R}^{6\times6}$ and $Y \in \mathbb{R}^{3\times3}$ such that

$$XC_1^TY = \begin{bmatrix} X_1 \\ X_2 \end{bmatrix} C_1^TY = \begin{bmatrix} \Phi \\ 0 \end{bmatrix}, \tag{7.41}$$

where $X_1 \in \mathbb{R}^{3\times6}$, $X_2 \in \mathbb{R}^{3\times6}$, $\Phi = \text{diag}\{\alpha_1,\ \alpha_2,\ \alpha_3\}$, and $\alpha_1,\ \alpha_2,\ \alpha_3$ are nonzero singular values of C_1^T. By using Lemma 2 in [7], one can conclude that if the matrix J can be written as

$$J = X^T \begin{bmatrix} J_{11} & 0 \\ 0 & J_{22} \end{bmatrix} X = X_1^TJ_{11}X_1 + X_2^TJ_{22}X_2, \tag{7.42}$$

where J_{11} and J_{22} are symmetric positive definite matrices with appropriate dimensions, then there exists a nonsingular matrix $\bar{J}$ such that $JC_1^T = C_1^T\bar{J}$.

Based on Theorem 7.3 and the statement presented above, one can derive the following observer-based controller design criterion.

Corollary 7.1 *For given scalars* $\tau_m^{ca} \geq 0$, $\tau_m^{sc} \geq 0$, $\tau_M > 0$, $h > 0$, $\varpi > 0$, $\gamma > 0$, *the network-based T-S fuzzy DPS (7.19) is asymptotically stable with an* H_∞ *norm bound* γ *and the controller gain* $K = V_1^T(X_1^TJ_{11}X_1 + X_2^TJ_{22}X_2)^{-1}$*, and the observer gain* $L_j = V_{2j}^TY\Phi J_{11}^{-1}\Phi^{-1}Y^T$*, if there exist symmetric positive definite matrices* J_{11}, J_{22}, $\widetilde{Q}_1$, $\widetilde{Q}_2$, $\widetilde{R}_1$, $\widetilde{R}_2$, $\widetilde{R}_3$, $\widetilde{R}_4$, $\widetilde{S}_1$, *and* $\widetilde{S}_2$*, matrices* V_1*, and* V_{2j} *such that*

$$\begin{bmatrix} \widetilde{\Gamma}_{ii,J} & \widetilde{\Gamma}_{4,ii,J} \\ * & \widetilde{\Gamma}_{5,J} \end{bmatrix} < 0, \tag{7.43}$$

$$\begin{bmatrix} \widetilde{\Gamma}_{ij,J}^{com} & \widetilde{\Gamma}_{4,ii,J} & \widetilde{\Gamma}_{4,ij,J} & \widetilde{\Gamma}_{4,ji,J} \\ * & 3\widetilde{\Gamma}_{5,J} & 0 & 0 \\ * & * & 2\widetilde{\Gamma}_{5,J} & 0 \\ * & * & * & 2\widetilde{\Gamma}_{5,J} \end{bmatrix} < 0, \tag{7.44}$$

where $\widetilde{\Gamma}_{ii,J}$, $\widetilde{\Gamma}_{4,ii,J}$, $\widetilde{\Gamma}_{5,J}$, $\widetilde{\Gamma}_{ij,J}^{com}$, $\widetilde{\Gamma}_{4,ij,J}$, *and* $\widetilde{\Gamma}_{4,ji,J}$ *are derived from* $\widetilde{\Gamma}_{ii}$, $\widetilde{\Gamma}_{4,ii}$, $\widetilde{\Gamma}_{5}$, $\widetilde{\Gamma}_{ij}^{com}$, $\widetilde{\Gamma}_{4,ij}$, *and* $\widetilde{\Gamma}_{4,ji}$ *in (7.38)–(7.39), respectively, by substituting J with* $X_1^T J_{11} X_1 + X_2^T J_{22} X_2$.

If the networked closed-loop system (7.21) is investigated, the controller design criterion in Theorem 7.4 is followed readily.

Theorem 7.4 *For given scalars* $\tau_m^{ca} \geq 0$, $\tau_M^{ca} > 0$, $h > 0$, $\varpi > 0$, $\gamma > 0$, *the network-based T-S fuzzy DPS (7.21) is asymptotically stable with an* H_∞ *norm bound* γ *and the controller gain* $K = V_1^T (X_1^T J_{11} X_1 + X_2^T J_{22} X_2)^{-1}$, *and the observer gain* $L_j = V_{2j}^T Y \Phi J_{11}^{-1} \Phi^{-1} Y^T$, *if there exist symmetric positive definite matrices* J_{11}, J_{22}, $\widetilde{Q}_1$, $\widetilde{R}_1$, $\widetilde{R}_2$, *and* $\widetilde{S}_1$, *matrices* V_1, *and* V_{2j} *such that*

$$\begin{bmatrix} \hat{\Gamma}_{ii} & \widetilde{\Gamma}_{6,ii} \\ * & \widetilde{\Gamma}_7 \end{bmatrix} < 0, \tag{7.45}$$

$$\begin{bmatrix} \hat{\Gamma}_{ij}^{com} & \widetilde{\Gamma}_{6,ii} & \widetilde{\Gamma}_{6,ij} & \widetilde{\Gamma}_{6,ji} \\ * & 3\widetilde{\Gamma}_7 & 0 & 0 \\ * & * & 2\widetilde{\Gamma}_7 & 0 \\ * & * & * & 2\widetilde{\Gamma}_7 \end{bmatrix} < 0, \tag{7.46}$$

where

$$\hat{\Gamma}_{ij} = \begin{bmatrix} \hat{\Gamma}_{ij}^{11} & 0 & 0 & \hat{\Gamma}^{14} & \hat{\Gamma}^{15} & 0 & 0 & \bar{D} \\ * & \hat{\Gamma}^{22} & 0 & \hat{\Gamma}^{24} & 0 & \hat{\Gamma}^{26} & 0 & 0 \\ * & * & \hat{\Gamma}^{33} & \hat{\Gamma}^{34} & 0 & 0 & \hat{\Gamma}^{37} & 0 \\ * & * & * & \hat{\Gamma}^{44} & \hat{\Gamma}^{45} & \hat{\Gamma}^{46} & \hat{\Gamma}^{47} & 0 \\ * & * & * & * & \hat{\Gamma}^{55} & 0 & 0 & 0 \\ * & * & * & * & * & \hat{\Gamma}^{66} & 0 & 0 \\ * & * & * & * & * & * & \hat{\Gamma}^{77} & 0 \\ * & * & * & * & * & * & * & -\gamma I \end{bmatrix},$$

$$\hat{\Gamma}_{ij}^{11} = H_{5ij} + H_{5ij}^T - \frac{4\widetilde{Q}_1}{\tau^u} + \widetilde{R}_1,\ \hat{\Gamma}^{14} = H_2^T - \frac{2\widetilde{Q}_1}{\tau^u},\ \hat{\Gamma}^{15} = \frac{6\widetilde{Q}_1}{\tau^u},$$

$$\hat{\Gamma}^{22} = \widetilde{R}_2 - \widetilde{R}_1 - \frac{4\widetilde{S}_1}{\tau^u - \tau_l},\ \hat{\Gamma}^{24} = -\frac{2\widetilde{S}_1}{\tau^u - \tau_l},\ \hat{\Gamma}^{26} = \frac{6\widetilde{S}_1}{\tau^u - \tau_l},$$

$$\hat{\Gamma}^{33} = -\widetilde{R}_2 - \frac{4\widetilde{S}_1}{\tau^u - \tau_l},\ \hat{\Gamma}^{34} = -\frac{2\widetilde{S}_1}{\tau^u - \tau_l},\ \hat{\Gamma}^{37} = \frac{6\widetilde{S}_1}{\tau^u - \tau_l},$$
$$\hat{\Gamma}^{44} = -\frac{4\widetilde{Q}_1}{\tau^u} - \frac{8\widetilde{S}_1}{\tau^u - \tau_l},\ \hat{\Gamma}^{45} = \frac{6\widetilde{Q}_1}{\tau^u},\ \hat{\Gamma}^{46} = \hat{\Gamma}^{47} = \frac{6\widetilde{S}_1}{\tau^u - \tau_l},$$
$$\hat{\Gamma}^{55} = -\frac{12\widetilde{Q}_1}{\tau^u},\ \hat{\Gamma}^{66} = \hat{\Gamma}^{77} = -\frac{12\widetilde{S}_1}{\tau^u - \tau_l},\ \widetilde{\Gamma}_{6,ij} = \left[\, \widetilde{U}_{4ij}\ \widetilde{U}_{4ij}\ \widetilde{U}_{5i} \,\right],$$
$$\widetilde{\Gamma}_7 = \mathrm{diag}\{(\tau^u - \tau_l)^{-1}(\widetilde{Q}_1 - 2\Upsilon), (\tau^u - \tau_l)^{-1}(\widetilde{S}_1 - 2\Upsilon),\ -\gamma I\}, \tag{7.47}$$
$$\widetilde{U}_{4ij} = \left[\, H_{5ij}^T\ 0\ 0\ H_2^T\ 0\ 0\ 0\ \bar{D} \,\right]^T,\ \widetilde{U}_{5i} = \left[\, H_{4i}^T\ \underbrace{0, \ldots, 0}_{6}\ F_i \,\right]^T,$$
$$\hat{\Gamma}_{ij}^{com} = \frac{1}{3}\hat{\Gamma}_{ii} + \frac{1}{2}\hat{\Gamma}_{ij} + \frac{1}{2}\hat{\Gamma}_{ji},\ \Upsilon = \mathrm{diag}\{\widetilde{J},\ \widetilde{J}\},\ \widetilde{J} = X_1^T J_{11} X_1 + X_2^T J_{22} X_2,$$
$$H_{5ij} = \begin{bmatrix} \widetilde{J}\widetilde{A}_j^T + V_1\widetilde{D}^T & \widetilde{J}\widetilde{A}_i^T - \widetilde{J}\widetilde{A}_j^T - V_1\widetilde{D}^T \\ C_1^T V_{2j} & \widetilde{J}\widetilde{A}_i^T - C_1^T V_{2j} \end{bmatrix},$$

while H_2 and H_{4i} are the same as the corresponding items in Theorem 7.3; $\hat{\Gamma}_{ii}$ and $\widetilde{\Gamma}_{6,ii}$ are derived from $\hat{\Gamma}_{ij}$ and $\widetilde{\Gamma}_{6,ij}$, respectively, by substituting the subscript j with i; $\hat{\Gamma}_{ji}$ and $\widetilde{\Gamma}_{6,ji}$ are derived from $\hat{\Gamma}_{ij}$ and $\widetilde{\Gamma}_{6,ij}$, respectively, by interchanging the subscripts i and j.

Remark 7.7 Note that observer-based controller design, which can stabilize states of the UMV, for network-based T-S fuzzy DPSs (7.19) and (7.21) are studied in Corollary 7.1 and Theorem 7.4, respectively. The proposed design methods can be extended to investigate the system (7.16), and the corresponding result is omitted here for brevity.

7.4 Performance Analysis and Discussion

In this section, we show the effectiveness of the proposed observer-based dynamic positioning controller design for the network-based T-S fuzzy DPS (7.21).

Choose the matrices M, N, and G in the system (7.1) as follows (see also [2, 8])

$$M = \begin{bmatrix} 1.0852 & 0 & 0 \\ 0 & 2.0575 & -0.4087 \\ 0 & -0.4087 & 0.2153 \end{bmatrix},$$
$$N = \begin{bmatrix} 0.0865 & 0 & 0 \\ 0 & 0.0762 & 0.1510 \\ 0 & 0.0151 & 0.0031 \end{bmatrix},$$
$$G = \begin{bmatrix} 0.0389 & 0 & 0 \\ 0 & 0.0266 & 0 \\ 0 & 0 & 0 \end{bmatrix}.$$

Noting that $A = -M^{-1}G$, $B = -M^{-1}N$, $D = M^{-1}$, one has

$$
\begin{aligned}
A &= \begin{bmatrix} -0.0358 & 0 & 0 \\ 0 & -0.0208 & 0 \\ 0 & -0.0394 & 0 \end{bmatrix}, \\
B &= \begin{bmatrix} -0.0797 & 0 & 0 \\ 0 & -0.0818 & -0.1224 \\ 0 & -0.2254 & -0.2468 \end{bmatrix}, \\
D &= \begin{bmatrix} 0.9215 & 0 & 0 \\ 0 & 0.7802 & 1.4811 \\ 0 & 1.4811 & 7.4562 \end{bmatrix}.
\end{aligned}
\tag{7.48}
$$

Without loss of generality, suppose that

$$
\begin{aligned}
C_{21} &= \begin{bmatrix} 0.5 & 1 & 0 & 0.1 & 0 & -0.7 \end{bmatrix},\ C_{22} = \begin{bmatrix} 1 & 0 & 1 & 2.1 & -1.6 & 0 \end{bmatrix}, \\
C_{23} &= \begin{bmatrix} 0.2 & 0 & 2 & 1 & 0 & -0.8 \end{bmatrix},\ C_{24} = \begin{bmatrix} -1 & 0 & -0.7 & 0 & 1 & 0.6 \end{bmatrix}, \\
F_1 &= \begin{bmatrix} 0.8 & 1 & 0.3 \end{bmatrix},\ F_2 = \begin{bmatrix} 1 & -1 & 2 \end{bmatrix},\ F_3 = \begin{bmatrix} 0.6 & 1 & -0.5 \end{bmatrix}, \\
F_4 &= \begin{bmatrix} 1 & 0.9 & -2 \end{bmatrix},\ C_1 = \begin{bmatrix} 1 & 0 & 0 & 1 & 0 & 0 \\ 0 & 1 & 0 & 0 & 1 & 0 \\ 0 & 0 & 1 & 0 & 0 & 1 \end{bmatrix}.
\end{aligned}
\tag{7.49}
$$

For Theorem 7.4, let $\tau_m^{ca} = 0\text{s}$, $\tau_M^{ca} = 0.03\text{s}$, $h = 0.02\text{s}$, $\varpi = 2$.
By using matrix singular value decomposition and from (7.41), one has

$$
\begin{aligned}
X_1 &= \begin{bmatrix} -0.7071 & 0 & 0 & -0.7071 & 0 & 0 \\ 0 & -0.7071 & 0 & 0 & -0.7071 & 0 \\ 0 & 0 & -0.7071 & 0 & 0 & -0.7071 \end{bmatrix}, \\
X_2 &= \begin{bmatrix} -0.7071 & 0 & 0 & 0.7071 & 0 & 0 \\ 0 & -0.7071 & 0 & 0 & 0.7071 & 0 \\ 0 & 0 & -0.7071 & 0 & 0 & 0.7071 \end{bmatrix}, \\
Y &= \begin{bmatrix} -1 & 0 & 0 \\ 0 & -1 & 0 \\ 0 & 0 & -1 \end{bmatrix},\ \Phi = \begin{bmatrix} 1.4142 & 0 & 0 \\ 0 & 1.4142 & 0 \\ 0 & 0 & 1.4142 \end{bmatrix}.
\end{aligned}
$$

Suppose that the initial state of the closed-loop system (7.21) is $\bar{\xi}(t) = [0.1\ -0.1\ 0.2\ -0.2\ 0.3\ -0.3\ 0\ 0\ 0\ 0\ 0\ 0]^T$. The disturbances of surge, sway and yaw motions $\omega_1(t)$, $\omega_2(t)$, and $\omega_3(t)$ are given as

$$
\begin{cases}
\omega_1(t) = 0.27F_1(s)N_1(t)N_2(t), \\
\omega_2(t) = -0.6\cos(1.6t)e^{-0.12t}, \\
\omega_3(t) = 0.58F_2(s)N_3(t)N_4(t),
\end{cases}
\tag{7.50}
$$

where $F_1(s)$ and $F_2(s)$ are shaping filters described by $\frac{K_{\omega 1} s}{s^2+2\varepsilon_1\sigma_1 s+\sigma_1^2}$ and $\frac{K_{\omega 2} s}{s^2+2\varepsilon_2\sigma_2 s+\sigma_2^2}$, respectively; $K_{\omega 1}$ and $K_{\omega 2}$ denote the dominate wave strength coefficients with $K_{\omega 1} = 0.26$ and $K_{\omega 2} = 0.8$; ε_1 and ε_2 denote the damping coefficients with $\varepsilon_1 = 0.2$ and $\varepsilon_2 = 1.7$; σ_1 and σ_2 denote the encountering wave frequencies with $\sigma_1 = 1.3$ and $\sigma_2 = 0.9$; $N_1(t)$ and $N_3(t)$ are band-limited white noise with noise powers 2.69 and 1.56, respectively; while

$$N_2(t) = \begin{cases} 1, & t \in [0s,\ 6s], \\ 0, & otherwise, \end{cases}$$
$$N_4(t) = \begin{cases} 1, & t \in [0s,\ 5.5s], \\ 0, & otherwise. \end{cases}$$

The number of packet dropouts and the network-induced delays τ_k vary stochastically.

The responses of the UMV state are given in Figs. 7.3, 7.4, and 7.5, from which figures one can see that the proposed dynamic positioning scheme can guarantee

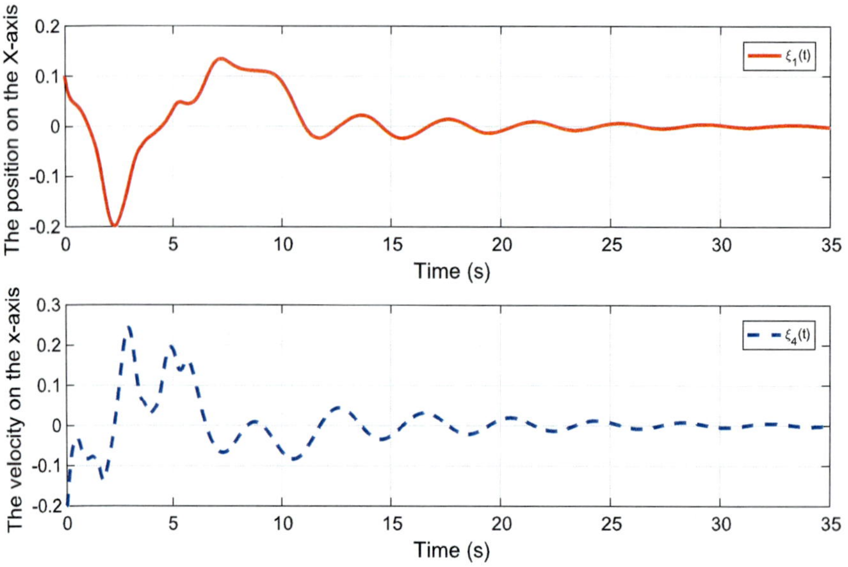

Fig. 7.3 The responses of the earth-fixed position on the X-axis and the body-fixed velocity on the x-axis

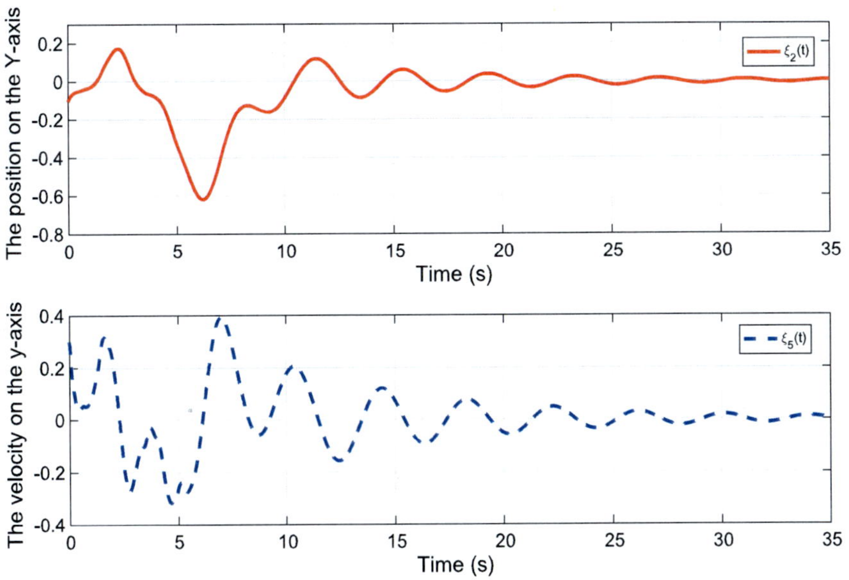

Fig. 7.4 The responses of the earth-fixed position on the Y-axis and the body-fixed velocity on the y-axis

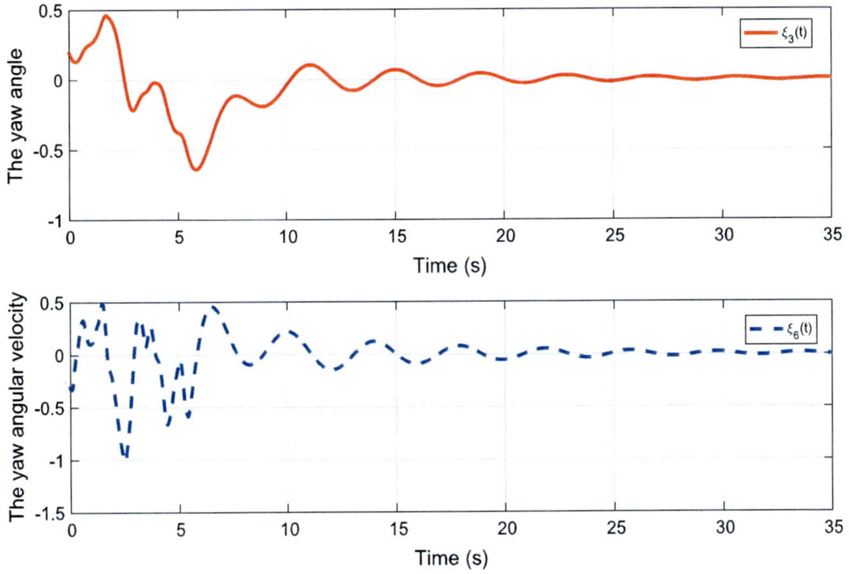

Fig. 7.5 The responses of the yaw angle and yaw angular velocity

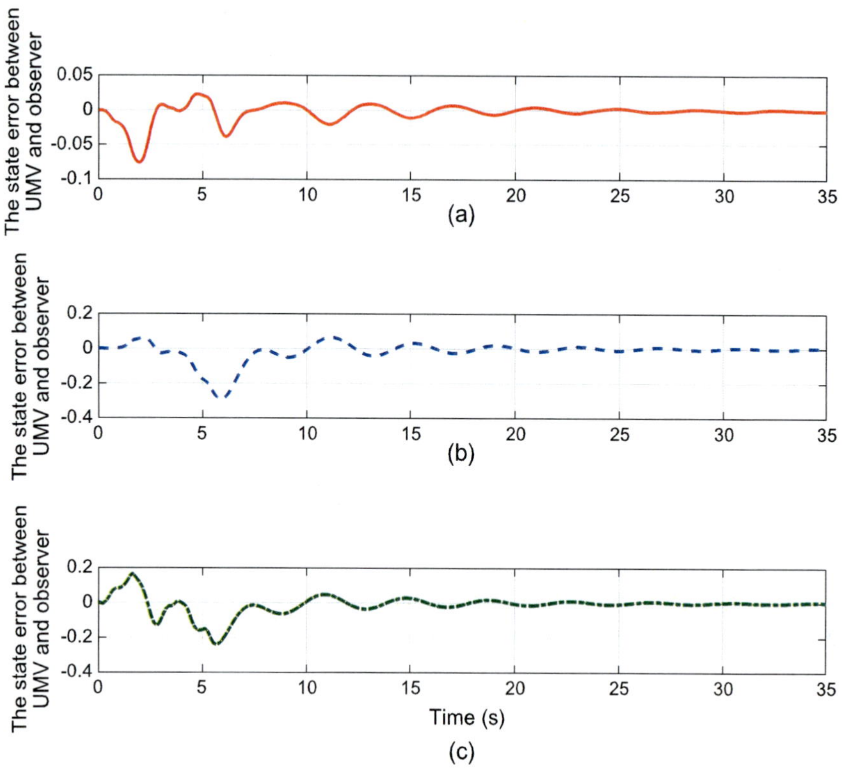

Fig. 7.6 The responses of state error between the UMV and the observer. **a**, **b**, and **c** are corresponding to $\tilde{\xi}_1(t)$, $\tilde{\xi}_2(t)$, and $\tilde{\xi}_3(t)$, respectively

satisfying performance for the UMV. The responses of state error between the UMV and the observer are presented in Figs. 7.6 and 7.7. In fact, the observer-based dynamic positioning controller design scheme can provide a small state error between the UMV and the observer, which is verified by Figs. 7.6 and 7.7. The responses of observer state are described by Figs. 7.8 and 7.9, while Figs. 7.10, and 7.11 present the responses of the control forces and moment provided by the thruster system, and the wave-induced disturbances, respectively. Even if the wave-induced disturbances are imposed on the UMV, the control cost is still acceptable. This statement is verified by Figs. 7.10 and 7.11.

It should be mentioned that the dynamic positioning controller design scheme in Corollary 7.1 is also applicable. The detailed performance analysis corresponding to the design scheme in Corollary 7.1 is omitted here.

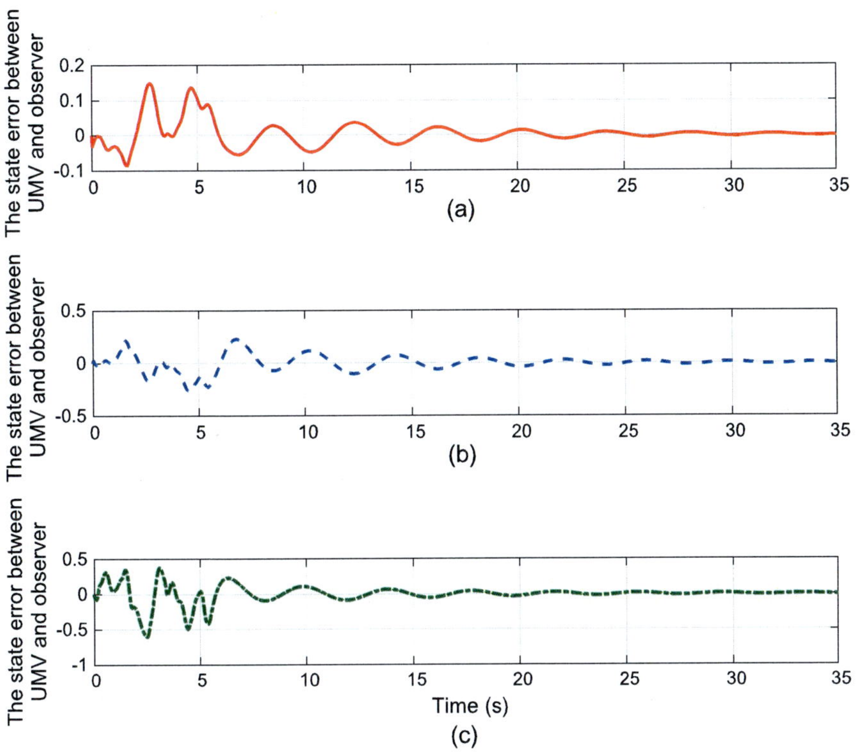

Fig. 7.7 The responses of state error between the UMV and the observer. **a**, **b**, and **c** are corresponding to $\tilde{\xi}_4(t)$, $\tilde{\xi}_5(t)$, and $\tilde{\xi}_6(t)$, respectively

7.5 Conclusions

The networked modeling, stability analysis, and observer-based controller design for the T-S fuzzy DPS of a UMV subject to wave-induced disturbances have been investigated. Network-based T-S fuzzy models for the DPS have been established by making full use of the variation scope of the yaw angle, and the sampler-to-controller and controller-to-actuator network-induced characteristics. A novel stability criterion has been derived by taking into account the asynchronous difference between the normalized membership function of the UMV and that of the controller. The proposed observer-based controller design can provide satisfying dynamic positioning performance.

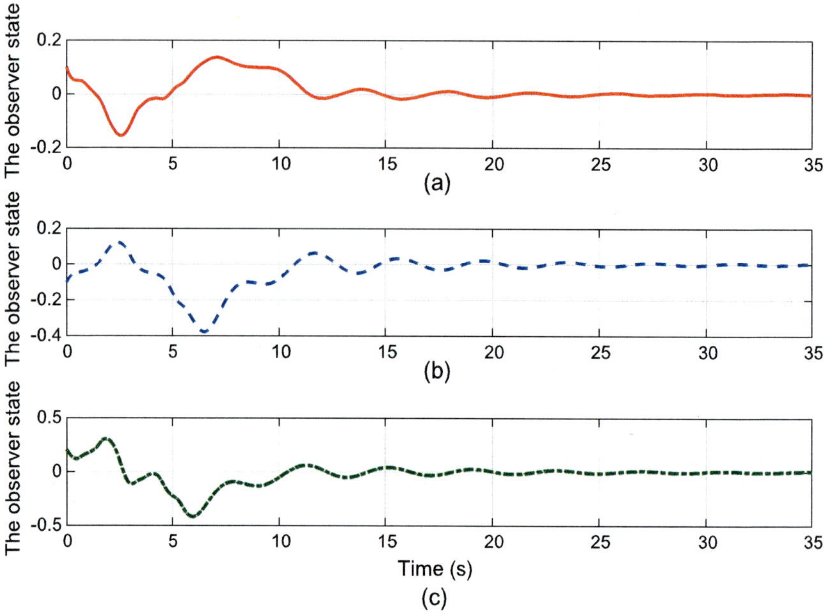

Fig. 7.8 The responses of observer state. **a**, **b**, and **c** are corresponding to $\hat{\xi}_1(t)$, $\hat{\xi}_2(t)$, and $\hat{\xi}_3(t)$, respectively

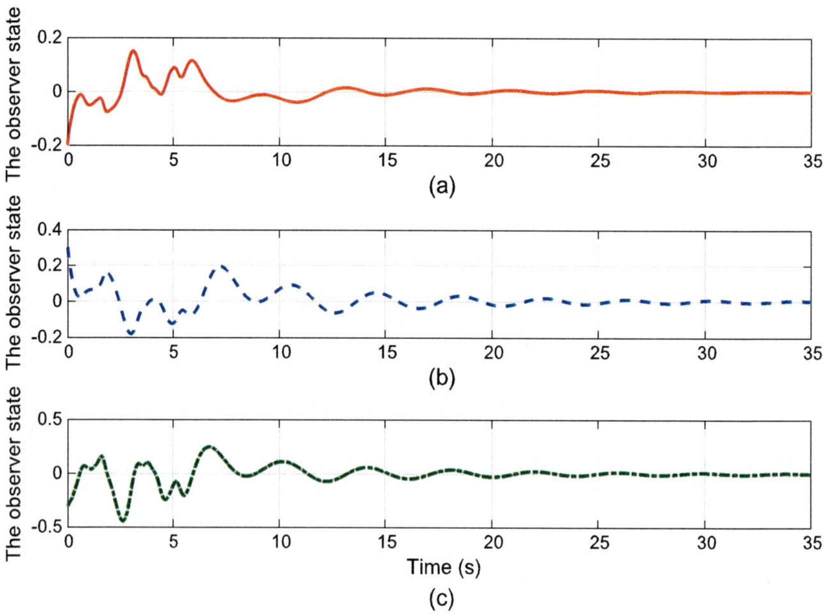

Fig. 7.9 The responses of observer state. **a**, **b**, and **c** are corresponding to $\hat{\xi}_4(t)$, $\hat{\xi}_5(t)$, and $\hat{\xi}_6(t)$, respectively

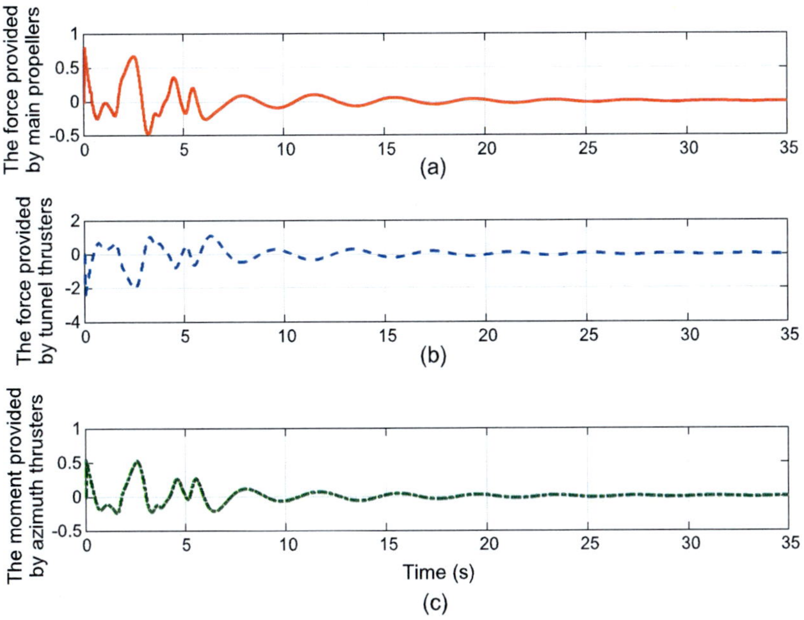

Fig. 7.10 The responses of the control forces and moment provided by the thruster system. **a**, **b**, and **c** are corresponding to the force $u_1(t)$ provided by main propellers, the force $u_2(t)$ provided by tunnel thrusters, and the moment $u_3(t)$ provided by azimuth thrusters, respectively

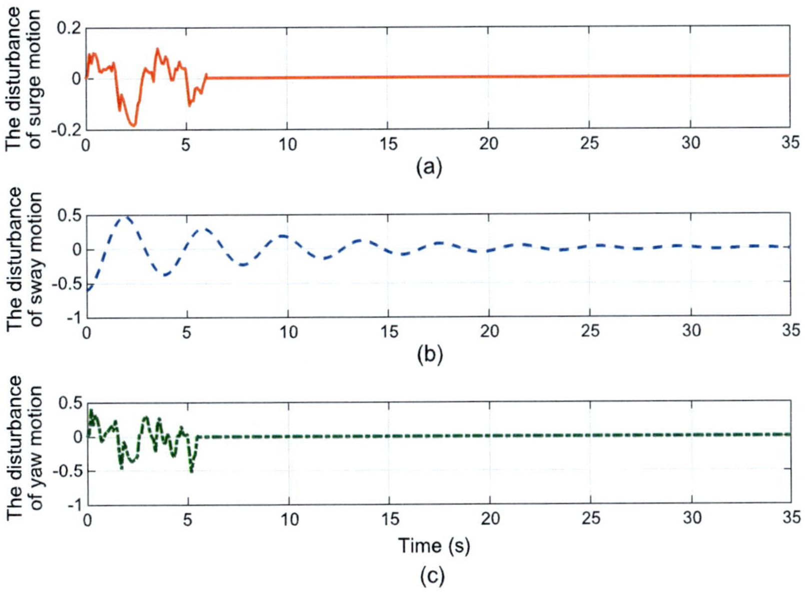

Fig. 7.11 The responses of the disturbances. **a**, **b**, and **c** are corresponding to the disturbance of surge motion $\omega_1(t)$, the disturbance of sway motion $\omega_2(t)$, and the disturbance of yaw motion $\omega_3(t)$, respectively

7.6 Notes

For the networked T-S fuzzy DPS of a UMV, how to take into account the asynchronous difference between the normalized membership function of the UMV and that of the controller is significant [5, 9]. Moreover, in practical situations, the UMV states are not always measurable. Then it is of paramount importance to introduce an observer for such DPS of a UMV. Motivated by these facts, the networked modeling, stability analysis, and observer-based controller design for the T-S fuzzy DPS of a UMV are investigated. The results in this chapter are based mainly on [10]. Future research includes network-based filtering [11, 12] for the T-S fuzzy DPS of a UMV.

References

1. T.I. Fossen, *Handbook of Marine Craft Hydrodynamics and Motion Control* (Wiley, Chichester, U.K., 2011)
2. W.-H. Ho, S.-H. Chen, J.-H. Chou, Optimal control of Takagi–Sugeno fuzzy-model-based systems representing dynamic ship positioning systems. Appl. Soft Comput. **13**(7), 3197–3210 (2013)
3. A. Seuret, F. Gouaisbaut, Wirtinger-based integral inequality: application to time-delay systems. Automatica **49**(9), 2860–2866 (2013)
4. X.-M. Zhang, Q.-L. Han, A. Seuret, F. Gouaisbaut, An improved reciprocally convex inequality and an augmented Lyapunov-Krasovskii functional for stability of linear systems with time-varying delay. Automatica **84**, 221–226 (2017)
5. D. Zhang, Q.-L. Han, X. Jia, Network-based output tracking control for a class of T-S fuzzy systems that can not be stabilized by nondelayed output feedback controllers. IEEE Trans. Cybern. **45**(8), 1511–1524 (2015)
6. H.D. Tuan, P. Apkarian, T. Narikiyo, Y. Yamamoto, Parameterized linear matrix inequality techniques in fuzzy control system design. IEEE Trans. Fuzzy Syst. **9**(2), 324–332 (2001)
7. F. Yang, Z. Wang, Y.S. Hung, M. Gani, H_∞ control for networked systems with random communication delays. IEEE Trans. Autom. Control **51**(3), 511–518 (2006)
8. N.E. Kahveci, P.A. Ioannou, Adaptive steering control for uncertain ship dynamics and stability analysis. Automatica **49**(3), 685–697 (2013)
9. D. Zhang, Q.-L. Han, X. Jia, Network-based output tracking control for T-S fuzzy systems using an event-triggered communication scheme. Fuzzy Sets Syst. **273**, 26–48 (2015)
10. Y.-L. Wang, Q.-L. Han, M.-R. Fei, C. Peng, Network-based T-S fuzzy dynamic positioning controller design for unmanned marine vehicles. IEEE Trans. Cybern. **48**(9), 2750–2763 (2018)
11. X.-M. Zhang, Q.-L. Han, Network-based H_∞ filtering using a logic jumping-like trigger. Automatica **49**(5), 1428–1435 (2013)
12. S. Zhu, Q.-L. Han, C. Zhang, l_1-gain performance analysis and positive filter design for positive discrete-time Markov jump linear systems: a linear programming approach. Automatica **50**(8), 2098–2107 (2014)

Chapter 8
Network-Based Dynamic Output Feedback Control of UMVs

This chapter is concerned with network-based modeling and dynamic output feedback control for a UMV in network environments. A network-based model for the UMV is established by taking sampler-to-control station packet dropouts, network-induced delays, and packet disordering into account. This model is then extended to the UMV system subject to control station-to-actuator, and both sampler-to-control station and control station-to-actuator packet dropouts, network-induced delays, and packet disordering. Based on these models, DOFCs are designed to attenuate the oscillation amplitudes of the yaw velocity error and the yaw angle. A benchmark example demonstrates that (i) compared with the UMV without control, the designed DOFCs can attenuate the oscillation amplitudes of the yaw velocity error and the yaw angle; and (ii) compared with the proportional-integral (PI) controller, the designed DOFCs can provide much smaller oscillation amplitudes of the yaw velocity error and the yaw angle.

8.1 Network-Based Modeling

Note that the first order model of Nomoto is the simplest model to describe the dynamics of the marine vehicle. Due to the nominal high order state-space model's resemblance of the Nomoto model, this chapter investigates an anchored marine vehicle, which is equipped with thrusters [1, 2]. Consider the body-fixed and earth-fixed reference frames presented in Fig. 7.1, where x, y, and z denote the longitudinal axis, transverse axis, and normal axis, respectively; X, Y, and Z denote earth-fixed reference frames.

The body-fixed equations of motion in surge, sway, and yaw are described as

$$M\dot{\nu}(t) + N\nu(t) + G\eta(t) = u(t), \tag{8.1}$$

Y.-L. Wang et al., *Network-Based Control of Unmanned Marine Vehicles*,
https://doi.org/10.1007/978-3-031-28605-6_8

where $\nu(t) = [\rho(t)\ \upsilon(t)\ r(t)]^T$ with $\rho(t)$, $\upsilon(t)$, and $r(t)$ denoting the surge velocity, sway velocity, and yaw velocity, respectively; $\eta(t) = [x_p(t)\ y_p(t)\ \psi(t)]^T$ with $x_p(t)$ and $y_p(t)$ denoting positions and $\psi(t)$ denoting the yaw angle. The control input vector $u(t) = [u_1(t)\ u_2(t)\ u_3(t)]^T$ with $u_1(t)$ and $u_2(t)$ denoting the forces in surge and sway, respectively, and $u_3(t)$ denoting the moment in yaw provided by the thruster system; M denotes the matrix of inertia which is invertible with $M = M^T > 0$; N introduces damping; the matrix G represents mooring forces; and

$$\dot{\eta}(t) = J(\psi(t))\nu(t), \tag{8.2}$$

where

$$J(\psi(t)) = \begin{bmatrix} \cos(\psi(t)) & -\sin(\psi(t)) & 0 \\ \sin(\psi(t)) & \cos(\psi(t)) & 0 \\ 0 & 0 & 1 \end{bmatrix}.$$

Let $A_1 = M^{-1}G$, $A = -M^{-1}N$, $B = M^{-1}$, and $x(t) = \nu(t)$. Then the system (8.1) can be expressed as

$$\dot{x}(t) = Ax(t) + Bu(t) - A_1 f(t, x(t)), \tag{8.3}$$

where $f(t, x(t)) = \eta(t)$ with $f(t, x(t))$ denoting a time-varying, nonlinear vector-valued function of $x(t)$. If the disturbance, denoted as $\tilde{D}(t)$, induced by waves, wind, and current, is taken into account, the system in (8.3) is converted to

$$\dot{x}(t) = Ax(t) + Bu(t) + \tilde{D}(t) - A_1 f(t, x(t)). \tag{8.4}$$

In practical applications, a marine vehicle may be stopped and anchored when carrying out tasks such as scientific characterization and exploration. Whenever the yaw angle $\psi(t)$ is small enough, one has that $\cos(\psi(t)) \approx 1$, $\sin(\psi(t)) \approx 0$, and $J(\psi(t)) \approx I$. For a marine vehicle, the disturbances induced by waves, wind, and current are unavoidable. Since this chapter aims to attenuate the oscillation amplitude of the yaw velocity error, which can be achieved by making the actual $x(t)$ accurately track a desired constant or piecewise constant state reference x_{ref}, one can combine the items $\tilde{D}(t)$ and $-A_1 f(t, x(t))$ as one item $\bar{D}(t)$ to be rejected. Thus, (8.4) is converted to

$$\dot{x}(t) = Ax(t) + Bu(t) + \bar{D}(t). \tag{8.5}$$

For the given x_{ref}, the control objective is to regulate the tracking error $e(t) \triangleq x(t) - x_{ref}$ to as small as possible. From (8.5) and the definition for $e(t)$, one has

$$\dot{e}(t) = Ae(t) + Bu(t) + \omega(t), \tag{8.6}$$

where $\omega(t) = [\omega_1(t)\ \omega_2(t)\ \omega_3(t)]^T = Ax_{ref} + \bar{D}(t)$.

To attenuate the oscillation amplitudes of the yaw velocity error and the yaw angle, one can choose the controlled output $z(t)$ as

$$z(t) = C_1 e(t) + Du(t), \tag{8.7}$$

with $C_1 = D = [0\ 0\ 1]$.

From $e(t)$ in (8.6) and $z(t)$ in (8.7), one has

$$\begin{cases} \dot{e}(t) = Ae(t) + Bu(t) + \omega(t), \\ z(t) = C_1 e(t) + Du(t),\ e(t_0) = e_0, \end{cases} \tag{8.8}$$

where $\omega(t)$ is assumed to belong to $L_2[t_0, \infty)$; $e_0 \in \mathbb{R}^n$ denotes the initial condition; A, B, C_1, and D are known constant matrices of appropriate dimensions.

In what follows, we turn to network-based modeling for the UMV.

Throughout this chapter, we consider an anchored UMV which is controlled by a remote land-based control station and equipped with thrusters. The UMV and the control station are connected through communication networks. The control station is composed of the controller and the state error calculator. The actuator is chosen as the zero order holder which is connected to the thruster system.

It should be mentioned that the surge velocity, the sway velocity, and the yaw velocity are not always measurable. In such a case, dynamic output feedback control is an ideal choice for attenuating the oscillation amplitudes of the yaw velocity error and the yaw angle. For the UMV, the measurement output $y(t)$ is constructed as $y(t) = C_2 e(t)$. If the measurement output $y(t)$ can accurately reflect the variation of the state error $e(t)$, one can choose $C_2 = I$; otherwise, the matrix C_2 can be chosen as $C_2 = \text{diag}\{1.02,\ 0.97,\ 1.1\}$, $C_2 = \text{diag}\{0.96,\ 0.99,\ 1.15\}$, and so on. Without loss of generality, we choose $C_2 = I$ in this chapter.

The UMV and the control station are connected through communication networks. The sampler is time-driven, while the control station and the actuator are event-driven. If there exists packet disordering, the latest available measurement outputs or control inputs will be utilized by the control station or the actuator, respectively, and disordered packets will be dropped purposely. Figure 8.1 depicts the signal transmission for the UMV subject to sampler-to-control station packet dropouts, network-induced delays, and packet disordering, where the solid lines denote that data packets are transmitted successfully, while the dashed lines and the dotted lines denote packet dropouts induced by communication networks unreliability and packet disordering, respectively. As observed from Fig. 8.1, the sampled data based on the measured outputs at the instants $t_k,\ t_{k+1},\ \ldots (k = 0,\ 1,\ 2,\ \ldots)$ are transmitted to the control station successfully, while the sampled data based on the measured outputs at the instants $\tilde{t}_{k1}$ and $\tilde{t}_{k2}$ are dropped due to communication networks unreliability and packet disordering, respectively. h denotes the length of the sampling period, τ_k denotes the time from the instant t_k when the sampler samples data from the UMV to the instant when the actuator receives the data, and $\tau_m \le \tau_k \le \tau_M$, where $\tau_m \ge 0$ and $\tau_M > 0$.

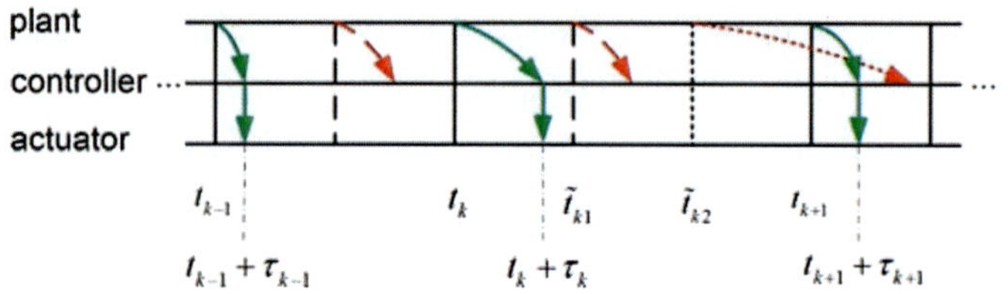

Fig. 8.1 Signal transmission for the UMV

For $t \in [t_k + \tau_k,\ t_{k+1} + \tau_{k+1})$, we seek a DOFC in the following form

$$\begin{cases} \dot{x_c}(t) = A_c x_c(t) + B_c y(t) + D_c x_c(t_k), \\ u(t) = C_c x_c(t), \end{cases} \tag{8.9}$$

where $x_c(t)$ is the controller state vector; A_c, B_c, C_c, and D_c are real matrices to be determined. Thus, the measurement output utilized by the control station is described as

$$y(t) = y(t_k). \tag{8.10}$$

For $t \in [t_k + \tau_k,\ t_{k+1} + \tau_{k+1})$, define the number of consecutive packet dropouts as δ_k, and $\tau(t) = t - t_k$ with $\tau(t)$ denoting an interval time-varying delay. Define the upper bound of δ_k as δ. Combining (8.8), (8.9), and (8.10), one can obtain the following augmented closed-loop system

$$\begin{cases} \dot{\xi}(t) = \phi_{1\xi}\xi(t) + \phi_{2\xi}\xi(t-\tau(t)) + \phi_{3\xi}\omega(t), \\ z(t) = \phi_{1z}\xi(t), \end{cases} \tag{8.11}$$

where

$$\xi(t) = \begin{bmatrix} e(t) \\ x_c(t) \end{bmatrix}, \phi_{1\xi} = \begin{bmatrix} A & BC_c \\ 0 & A_c \end{bmatrix}, \phi_{2\xi} = \begin{bmatrix} 0 & 0 \\ B_c C_2 & D_c \end{bmatrix}, \phi_{3\xi} = \begin{bmatrix} I \\ 0 \end{bmatrix}, \phi_{1z} = \begin{bmatrix} C_1 & DC_c \end{bmatrix}.$$

For $t \in [t_k + \tau_k,\ t_{k+1} + \tau_{k+1})$, by considering that $\tau(t) = t - t_k$ and $\tau_m \le \tau_k \le \tau_M$, one has $\tau(t) \in [\tau_m,\ \mu)$, where $\mu = (\delta + 1)h + \tau_M$. To make full use of the non-uniform distribution characteristic of the interval time-varying delay $\tau(t)$, we divide $[\tau_m,\ \mu)$ into two subintervals Φ_1 and Φ_2, where $\Phi_1 = [\tau_m,\ \beta)$, $\Phi_2 = [\beta,\ \mu)$, and $\beta = \frac{\tau_m + \mu}{2}$.

Consider the case that the interval time-varying delay $\tau(t)$ is non-uniformly distributed, and the statistic characteristic of $\tau(t)$ can be described by the following formula

$$\begin{cases} \text{Prob}\{\tau(t) \in \Phi_1\} = \bar{\lambda}, \\ \text{Prob}\{\tau(t) \in \Phi_2\} = 1 - \bar{\lambda}, \end{cases} \tag{8.12}$$

where $\bar{\lambda} \in [0,\ 1]$.

Define a stochastic variable $\lambda(t)$

$$\lambda(t)=\begin{cases}1, & \tau(t)\in\Phi_1,\\ 0, & \tau(t)\in\Phi_2.\end{cases} \tag{8.13}$$

By using the Bernoulli distributed white sequence to describe the stochastic variable $\lambda(t)$, one has

$$\begin{cases}\text{Prob}\{\lambda(t)=1\}=E\{\lambda(t)\}=\bar{\lambda},\\ \text{Prob}\{\lambda(t)=0\}=1-E\{\lambda(t)\}=1-\bar{\lambda}.\end{cases} \tag{8.14}$$

Taking into account the non-uniform distribution characteristic of $\tau(t)$, one can describe the measurement output utilized by the control station as

$$y(t)=\lambda(t)y(t-\tau_1(t))+(1-\lambda(t))y(t-\tau_2(t)), \tag{8.15}$$

where

$$\tau_1(t)=\begin{cases}\tau(t), & \tau(t)\in\Phi_1\\ \bar{\tau}_1, & \tau(t)\in\Phi_2\end{cases},\quad \tau_2(t)=\begin{cases}\tau(t), & \tau(t)\in\Phi_2\\ \bar{\tau}_2, & \tau(t)\in\Phi_1\end{cases}, \tag{8.16}$$

with $\bar{\tau}_1$ and $\bar{\tau}_2$ being constants, and $\bar{\tau}_1\in\Phi_1$, $\bar{\tau}_2\in\Phi_2$. Thus, the system (8.11) is converted to

$$\begin{cases}\dot{\xi}(t)=\hat{\psi}_1(t)+(\lambda(t)-\bar{\lambda})\hat{\psi}_2(t)+\phi_{3\xi}\omega(t),\\ z(t)=\phi_{1z}\xi(t),\end{cases} \tag{8.17}$$

where

$$\begin{aligned}\hat{\psi}_1(t)&=\phi_{1\xi}\xi(t)+\bar{\lambda}\phi_{2\xi}\xi(t-\tau_1(t))+(1-\bar{\lambda})\phi_{2\xi}\xi(t-\tau_2(t)),\\ \hat{\psi}_2(t)&=\phi_{2\xi}[\xi(t-\tau_1(t))-\xi(t-\tau_2(t))].\end{aligned}$$

For the system (8.17), the initial condition of the augmented state vector $\xi(t)$ on $[t_0-\mu,\ t_0]$ is supplemented as

$$\xi(\varsigma)=\varpi(\varsigma),\ \varsigma\in[t_0-\mu,\ t_0],$$

where $\varpi\in\mathbb{W}$ and $\mathbb{W}$ denotes the Banach space of absolutely continuous functions $[-\mu,0]\to\mathbb{R}^{2n}$ with square-integrable derivative and with the norm

$$\|\varpi\|_W^2=\|\varpi(0)\|^2+\int_{-\mu}^{0}\|\varpi(\varsigma)\|^2d\varsigma+\int_{-\mu}^{0}\|\dot{\varpi}(\varsigma)\|^2d\varsigma,$$

while the vector norm $\|\cdot\|$ represents the Euclidean norm.

Note that the disturbance $\tilde{D}(t)$ and the nonlinear item $-A_1 f(t, x(t))$ are combined as $\bar{D}(t)$ in (8.5). If the nonlinear item $-A_1 f(t, x(t))$ is investigated separately, the tracking error system in (8.6) is converted to

$$\dot{e}(t) = Ae(t) + Bu(t) + \tilde{\omega}(t) - A_1 f(t, e(t)), \tag{8.18}$$

where $\tilde{\omega}(t) = Ax_{ref} + \tilde{D}(t) = [\tilde{\omega}_1(t)\ \tilde{\omega}_2(t)\ \tilde{\omega}_3(t)]^T$, and $f(t, e(t))$ denotes a time-varying, nonlinear vector-valued function of $e(t)$. Then the system (8.17) is converted to

$$\begin{cases} \dot{\xi}(t) = \hat{\psi}_1(t) + (\lambda(t) - \bar{\lambda})\hat{\psi}_2(t) + \phi_{3\xi}\tilde{\omega}(t) - \bar{A}_1 f(t, e(t)), \\ z(t) = \phi_{1z}\xi(t), \end{cases} \tag{8.19}$$

where $\xi(t)$, $\hat{\psi}_1(t)$, $\hat{\psi}_2(t)$, $\phi_{3\xi}$, and ϕ_{1z} are the same as the corresponding items in (8.17), and $\bar{A}_1 = \begin{bmatrix} A_1 \\ 0 \end{bmatrix}$. For the nonlinear function $f(t, e(t))$, we consider the following sector condition [3]

$$[f(t, e(t)) - M_1 e(t)]^T [f(t, e(t)) - M_2 e(t)] \le 0, \tag{8.20}$$

where M_1 and M_2 are constant real matrices and $M = M_2 - M_1$ is a symmetric positive definite matrix, which means that the nonlinear function $f(t, e(t))$ belongs to a sector $[M_1,\ M_2]$.

As one can see, Fig. 8.1 takes sampler-to-control station packet dropouts, network-induced delays, and packet disordering into account. If the control station-to-actuator network-induced characteristics are considered, the signal transmission for the UMV in Fig. 8.1 is converted to the one in Fig. 8.2.

Similar to the modeling of the closed-loop system (8.11), one can derive the following augmented closed-loop system

$$\begin{cases} \dot{\xi}(t) = \hat{\phi}_{1\xi}\xi(t) + \hat{\phi}_{2\xi}\xi(t - \tau(t)) + \phi_{3\xi}\omega(t), \\ z(t) = \hat{\phi}_{1z}\xi(t) + \hat{\phi}_{2z}\xi(t - \tau(t)), \end{cases} \tag{8.21}$$

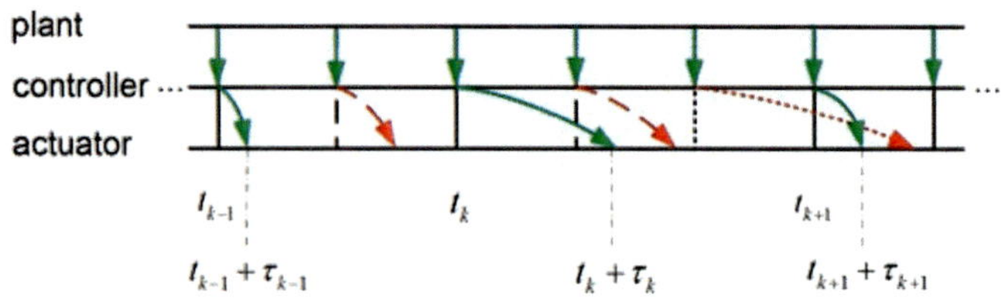

Fig. 8.2 Signal transmission for the UMV

where

$$\xi(t)=\begin{bmatrix} e(t) \\ x_c(t) \end{bmatrix}, \hat{\phi}_{1\xi}=\begin{bmatrix} A & 0 \\ B_cC_2 & A_c \end{bmatrix}, \hat{\phi}_{2\xi}=\begin{bmatrix} 0 & BC_c \\ 0 & D_c \end{bmatrix}, \phi_{3\xi}=\begin{bmatrix} I \\ 0 \end{bmatrix}, \hat{\phi}_{1z}=\begin{bmatrix} C_1 & 0 \end{bmatrix},$$

and $\hat{\phi}_{2z}=\begin{bmatrix} 0 & DC_c \end{bmatrix}$. Then one can take into account the non-uniform distribution characteristic of $\tau(t)$ to establish a closed-loop system accordingly. The corresponding modeling is omitted here for brevity.

The sampler-to-control station and the control station-to-actuator network-induced characteristics are considered in Figs. 8.1 and 8.2, respectively. If sampler-to-control station and control station-to-actuator (that is the dual-channel-network) packet dropouts, network-induced delays, and packet disordering are considered simultaneously, the signal transmission in Figs. 8.1 and 8.2 is converted to Fig. 8.3. As observed from Fig. 8.3, if sampled measurement outputs are transmitted to the control station successfully, then the control station will generate control inputs immediately, otherwise, no action is taken. τ_k^{sc} denotes the time from the instant t_k when the sampler samples data from the UMV to the instant when the control station receives the data, τ_k^{ca} denotes the time from the instant $t_k+\tau_k^{sc}$ when the control station begins to transmit the control input to the instant when the actuator receives the control input. Let $\tau_m^{sc}\leq\tau_k^{sc}\leq\tau_M^{sc}$, $\tau_m^{ca}\leq\tau_k^{ca}\leq\tau_M^{ca}$, where $\tau_m^{sc}\geq 0$ and $\tau_m^{ca}\geq 0$, while τ_M^{sc} and τ_M^{ca} may be larger than h.

From Fig. 8.3 and for $t\in[t_k+\tau_k,\ t_{k+1}+\tau_{k+1})$, one can see that the control input utilized by the UMV is

$$u(t)=u(t_k+\tau_k^{sc}). \tag{8.22}$$

However, for $t\in[t_k+\tau_k,\ t_{k+1}+\tau_{k+1})$, the measurement outputs available to the control station are variable. At the instant $t_k+\tau_k$, the available latest and oldest measurement outputs to the control station are sampled at instants $t_k+\lfloor\frac{\tau_k}{h}\rfloor h$ and t_k, respectively, where $\lfloor\frac{\tau_k}{h}\rfloor$ is the largest integer smaller than or equal to $\frac{\tau_k}{h}$. Introduce a stochastic variable $\alpha(t)$. At the instant $t_k+\tau_k$, if the latest available measurement output to the control station is sampled at the instant $t_k+\lfloor\frac{\tau_k}{h}\rfloor h$, set $\alpha(t)=1$, otherwise, set $\alpha(t)=0$. Then one has

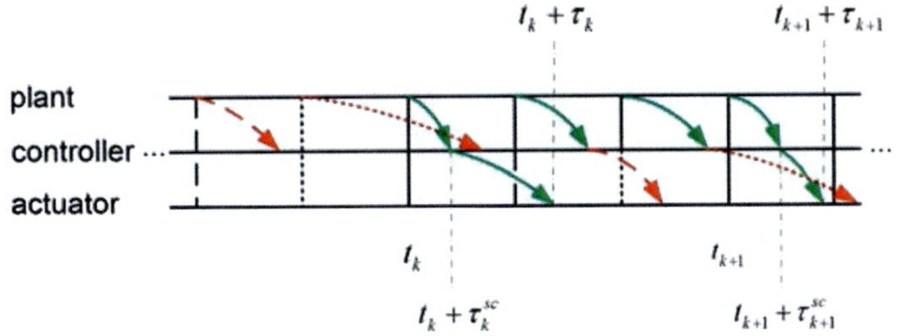

Fig. 8.3 Signal transmission for the dual-channel-network UMV

$$y(t) = \alpha(t)y(t_k + \lfloor \frac{\tau_k}{h} \rfloor h) + (1 - \alpha(t))y(t_k). \tag{8.23}$$

Remark 8.1 When the control input $u(t_k + \tau_k^{sc})$ is received by the actuator at the instant $t_k + \tau_k$, the actuator sends an acknowledgement signal to the control station. The acknowledgement signal is assigned the highest transmission priority and its network-induced delays are negligible. Thus, for $t \in [t_k + \tau_k, \ t_{k+1} + \tau_{k+1})$, the measurement output $y(t)$ utilized by the control station may be $y(t_k + \lfloor \frac{\tau_k}{h} \rfloor h)$ or $y(t_k)$.

By taking into consideration the system (8.8), the DOFC (8.9), the control input (8.22), and the measurement output (8.23), one can establish the following network-based closed-loop system

$$\begin{cases} \dot{\xi}(t) = \vartheta_{1\xi}\xi(t) + \vartheta_{2\xi}\xi(t - d_1(t)) + \vartheta_{3\xi}\xi(t - d_2(t)) \\ \qquad + \vartheta_{4\xi}\xi(t - d_3(t)) + \vartheta_{5\xi}\omega(t), \\ z(t) = \tilde{\vartheta}_{1z}\xi(t) + \tilde{\vartheta}_{2z}\xi(t - d_1(t)), \end{cases} \tag{8.24}$$

where $d_1(t) = t - (t_k + \tau_k^{sc})$, $d_2(t) = t - t_k$, $d_3(t) = t - (t_k + \lfloor \frac{\tau_k}{h} \rfloor h)$, $\vartheta_{1\xi} = \begin{bmatrix} A & 0 \\ 0 & A_c \end{bmatrix}$, $\vartheta_{2\xi} = \begin{bmatrix} 0 & BC_c \\ 0 & 0 \end{bmatrix}$, $\vartheta_{3\xi} = \begin{bmatrix} 0 & 0 \\ (1 - \alpha(t))B_c C_2 & D_c \end{bmatrix}$, $\vartheta_{4\xi} = \begin{bmatrix} 0 & 0 \\ \alpha(t)B_c C_2 & 0 \end{bmatrix}$, $\vartheta_{5\xi} = \begin{bmatrix} I \\ 0 \end{bmatrix}$, $\tilde{\vartheta}_{1z} = \begin{bmatrix} C_1 & 0 \end{bmatrix}$, $\tilde{\vartheta}_{2z} = \begin{bmatrix} 0 & DC_c \end{bmatrix}$.

From the definition of $d_1(t)$, $d_2(t)$, and $d_3(t)$, and by considering that $\tau_m^{ca} \leq \tau_k^{ca} \leq \tau_M^{ca}$, $\tau_m^{sc} \leq \tau_k^{sc} \leq \tau_M^{sc}$, $\tau_m \leq \tau_k \leq \tau_M$, one has that $d_1(t) \in [\tau_m^{ca}, \ (\delta + 1)h + \tau_M - \tau_m^{sc})$, $d_2(t) \in [\tau_m, \ (\delta + 1)h + \tau_M)$, $d_3(t) \in [0, \ (\delta + 1)h + \tau_M)$. Based on the upper-bounds and the lower-bounds of $d_1(t)$, $d_2(t)$, and $d_3(t)$, one can study DOFC design for the closed-loop system (8.24).

If there is no communication network between the UMV and the land-based control station, the augmented closed-loop system for the UMV can be described as

$$\begin{cases} \dot{\xi}(t) = \zeta_{1\xi}\xi(t) + \phi_{3\xi}\omega(t), \\ z(t) = \phi_{1z}\xi(t), \end{cases} \tag{8.25}$$

where $\zeta_{1\xi} = \begin{bmatrix} A & BC_c \\ B_c C_2 & A_c \end{bmatrix}$, while $\phi_{3\xi}$ and ϕ_{1z} are the same as the corresponding items in (8.11).

8.2 DOFC Design

This section is concerned with DOFC design for the UMV. In doing so, we construct the following Lyapunov functional [4]

$$V(t,\xi_t) = \xi^T(t)P\xi(t) + \int_{-\tau_m}^{0}\int_{t+s}^{t}\dot{\xi}^T(\theta)Q_0\dot{\xi}(\theta)d\theta ds \\ + \int_{-\beta}^{-\tau_m}\int_{t+s}^{t}\dot{\xi}^T(\theta)Q_1\dot{\xi}(\theta)d\theta ds + \int_{-\mu}^{-\beta}\int_{t+s}^{t}\dot{\xi}^T(\theta)Q_2\dot{\xi}(\theta)d\theta ds, \tag{8.26}$$

where $\xi_t = \xi(t+\varsigma)$, $\varsigma \in [t_0+\tau_0-\mu,\ t_0+\tau_0]$, the matrices P, Q_0, Q_1, and Q_2 are symmetric positive definite with appropriate dimensions. We now establish the following result for the network-based system (8.17).

Theorem 8.1 *For given positive scalars ε_0, ε_1, ε_2, h, δ, τ_m, τ_M, γ, and a scalar $\bar{\lambda} \in [0,\ 1]$, the closed-loop system described by (8.17) is asymptotically stable in the sense of mean-square with an H_∞ norm bound γ, and the gains for the DOFC in (8.9) are given by*

$$\begin{aligned} &A_c = S^{-1}(\hat{A}^T - XAY - XB\hat{C}^T)W^{-T}, \quad B_c = S^{-1}\hat{B}, \\ &C_c = \hat{C}^T W^{-T}, \quad D_c = S^{-1}(\hat{D} - SB_cC_2Y)W^{-T}, \end{aligned} \tag{8.27}$$

where S and W are nonsingular matrices satisfying $SW^T = I - XY$, if there exist symmetric positive definite matrices X, Y, and matrices $\hat{A}$, $\hat{B}$, $\hat{C}$, $\hat{D}$ such that

$$\begin{bmatrix} \widetilde{\Pi}_{11} & \widetilde{\Pi}_{12} & \widetilde{\Pi}_{13} & \widetilde{\Pi}_{14} \\ * & \widetilde{\Pi}_{22} & 0 & 0 \\ * & * & \widetilde{\Pi}_{33} & 0 \\ * & * & * & -\gamma I \end{bmatrix} < 0, \tag{8.28}$$

where

$$\widetilde{\Pi}_{11} = \begin{bmatrix} \widetilde{\Delta}_1 & \widetilde{\Delta}_2 \\ * & \widetilde{\Delta}_3 \end{bmatrix}, \ \widetilde{\Pi}_{12} = \begin{bmatrix} \widetilde{\Upsilon}_1 & \widetilde{\Upsilon}_1 & \widetilde{\Upsilon}_1 \end{bmatrix}, \ \widetilde{\Pi}_{13} = \begin{bmatrix} \widetilde{\Upsilon}_2 & \widetilde{\Upsilon}_2 & \widetilde{\Upsilon}_2 \end{bmatrix}, \tag{8.29}$$

$$\widetilde{\Pi}_{14} = \begin{bmatrix} H_5^T & \underbrace{0,\ \ldots,\ 0}_{13} \end{bmatrix}^T, \ \widetilde{\Pi}_{33} = \hat{\lambda}^{-1}\widetilde{\Pi}_{22},$$

$$\widetilde{\Pi}_{22} = \mathrm{diag}\{-\tau_m^{-1}\varepsilon_0^{-1}H_3,\ -(\beta-\tau_m)^{-1}\varepsilon_1^{-1}H_3,\ -(\mu-\beta)^{-1}\varepsilon_2^{-1}H_3\},$$

$$\hat{\lambda} = \bar{\lambda}(1-\bar{\lambda}),\ \mu = (\delta+1)h + \tau_M,\ \beta = \frac{\tau_m+\mu}{2},$$

$$\widetilde{\Delta}_1 = \begin{bmatrix} \widetilde{\Omega}_{11} & \widetilde{\Omega}_{12} & \widetilde{\Omega}_{13} & 0 & \widetilde{\Omega}_{15} & 0 & \widetilde{\Omega}_{17} \\ * & \widetilde{\Omega}_{22} & \widetilde{\Omega}_{23} & \widetilde{\Omega}_{24} & 0 & 0 & \widetilde{\Omega}_{27} \\ * & * & \widetilde{\Omega}_{33} & \widetilde{\Omega}_{34} & 0 & 0 & 0 \\ * & * & * & \widetilde{\Omega}_{44} & \widetilde{\Omega}_{45} & \widetilde{\Omega}_{46} & 0 \\ * & * & * & * & \widetilde{\Omega}_{55} & \widetilde{\Omega}_{56} & 0 \\ * & * & * & * & * & \widetilde{\Omega}_{66} & 0 \end{bmatrix},$$

$$\widetilde{\Delta}_2 = \begin{bmatrix} 0 & 0 & 0 & 0 & 0 & 0 & H_2 \\ \widetilde{\Omega}_{28} & 0 & 0 & \widetilde{\Omega}_{2,11} & 0 & 0 & 0 \\ \widetilde{\Omega}_{38} & \widetilde{\Omega}_{39} & 0 & 0 & 0 & 0 & 0 \\ 0 & \widetilde{\Omega}_{49} & \widetilde{\Omega}_{4,10} & \widetilde{\Omega}_{4,11} & \widetilde{\Omega}_{4,12} & 0 & 0 \\ 0 & 0 & 0 & 0 & \widetilde{\Omega}_{5,12} & \widetilde{\Omega}_{5,13} & 0 \\ 0 & 0 & \widetilde{\Omega}_{6,10} & 0 & 0 & \widetilde{\Omega}_{6,13} & 0 \end{bmatrix},$$

$$\begin{aligned}
\tilde{\Delta}_3 &= \operatorname{diag}\{\tilde{\Omega}_{77}, \tilde{\Omega}_{88}, \tilde{\Omega}_{99}, \tilde{\Omega}_{10,10}, \tilde{\Omega}_{11,11}, \tilde{\Omega}_{12,12}, \tilde{\Omega}_{13,13}, -\gamma I\},\\
\tilde{\Upsilon}_1 &= \begin{bmatrix} H_4^T\ 0\ \bar{\lambda}H_1\ 0\ (1-\bar{\lambda})H_1\ \underbrace{0, \ \ldots, \ 0}_{8}\ H_2 \end{bmatrix}^T,\\
\tilde{\Upsilon}_2 &= \begin{bmatrix} 0\ 0\ H_1\ 0\ -H_1\ \underbrace{0, \ \ldots, \ 0}_{9} \end{bmatrix}^T,\\
\tilde{\Omega}_{11} &= H_4 + H_4^T - \frac{4}{\tau_m}\varepsilon_0 H_3, \ \tilde{\Omega}_{12} = -\frac{2}{\tau_m}\varepsilon_0 H_3, \ \tilde{\Omega}_{13} = \bar{\lambda}H_1,\\
\tilde{\Omega}_{15} &= (1-\bar{\lambda})H_1, \ \tilde{\Omega}_{17} = \tilde{\Omega}_{27} = \frac{6}{\tau_m}\varepsilon_0 H_3,\\
\tilde{\Omega}_{22} &= -\frac{4}{\tau_m}\varepsilon_0 H_3 - \frac{4}{\beta-\tau_m}\varepsilon_1 H_3, \ \tilde{\Omega}_{23} = \tilde{\Omega}_{34} = -\frac{2\bar{\lambda}}{\beta-\tau_m}\varepsilon_1 H_3,\\
\tilde{\Omega}_{24} &= -\frac{2(1-\bar{\lambda})}{\beta-\tau_m}\varepsilon_1 H_3, \ \tilde{\Omega}_{66} = -\frac{4}{\mu-\beta}\varepsilon_2 H_3,\\
\tilde{\Omega}_{28} &= \tilde{\Omega}_{38} = \tilde{\Omega}_{39} = \tilde{\Omega}_{49} = \frac{6\bar{\lambda}}{\beta-\tau_m}\varepsilon_1 H_3,\\
\tilde{\Omega}_{2,11} &= \tilde{\Omega}_{4,11} = \frac{6(1-\bar{\lambda})}{\beta-\tau_m}\varepsilon_1 H_3, \ \tilde{\Omega}_{33} = -\frac{8\bar{\lambda}}{\beta-\tau_m}\varepsilon_1 H_3,\\
\tilde{\Omega}_{44} &= -\frac{4}{\beta-\tau_m}\varepsilon_1 H_3 - \frac{4}{\mu-\beta}\varepsilon_2 H_3, \ \tilde{\Omega}_{45} = \tilde{\Omega}_{56} = -\frac{2(1-\bar{\lambda})}{\mu-\beta}\varepsilon_2 H_3,\\
\tilde{\Omega}_{46} &= -\frac{2\bar{\lambda}}{\mu-\beta}\varepsilon_2 H_3, \ \tilde{\Omega}_{4,10} = \tilde{\Omega}_{6,10} = \frac{6\bar{\lambda}}{\mu-\beta}\varepsilon_2 H_3,\\
\tilde{\Omega}_{4,12} &= \tilde{\Omega}_{5,12} = \tilde{\Omega}_{5,13} = \tilde{\Omega}_{6,13} = \frac{6(1-\bar{\lambda})}{\mu-\beta}\varepsilon_2 H_3,\\
\tilde{\Omega}_{55} &= -\frac{8(1-\bar{\lambda})}{\mu-\beta}\varepsilon_2 H_3, \ \tilde{\Omega}_{77} = -\frac{12}{\tau_m}\varepsilon_0 H_3, \ \tilde{\Omega}_{88} = \tilde{\Omega}_{99} = -\frac{12\bar{\lambda}}{\beta-\tau_m}\varepsilon_1 H_3,\\
\tilde{\Omega}_{10,10} &= -\frac{12\bar{\lambda}}{\mu-\beta}\varepsilon_2 H_3, \ \tilde{\Omega}_{11,11} = -\frac{12(1-\bar{\lambda})}{\beta-\tau_m}\varepsilon_1 H_3,\\
\tilde{\Omega}_{12,12} &= \tilde{\Omega}_{13,13} = -\frac{12(1-\bar{\lambda})}{\mu-\beta}\varepsilon_2 H_3, \ H_1 = \begin{bmatrix} 0 & 0 \\ \hat{D} & \hat{B}C_2 \end{bmatrix},\\
H_2 &= \begin{bmatrix} I \\ X \end{bmatrix}, \ H_3 = \begin{bmatrix} Y & I \\ I & X \end{bmatrix}, \ H_4 = \begin{bmatrix} YA^T + \hat{C}B^T & \hat{A} \\ A^T & A^T X \end{bmatrix}, \ H_5 = \begin{bmatrix} YC_1^T + \hat{C}D^T \\ C_1^T \end{bmatrix}.
\end{aligned}$$

Proof Considering the non-uniform distribution characteristic of $\tau(t)$ and using the Wirtinger-based integral inequality [5], and adopting appropriate matrix manipulations, one can obtain the above given DOFC design criterion. This completes the proof. □

Remark 8.2 Theorem 8.1 addresses the DOFC design issue for the UMV subject to sampler-to-control station packet dropouts, network-induced delays, and packet disordering. The DOFC design in Theorem 8.1 can be extended to deal with the DOFC design for the closed-loop systems (8.21) and (8.24).

The non-uniform distribution and the Wirtinger-based integral inequality approach are adopted in the proof of Theorem 8.1 to deal with integral inequalities for products of vectors. If the non-uniform distribution characteristic is neglected, the following DOFC design criterion can be derived.

Corollary 8.1 *For given positive scalars ε_0, ε_1, h, δ, τ_m, τ_M, and γ, the closed-loop system described by (8.11) is asymptotically stable with an H_∞ norm bound γ, and the gains for the DOFC in (8.9) are the same as the corresponding items in (8.27), if there exist symmetric positive definite matrices X, Y, and matrices $\hat{A}$, $\hat{B}$, $\hat{C}$, $\hat{D}$ such that*

$$\begin{bmatrix} \breve{\Pi}_{11} & \breve{\Pi}_{12} & \breve{\Pi}_{13} \\ * & \breve{\Pi}_{22} & 0 \\ * & * & -\gamma I \end{bmatrix} < 0, \tag{8.30}$$

where

$$\breve{\Pi}_{11} = \begin{bmatrix} \widetilde{\Omega}_{11} & \widetilde{\Omega}_{12} & H_1 & 0 & \breve{\Omega}_{15} & 0 & 0 & H_2 \\ * & \breve{\Omega}_{22} & \breve{\Omega}_{23} & 0 & \breve{\Omega}_{25} & \breve{\Omega}_{26} & 0 & 0 \\ * & * & \breve{\Omega}_{33} & \breve{\Omega}_{34} & 0 & \breve{\Omega}_{36} & \breve{\Omega}_{37} & 0 \\ * & * & * & \breve{\Omega}_{44} & 0 & 0 & \breve{\Omega}_{47} & 0 \\ * & * & * & * & \breve{\Omega}_{55} & 0 & 0 & 0 \\ * & * & * & * & * & \breve{\Omega}_{66} & 0 & 0 \\ * & * & * & * & * & * & \breve{\Omega}_{77} & 0 \\ * & * & * & * & * & * & * & -\gamma I \end{bmatrix},$$

$$\breve{\Pi}_{12} = \begin{bmatrix} \breve{\Upsilon} & \breve{\Upsilon} \end{bmatrix}, \ \breve{\Pi}_{13} = \begin{bmatrix} H_5^T & \underbrace{0, \ldots, 0}_{7} \end{bmatrix}^T,$$

$$\breve{\Pi}_{22} = \mathrm{diag}\{-\tau_m^{-1}\varepsilon_0^{-1}H_3, \ -(\mu-\tau_m)^{-1}\varepsilon_1^{-1}H_3\},$$

$$\breve{\Upsilon} = \begin{bmatrix} H_4^T & 0 & H_1 & 0 & 0 & 0 & 0 & H_2 \end{bmatrix}^T,$$

$$\breve{\Omega}_{15} = \breve{\Omega}_{25} = \frac{6}{\tau_m}\varepsilon_0 H_3, \ \breve{\Omega}_{22} = -\frac{4}{\tau_m}\varepsilon_0 H_3 - \frac{4}{\mu-\tau_m}\varepsilon_1 H_3,$$

$$\breve{\Omega}_{23} = -\frac{2}{\mu-\tau_m}\varepsilon_1 H_3, \ \breve{\Omega}_{26} = \breve{\Omega}_{36} = \breve{\Omega}_{37} = \breve{\Omega}_{47} = \frac{6}{\mu-\tau_m}\varepsilon_1 H_3,$$

$$\breve{\Omega}_{33} = -\frac{8}{\mu-\tau_m}\varepsilon_1 H_3, \ \breve{\Omega}_{34} = -\frac{2}{\mu-\tau_m}\varepsilon_1 H_3, \ \breve{\Omega}_{44} = -\frac{4\varepsilon_1 H_3}{\mu-\tau_m},$$

$$\breve{\Omega}_{55} = -\frac{12}{\tau_m}\varepsilon_0 H_3, \ \breve{\Omega}_{66} = \breve{\Omega}_{77} = -\frac{12}{\mu-\tau_m}\varepsilon_1 H_3,$$

while μ, $\widetilde{\Omega}_{11}$, $\widetilde{\Omega}_{12}$, H_1, H_2, H_3, H_4, and H_5 are the same as the corresponding items in Theorem 8.1.

Remark 8.3 Theorem 8.1 provides a DOFC design criterion for the network-based system (8.17). If the nonlinear vector-valued function $f(t, e(t))$ and the sector condition (8.20) are taken into account, one can obtain the DOFC design criterion for the system (8.19) immediately. It is omitted here for brevity.

Theorem 8.1 presents the DOFC design for the closed-loop system (8.17). If no communication network is considered, the following DOFC design criterion is obtained.

Corollary 8.2 *For a given scalar $\gamma > 0$, the closed-loop system (8.25) is asymptotically stable with an H_∞ norm bound γ, and the parameters for the DOFC in (8.9) are given by*

$$\begin{aligned} A_c &= S^{-1}(\hat{A}^T - XAY - XB\hat{C}^T - \hat{B}C_2Y)W^{-T}, \\ B_c &= S^{-1}\hat{B},\ C_c = \hat{C}^T W^{-T}, \end{aligned} \tag{8.31}$$

where S and W are nonsingular matrices satisfying $SW^T = I - XY$, if there exist symmetric positive definite matrices X, Y, and matrices $\hat{A}$, $\hat{B}$, and $\hat{C}$ such that

$$\begin{bmatrix} \hat{H}_0 & H_2 & H_5 \\ * & -\gamma I & 0 \\ * & * & -\gamma I \end{bmatrix} < 0, \tag{8.32}$$

where $\hat{H}_0 = \begin{bmatrix} YA^T + AY + \hat{C}B^T + B\hat{C}^T & \hat{A} + A \\ \hat{A}^T + A^T & \Psi \end{bmatrix}$, $\Psi = A^T X + XA + C_2^T \hat{B}^T + \hat{B}C_2$, *while H_2 and H_5 are the same as the corresponding items in (8.28).*

8.3 Performance Analysis and Discussion

In this section, we show the effectiveness of the proposed DOFC design for the network-based closed-loop system (8.17) in attenuating the oscillation amplitudes of the yaw velocity error and the yaw angle. Choose the matrices M, N and G in the system (8.1) as

$$M = \begin{bmatrix} 1.0852 & 0 & 0 \\ 0 & 2.0575 & -0.4087 \\ 0 & -0.4087 & 0.2153 \end{bmatrix},$$

$$N = \begin{bmatrix} 0.0865 & 0 & 0 \\ 0 & 0.0762 & 0.0151 \\ 0 & 0.0151 & 0.031 \end{bmatrix},\ G = \begin{bmatrix} 0.0389 & 0 & 0 \\ 0 & 0.0266 & 0 \\ 0 & 0 & 0 \end{bmatrix}.$$

Noting that $A = -M^{-1}N$, $B = M^{-1}$, one can get matrices A and B. It is clearly seen in Sect. 8.1 that C_1 and D are chosen as $C_1 = [0\ 0\ 1]$ and $D = [0\ 0\ 1]$, while the matrix $C_2 = I$.

8.3.1 The Merits of the DOFC Design

The merits of the non-uniform distribution and Wirtinger integral inequality-based DOFC design are verified by the following analysis.

For Theorem 8.1 and Corollary 8.1, suppose that $h = 0.2$, $\tau_m = 0.02$, $\tau_M = 0.25$, and $\delta = 2$. It should be mentioned that if an algorithm similar to Algorithm 2 in [6] is adopted to deal with Theorem 8.1 and Corollary 8.1, locally optimal parameters ε_0, ε_1, and ε_2 can be achieved. However, Algorithm 2 in [6] improves the performance at the cost of increased computational complexity. Without loss of generality, choose $\varepsilon_0 = 2$ and $\varepsilon_1 = 0.6$ in Theorem 8.1 and Corollary 8.1, while $\varepsilon_2 = 0.9$ in Theorem 8.1.

Then the H_∞ norm bound derived by Corollary 8.1 is $\gamma = 2.0675$. For Theorem 8.1, H_∞ norm bounds corresponding to different $\bar{\lambda}$ are given in Table 8.1. From the formula (8.12), one can conclude that the larger the $\bar{\lambda}$, the lower the probability that $\tau(t)$ falls into the subinterval Φ_2 and the smaller the H_∞ norm bound γ, which conclusion is verified by Table 8.1. Note that H_∞ norm bounds derived by Theorem 8.1 are much smaller than the one derived by Corollary 8.1, while the H_∞ norm bound under $\bar{\lambda} = 0.99$ is only 61.6% of that derived by Corollary 8.1, which facts illustrate the merits of the non-uniform distribution and Wirtinger integral inequality-based DOFC design.

8.3.2 The Effectiveness of the DOFC Design

In this subsection, the effectiveness of the proposed DOFC design, and the effects of packet dropouts and network-induced delays on the UMV controlled through communication networks are presented and discussed.

For the UMV, define the yaw velocity error as $r(t) - r_{ref}$ with r_{ref} denoting the reference yaw velocity. The reference signal $\rho_{ref} = \upsilon_{ref} = 0$, and r_{ref} is a piecewise constant function described by

$$r_{ref} = \begin{cases} -0.2, & t \in [0,\ 5)\text{s},\ [10,\ 15)\text{s},\ \text{and } [20,\ 25)\text{s}, \\ 0.2, & t \in [5,\ 10)\text{s},\ [15,\ 20)\text{s},\ \text{and } [25,\ 30]\text{s}. \end{cases} \tag{8.33}$$

Define the reduction percentage of the yaw velocity error, and the reduction percentage of the yaw angle as the ratio between the oscillation amplitudes of the yaw velocity error with and without control, and the ratio between the oscillation

Table 8.1 H_∞ norm bounds corresponding to different $\bar{\lambda}$

$\bar{\lambda}$	0.01	0.1	0.5	0.9	0.99
γ	1.8216	1.7404	1.4813	1.3070	1.2741

amplitudes of the yaw angle with and without control, respectively. The discussed oscillation amplitudes in this chapter refer to peak to peak oscillation amplitudes. From the definitions presented above, one can see that the smaller the reduction percentage of the yaw velocity error and the yaw angle, the better the UMV performance.

For the purpose of comparison, we first give the oscillation amplitudes of the yaw velocity error and the yaw angle for the UMV without control input. Suppose that the initial state of the state error dynamics (8.6) is $e_0 = [0.1 \ -0.1 \ 0.25]^T$. The disturbances of surge, sway and yaw motions $\omega_1(t)$, $\omega_2(t)$, and $\omega_3(t)$ are given as

$$\begin{cases} \omega_1(t) = 2F_1(s)N_1(t) + I_1 A x_{ref}, \\ \omega_2(t) = -\cos(3t)e^{-0.3t} + I_2 A x_{ref}, \\ \omega_3(t) = 0.9F_2(s)N_2(t) + I_3 A x_{ref}, \end{cases} \tag{8.34}$$

where $F_1(s)$ and $F_2(s)$ are shaping filters described by $\frac{K_{\omega 1}s}{s^2+2\varepsilon_1\sigma_1 s+\sigma_1^2}$ and $\frac{K_{\omega 2}s}{s^2+2\varepsilon_2\sigma_2 s+\sigma_2^2}$, respectively; $K_{\omega 1}$ and $K_{\omega 2}$ denote the dominate wave strength coefficients with $K_{\omega 1} = 0.2$ and $K_{\omega 2} = 0.6$; ε_1 and ε_2 denote the damping coefficients with $\varepsilon_1 = 0.5$ and $\varepsilon_2 = 1.6$; σ_1 and σ_2 denote the encountering wave frequencies with $\sigma_1 = 0.7$ and $\sigma_2 = 1$; $N_1(t)$ and $N_2(t)$ are band-limited white noise with noise powers 2 and 1.8, respectively; $I_1 = [1\,0\,0]$, $I_2 = [0\,1\,0]$, $I_3 = [0\,0\,1]$. Thus, the oscillation amplitudes of the yaw velocity and the yaw velocity error are all 1.1945 deg/s; and the oscillation amplitude of the yaw angle is 2.265 deg.

Note that the oscillation amplitudes of the yaw velocity and the yaw velocity error are the same, which is induced by the definition for the yaw velocity error. Thus, we only consider the yaw velocity error and the yaw angle in this chapter.

8.3.2.1 The Effects of Constant Packet Dropouts and Network-Induced Delays

In this subsection, we investigate the effects of constant packet dropouts and network-induced delays on the UMV controlled through communication networks. For Theorem 8.1, let $\bar{\lambda} = 0.9$, while h, ε_0, ε_1, and ε_2 are the same as the corresponding items presented above. Consider the case that the network-induced delays τ_k and the number of packet dropouts δ_k are constant. By setting $\tau_k = 0.01s, \ \delta_k = 2$ (Case 1), $\tau_k = 0.1s, \ \delta_k = 1$ (Case 2), $\tau_k = 0.1s, \ \delta_k = 2$ (Case 3), and $\tau_k = 0.2s, \ \delta_k = 1$ (Case 4), respectively, one can get different DOFCs which are denoted as DOFC1, DOFC2, DOFC3, and DOFC4, respectively, for the system (8.17). The controller gains corresponding to DOFC1, DOFC2, DOFC3, and DOFC4 are omitted here for brevity. For the system (8.17), let $\xi_0 = [0.1 \ \ -0.1 \ \ 0.25 \ \ 0.05 \ \ -0.05 \ \ 0.02]^T$. Then Figs. 8.4 and 8.5 depict the responses of the yaw velocity error, the yaw angle, and the yaw moment under DOFC1 and DOFC4, respectively. For the UMV subject to constant packet dropouts and network-induced delays, the oscillation amplitudes of the yaw velocity error, the yaw angle, and the yaw moment; and the reduction

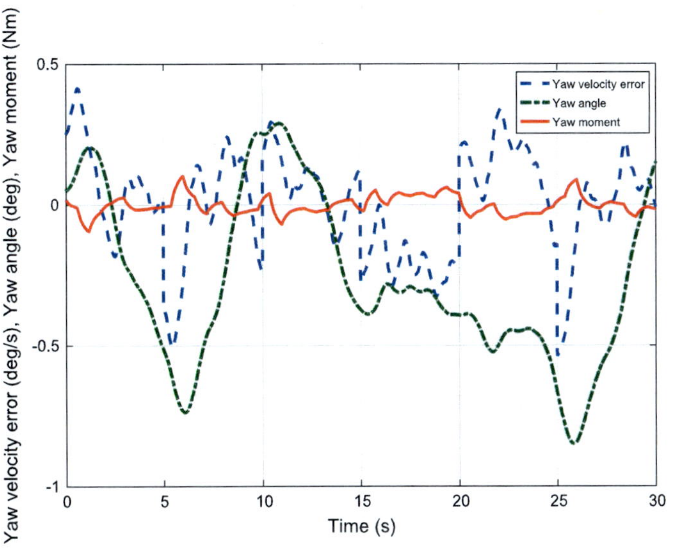

Fig. 8.4 The responses of the yaw velocity error, the yaw angle, and the yaw moment under DOFC1

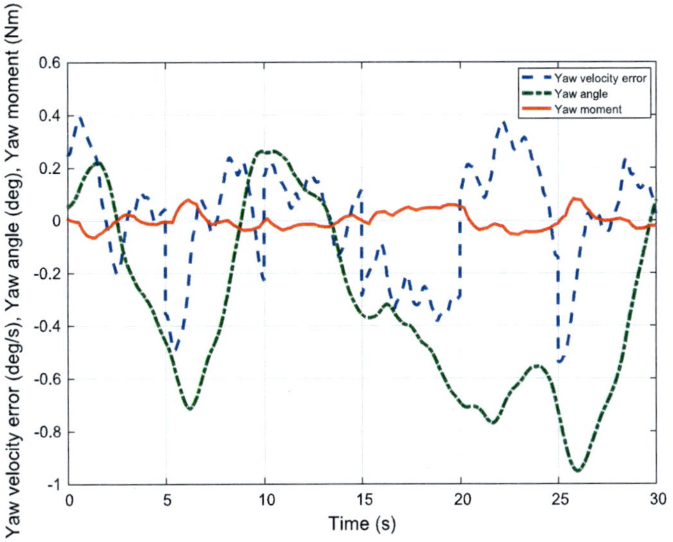

Fig. 8.5 The responses of the yaw velocity error, the yaw angle, and the yaw moment under DOFC4

percentage of the yaw velocity error and the yaw angle are listed in Table 8.2, where '–' in Table 8.2 denotes that the corresponding item does not exist.

From Table 8.2, Figs. 8.4, and 8.5, one can see that DOFC1, DOFC2, DOFC3, and DOFC4 are effective in reducing the oscillation amplitudes of the yaw velocity

Table 8.2 The oscillation amplitudes of the yaw velocity error, the yaw angle, and the yaw moment (denoted as OA_{yve}, OA_{ya}, and OA_{ym}, respectively), and the reduction percentage of the yaw velocity error and the yaw angle (denoted as RP_{yve} and RP_{ya}, respectively) under constant packet dropouts and network-induced delays

Controllers	OA_{yve}	RP_{yve} (%)	OA_{ya}	RP_{ya} (%)	OA_{ym}
No control	1.1945	–	2.265	–	–
DOFC1	0.9538	79.9	1.1389	50.3	0.1959
DOFC2	0.9520	79.7	1.1979	52.9	0.1834
DOFC3	0.9502	79.6	1.0612	46.9	0.1729
DOFC4	0.9319	78.0	1.2144	53.6	0.1457

error and the yaw angle. Note that DOFC3 can reduce the oscillation amplitude of the yaw angle to 46.9% of that of the system without control, while DOFC4 can reduce the oscillation amplitude of the yaw velocity error to 78.0% of that of the system without control. Moreover, the reduction percentage of the yaw velocity error derived by DOFC1, DOFC2, and DOFC3 is larger than or equal to 79.6%, and the reduction percentage of the yaw velocity error derived by DOFC4 is equal to 78.0%; the reduction percentage of the yaw angle derived by DOFC1 and DOFC3 is smaller than or equal to 50.3%, and the reduction percentage of the yaw angle derived by DOFC2 and DOFC4 is larger than or equal to 52.9%. It is also seen from Table 8.2 that the oscillation amplitudes of the yaw moment under DOFC1 and DOFC2 are larger than the corresponding items under DOFC3 and DOFC4. In addition, the oscillation amplitude of the yaw moment under DOFC1 is about 1.34 times of the oscillation amplitude of the yaw moment under DOFC4. Thus, DOFC4 provides the smallest oscillation amplitude of the yaw velocity error and the largest oscillation amplitude of the yaw angle at the smallest control cost.

The responses of the yaw velocity error under DOFC1, DOFC2, DOFC3, and DOFC4 are given in Fig. 8.6. The effectiveness of the DOFC design scheme is verified also by Fig. 8.6.

8.3.2.2 The Effects of Time-Varying Packet Dropouts and Network-Induced Delays

We investigate the effects of time-varying packet dropouts and network-induced delays on the UMV in this subsection. For Theorem 8.1, choose $\bar{\lambda}$, h, ε_0, ε_1, and ε_2 the same as the corresponding items presented above. Consider the case that the network-induced delays τ_k and the number of consecutive packet dropouts δ_k are time-varying, and the upper bound (denoted as δ) of δ_k is a given constant. By setting $\tau_m = 0.01s$, $\tau_M = 0.1s$, and $\delta = 1$ (Case 5), $\tau_m = 0.01s$, $\tau_M = 0.2s$, and $\delta = 1$ (Case 6), $\tau_m = 0.05s$, $\tau_M = 0.15s$, and $\delta = 1$ (Case 7), $\tau_m = 0.05s$, $\tau_M = 0.1s$, and $\delta = 2$ (Case 8), one can get different DOFCs which are denoted as DOFC5, DOFC6, DOFC7, and DOFC8, respectively, for the system (8.17). The controller gains corresponding to DOFC5, DOFC6, DOFC7, and DOFC8 are omitted here for

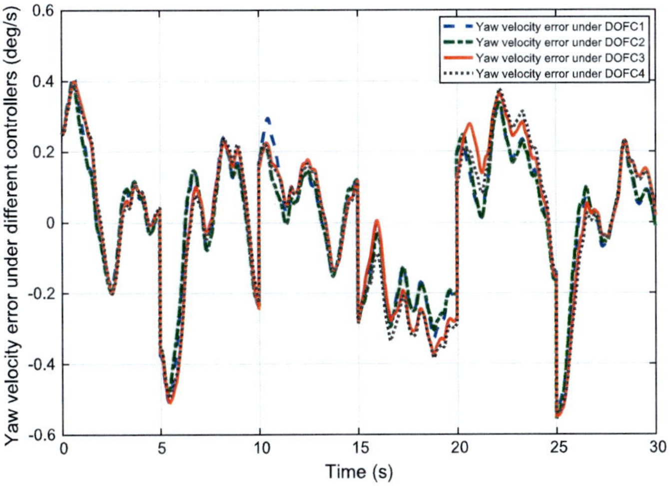

Fig. 8.6 The responses of the yaw velocity error under different controllers

brevity. For the system (8.17), suppose that ξ_0 is the same as the corresponding item in Sect. 8.3.2.1, while τ_k and δ_k vary stochastically under the constraint of the above given upper bounds and lower bounds. Then the responses of the yaw velocity error, the yaw angle, and the yaw moment under DOFC5 are presented in Fig. 8.7, while the responses of the yaw velocity error, the yaw angle, and the yaw moment under DOFC8 are presented in Fig. 8.8. The oscillation amplitudes of the yaw velocity error, the yaw angle, and the yaw moment; and the reduction percentage of the yaw velocity error and the yaw angle for the UMV subject to time-varying packet dropouts and network-induced delays are listed in Table 8.3.

It is seen clearly from Table 8.3 that DOFC7 can reduce the oscillation amplitude of the yaw angle to 47.2% of that of the system without control, while DOFC6 can reduce the oscillation amplitude of the yaw velocity error to 77.2% of that of the system without control. Note that the upper bounds of δ_k (that is δ) under DOFC5, DOFC6, and DOFC7 are all assumed to be 1, while the upper bounds of τ_k (denoted as τ_M) and the variation scopes of τ_k (denoted as $\tau_M - \tau_m$) under DOFC6 and DOFC7 are larger than the corresponding items under DOFC5. However, the oscillation amplitudes of the yaw velocity error and the yaw angle under DOFC6 and DOFC7 are smaller than the corresponding items under DOFC5, which phenomenon is induced by the fact that the control performance of the UMV is affected by both the upper bounds of τ_k, δ_k, and the actual values of τ_k, δ_k. For stochastic time-varying τ_k and δ_k, even if a large control input is imposed on the UMV, the control performance may not be the best, which statement is verified by the control performance under DOFC5. From Table 8.3, one can see that DOFC6 provides the smallest reduction percentage of the yaw velocity error, the second-smallest reduction percentage of the yaw angle, and the smallest control cost when compared with DOFC5, DOFC7, and DOFC8. Thus, if the control cost is specifically considered, DOFC6 is the best choice.

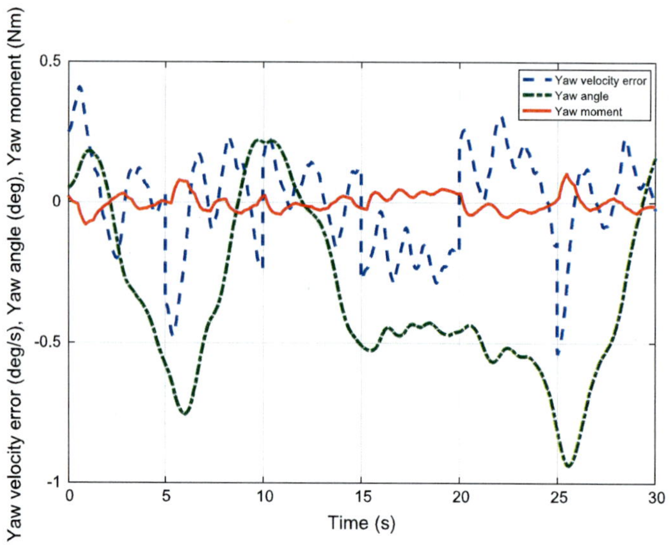

Fig. 8.7 The responses of the yaw velocity error, the yaw angle, and the yaw moment under DOFC5

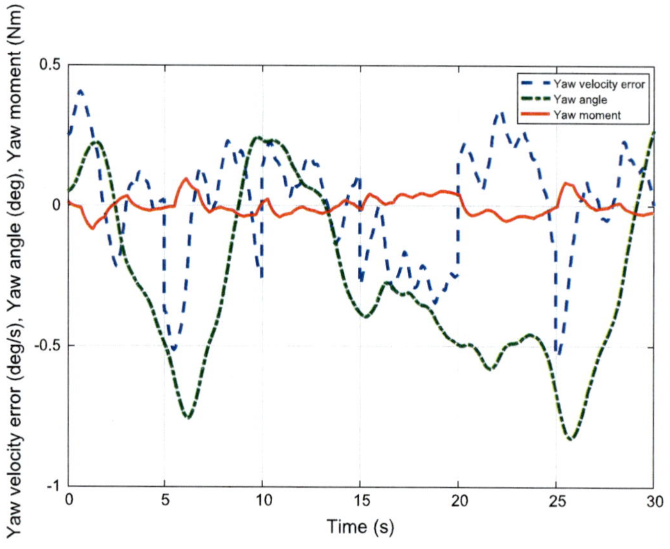

Fig. 8.8 The responses of the yaw velocity error, the yaw angle, and the yaw moment under DOFC8

The responses of the yaw velocity error under DOFC5, DOFC6, DOFC7, and DOFC8 are given in Fig. 8.9, which demonstrates the effectiveness of the proposed DOFC design.

Table 8.3 The oscillation amplitudes of the yaw velocity error, the yaw angle, and the yaw moment (denoted as OA_{yve}, OA_{ya}, and OA_{ym}, respectively), and the reduction percentage of the yaw velocity error and the yaw angle (denoted as RP_{yve} and RP_{ya}, respectively) under time-varying packet dropouts and network-induced delays

Controllers	OA_{yve}	RP_{yve} (%)	OA_{ya}	RP_{ya} (%)	OA_{ym}
No control	1.1945	–	2.265	–	–
DOFC5	0.9501	79.5	1.1575	51.1	0.1854
DOFC6	0.9226	77.2	1.0718	47.3	0.1655
DOFC7	0.9323	78.1	1.0679	47.2	0.1896
DOFC8	0.9427	78.9	1.0961	48.4	0.1805

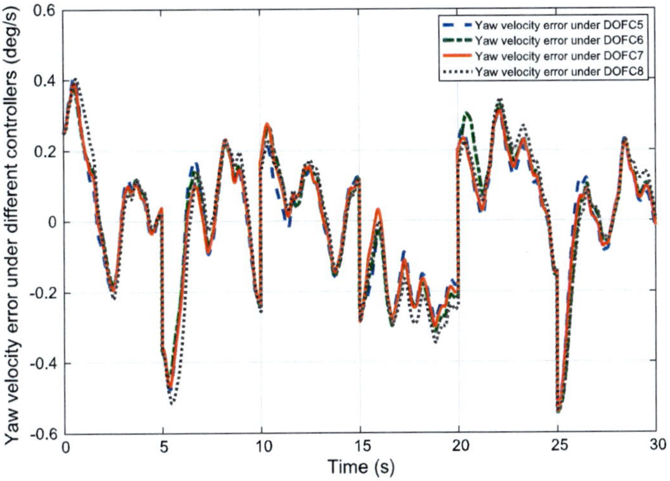

Fig. 8.9 The responses of the yaw velocity error under different controllers

8.4 Performance Comparison and Discussion

In this section, we compare the performance between the DOFC designed in this chapter and the PI controller designed in [1]. Suppose that system parameter matrices M, N, G, C_1, C_2, and D are the same as the ones in Sect. 8.3.

Based on an algebraic Riccati Equation (8) in [1], which is rewritten as (8.35) in this chapter, a PI controller for a linear plant without control input constraint is designed in [1], where

$$A_{aug}^T P + PA_{aug} + Q_z - PB_{aug}R_z^{-1}B_{aug}^T P = 0, \tag{8.35}$$

and $A_{aug} = \begin{bmatrix} A & 0 \\ C_p & 0 \end{bmatrix}$, $B_{aug} = \begin{bmatrix} B \\ 0 \end{bmatrix}$, $Q_z = \begin{bmatrix} 0 & 0 \\ 0 & I \end{bmatrix} Q$, $Q > 0$, $R_z > 0$, $C_p = [0 \ \ 0 \ \ 1]$. The PI controller gain $K = R_z^{-1} B_{aug}^T P$, $K = [K_1 \ \ K_2]$, $u(t) = -K_1 e(t) - K_2 \int_0^t C_p e(\tau) d\tau$. By choosing $Q = I$, $R_z = 3I$, and solving the algebraic Riccati Equation (8.35), one can get the controller gain K for the state error dynamics (7) in [1], which is rewritten as (8.36) in this chapter

$$\dot{e} = Ae + Ax_r + Bu + D'. \tag{8.36}$$

Moreover, $K = \begin{bmatrix} 0 & 0 & 0 & 0 \\ 0 & 0 & 0.0668 & 0.1119 \\ 0 & -0.0302 & 0.3522 & 0.5664 \end{bmatrix}$.

For the DOFC design scheme presented in Corollary 8.2, one can impose constraints on the elements of matrices $\hat{A}$, $\hat{B}$, and $\hat{C}$ to avoid large DOFC gains A_c, B_c, and C_c. By solving the controller design criterion in Corollary 8.2, one obtains

$$\begin{aligned} A_c &= \begin{bmatrix} -9.9040 & 0 & 0 \\ 0 & -11.3132 & 0.4082 \\ 0 & -4.2177 & -35.7590 \end{bmatrix}, \\ B_c &= \begin{bmatrix} -0.0004 & 0 & 0 \\ 0 & 0.4304 & 0.0174 \\ 0 & 0.0944 & -5.1325 \end{bmatrix}, \\ C_c &= \begin{bmatrix} 0 & 0 & 0 \\ 0 & -0.0715 & 0.5790 \\ 0 & 0.4794 & 4.8770 \end{bmatrix}. \end{aligned} \tag{8.37}$$

Choose the initial state of the system (8.36) as $e_0 = [0.2 \ \ 0 \ \ 0.5]^T$. For the system (8.25) in this chapter, the initial state is given as $\xi_0 = [e_0 \ x_{c0}]^T$, where x_{c0} denotes the initial state of the DOFC state vector $x_c(t)$. For the system (8.25), let $e_0 = [0.2 \ \ 0 \ \ 0.5]^T$ and $x_{c0} = [0.1 \ \ -0.1 \ \ 0.03]^T$.

The reference signal $\rho_{ref} = \upsilon_{ref} = 0$, and the reference yaw velocity r_{ref} is the same as the item presented in (8.33). Consider the case that the disturbances of surge, sway and yaw motions $\omega_1(t)$, $\omega_2(t)$, and $\omega_3(t)$ are the same as the items given in (8.34). Then the curves of the yaw velocity error for the systems in (8.25) of this chapter and in (7) of [1] are presented in Fig. 8.10. As observed from Fig. 8.10, the DOFC in this chapter provides a much smaller yaw velocity error than the PI controller in [1].

In fact, if the control performance instead of the control cost is specifically considered, the DOFC in this chapter can provide much better control performance than the PI controller in [1]. For Corollary 8.2 in this chapter, consider the case that there is no constraint on the matrices $\hat{A}$, $\hat{B}$, and $\hat{C}$. By solving the controller design criterion in Corollary 8.2, one obtains

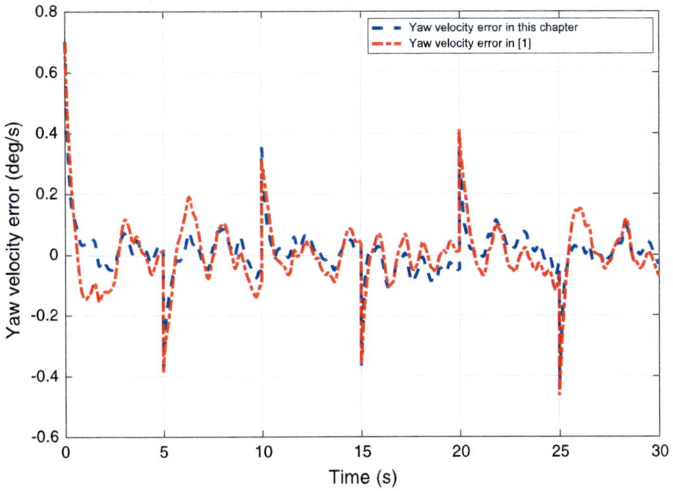

Fig. 8.10 The responses of the yaw velocity error under DOFC gains in (8.37) and the PI controller in [1]

$$
\begin{aligned}
A_c &= 10^4 \times \begin{bmatrix} -7.2225 & 0 & 0 \\ 0 & -7.2225 & 0 \\ 0 & 0 & -7.2232 \end{bmatrix}, \\
B_c &= \begin{bmatrix} -290.8473 & 0 & 0 \\ 0 & 285.3536 & -57.8999 \\ 0 & -60.2888 & -297.1282 \end{bmatrix}, \\
C_c &= \begin{bmatrix} 226.5573 & 0 & 0 \\ 0 & -223.7880 & 40.1344 \\ 0 & 49.3259 & 233.4622 \end{bmatrix}.
\end{aligned} \tag{8.38}
$$

Consider the case that the disturbances of surge, sway and yaw motions $\omega_1(t)$, $\omega_2(t)$, and $\omega_3(t)$ are the same as the items in (8.34). Then the curves of the yaw velocity error under DOFC gains in (8.38) in this chapter and the PI controller in [1] are given in Fig. 8.11.

Compared with the PI controller in [1], the DOFC gains in (8.37) and (8.38) provide much better control performance, which statement is verified by Figs. 8.10 and 8.11. To facilitate comparison, the curves of the yaw velocity error under DOFC gains in (8.37) and (8.38) are given in Fig. 8.12, from which figure one can see that the large DOFC gains in (8.38) provide better performance than the small DOFC gains in (8.37). Thus, there exists a tradeoff between the control cost and the control performance. If the control performance is specifically considered, the DOFC gains in (8.38) are preferred.

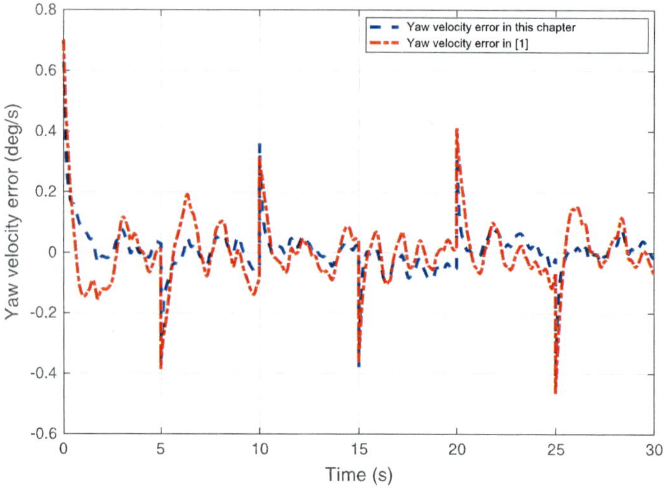

Fig. 8.11 The responses of the yaw velocity error under DOFC gains in (8.38) and the PI controller in [1]

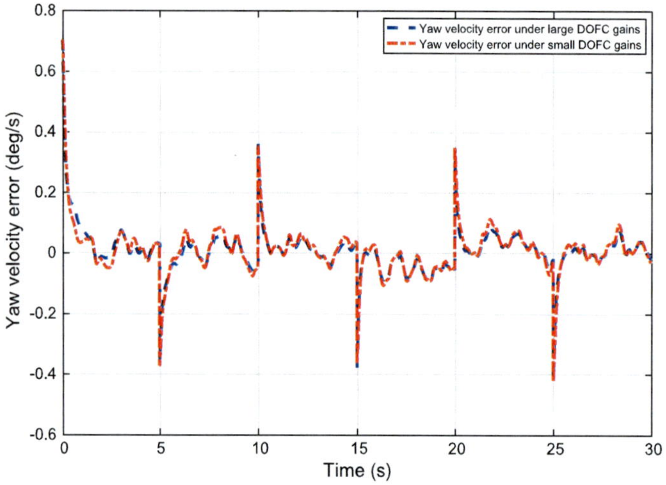

Fig. 8.12 The responses of the yaw velocity error under DOFC gains in (8.37) and (8.38)

8.5 Conclusions

The network-based modeling and DOFC design have been investigated to attenuate the oscillation amplitudes of the yaw velocity error and the yaw angle for a UMV. Network-based models for the UMV subject to packet dropouts, network-induced delays, and packet disordering have been established. Appropriate DOFC design schemes have been proposed. The performance analysis and discussion have demonstrated that the designed DOFCs can attenuate the oscillation amplitudes of the yaw velocity error and the yaw angle, and provide much smaller oscillation amplitudes than the PI controller.

8.6 Notes

In some practical situations, controlled plants' states may not be always measurable. Thus, observer-based control [7–9] and dynamic output feedback control [10, 11] have received much attention. For a UMV in network environments, if the surge velocity, sway velocity, and the yaw velocity are not measurable, how to propose an appropriate DOFC design scheme is of paramount importance and far from being resolved. Motivated by these facts, network-based modeling and DOFC design for a UMV are investigated in this chapter to attenuate the oscillation amplitudes of the yaw velocity error and the yaw angle. The results in this chapter are based mainly on [12]. For more results about output feedback control of systems, see also [13, 14], etc.

References

1. N.E. Kahveci, P.A. Ioannou, Adaptive steering control for uncertain ship dynamics and stability analysis. Automatica **49**(3), 685–697 (2013)
2. A. Grovlen, T.I. Fossen, Nonlinear control of dynamic positioned ships using only position feedback: an observer backstepping approach, in *Proceedings of the 35th IEEE Conference on Decision and Control*, Kobe, Japan, Dec. 1996, pp. 3388–3393
3. Q.-L. Han, Absolute stability of time-delay systems with sector-bounded nonlinearity. Automatica **41**(12), 2171–2176 (2005)
4. Q.-L. Han, Improved stability criteria and controller design for linear neutral systems. Automatica **45**(8), 1948–1952 (2009)
5. A. Seuret, F. Gouaisbaut, Wirtinger-based integral inequality: application to time-delay systems. Automatica **49**(9), 2860–2866 (2013)
6. Y.-L. Wang, Q.-L. Han, Modelling and observer-based H_∞ controller design for networked control systems. IET Control Theory Appl. **8**(15), 1478–1486 (2014)
7. S.-L. Du, X.-M. Sun, W. Wang, Guaranteed cost control for uncertain networked control systems with predictive scheme. IEEE Trans. Autom. Sci. Eng. **11**(3), 740–748 (2014)
8. Y.-L. Wang, Q.-L. Han, Network-based fault detection filter and controller coordinated design for unmanned surface vehicles in network environments. IEEE Trans. Ind. Inf. **12**(5), 1753–1765 (2016)
9. Y.-L. Wang, T.-B. Wang, Q.-L. Han, Fault detection filter design for data reconstruction-based continuous-time networked control systems. Inf. Sci. **328**, 577–594 (2016)
10. H. Gao, X. Meng, T. Chen, J. Lam, Stabilization of networked control systems via dynamic output-feedback controllers. SIAM J. Control Optim. **48**(5), 3643–3658 (2010)
11. M. Jungers, E.B. Castelan, V.M. Moraes, U.F. Moreno, A dynamic output feedback controller for NCS based on delay estimates. Automatica **49**(3), 788–792 (2013)

12. Y.-L. Wang, Q.-L. Han, Network-based modelling and dynamic output feedback control for unmanned marine vehicles in network environments. Automatica **91**, 43–53 (2018)
13. J. Zhang, C. Peng, X. Xie, D. Yue, Output feedback stabilization of networked control systems under a stochastic scheduling protocol. IEEE Trans. Cybern. **50**(6), 2851–2860 (2020)
14. D. Zhang, Q.-L. Han, X. Jia, Network-based output tracking control for T-S fuzzy systems using an event-triggered communication scheme. Fuzzy Sets Syst. **273**, 26–48 (2015)

Chapter 9
Cooperative Target Tracking of Multiple UMVs Under Switching Topologies

This chapter is concerned with the cooperative target tracking of multiple UMVs under switching network topologies. For the target to be tracked, only its position can be measured/received by some of the UMVs, and its velocity is unavailable to all the UMVs. A distributed extended state observer considering switching topologies is designed to integrally estimate unknown target dynamics and neighboring UMVs' dynamics. Accordingly, a novel kinematic controller is designed, which takes full advantage of known information and avoids the approximation of some virtual control vectors. Moreover, a disturbance observer is presented to estimate unknown time-varying environmental disturbances. Furthermore, a distributed dynamic controller is designed to regulate the involved UMVs to cooperatively track the target. It enables each UMV to adjust its forces and moments according to the received information from its neighbors. It is shown that the designed target tracking controller is effective.

9.1 Cooperative Target Tracking System Modeling

Consider a maritime target tracking system consisting of N UMVs, which aims to cooperatively track a marine target.

If each UMV is treated as a communication node, then the network topology among the N UMVs can be expressed by a weighted graph $\mathscr{G} = \{\mathscr{V}, \mathscr{E}, \mathscr{A}\}$, where $\mathscr{V} = \{1, 2, \ldots, N\}$ is the index set of N nodes (i.e. UMVs), $\mathscr{E} \subseteq \mathscr{V} \times \mathscr{V}$ is the edge set, and $\mathscr{A} = [a_{ij}]_{N\times N}$ is the adjacency matrix. The graph $\mathscr{G}$ is undirected, which means that the edges $(i,\ j)$ and $(j,\ i)$ $(\forall i, j \in \mathscr{V})$ are regarded as the same. In this chapter, the undirected edge $(i,\ j) \in \mathscr{E}$ indicates that UMVs i and j can receive position information from each other. That is, i and j are neighbors to each other. The neighbor set of node i is denoted by $\mathscr{N}_i = \{j \in \mathcal{V} : (j,\ i) \in \mathscr{E}\}$. Note that $a_{ij} > 0$ if $(j,\ i) \in \mathscr{E}$; otherwise, $a_{ij} = 0$. In the undirected graph $\mathscr{G}$, $a_{ij} = a_{ji},\ \forall i,\ j \in \mathscr{V}$.

Y.-L. Wang et al., *Network-Based Control of Unmanned Marine Vehicles*,
https://doi.org/10.1007/978-3-031-28605-6_9

The degree matrix $\mathscr{D}$ of $\mathscr{G}$ is a diagonal matrix whose ith diagonal entry is $\sum_{j=1}^{N} a_{ij}$. The Laplacian matrix of the undirected graph $\mathscr{G}$ is defined as $L = \mathscr{D} - \mathscr{A}$, which is a symmetric matrix, i.e., $L^T = L$.

This chapter considers the case that linear and angular velocities of the target are unknown. Some of the N UMVs can obtain the position vector information of the target, which means that the target is a neighbor of them. Treat the target as the node 0. The information exchange between the target and the N UMVs can be described by a weighted and augmented graph $\overline{\mathscr{G}}$ (with nodes 0, 1, 2, ..., N). The graph $\overline{\mathscr{G}}$ consists of graph $\mathscr{G}$, the node 0, and edges between the target and UMVs whose neighbors contain the target. Denote a_{i0} as the communication weight from the target to the UMV i (pinning weight), where $a_{i0} > 0$ if the target is a neighbor of the UMV i; otherwise, $a_{i0} = 0$. The pinning matrix is $D = \text{diag}\{a_{10}, a_{20}, \ldots, a_{N0}\}$.

Note that the graph associated with the interaction network topology of the target and the N UMVs may be switching. Denote the index set of all possible graphs as Ω with $\Omega = \{1, 2, \ldots, \varsigma\}$, where ς is a positive integer. Let $\sigma_t : [0, \infty) \to \Omega$ be the switching signal. For any $i, j \in \mathscr{V}$, $a_{ij}^{\sigma_t}$ and $a_{i0}^{\sigma_t}$ are communication weights from the UMV j to the UMV i and from the target to the UMV i at time t, respectively. The corresponding graph, Laplacian matrix, and pinning matrix are denoted by $\overline{\mathscr{G}}_{\sigma_t}$, L_{σ_t}, and D_{σ_t}, respectively. At switching mode σ_t, the neighbor set of node i is denoted by $\mathscr{N}_i^{\sigma_t}$. Without loss of generality, we assume that there exists at least one communication path from the target to every UMV for each graph $\overline{\mathscr{G}}_{\sigma_t}$. From [1], $L_{\sigma_t} + D_{\sigma_t}$ is a symmetric positive definite matrix.

The kinematical equation of the target is described by

$$\dot{\eta}_0 = R_0 \nu_0, \tag{9.1}$$

where $\eta_0 = [x_0 \ y_0 \ \psi_0]^T$ is the position vector with x_0, y_0 being positions and ψ_0 being the yaw angle in the earth-fixed frame; $\nu_0 = [u_0 \ v_0 \ r_0]^T$ is the velocity vector with u_0, v_0, and r_0 being the surge velocity, sway velocity, and yaw rate, respectively, in the body-fixed frame; $R_0 = R(\psi_0)$ with $R(\cdot)$ being the transformation matrix

$$R(\cdot) = \begin{bmatrix} cos(\cdot) & -sin(\cdot) & 0 \\ sin(\cdot) & cos(\cdot) & 0 \\ 0 & 0 & 1 \end{bmatrix}. \tag{9.2}$$

As mentioned above, η_0 is known to some UMVs and ν_0 is unknown to all the UMVs.

For surface ships, the motion of heave, pitch, and roll is usually neglected [2]. Then, the dynamical equation (surge-sway-yaw model) of the UMV i can be expressed as

$$\begin{cases} \dot{\eta}_i &= R_i \nu_i, \\ M_i \dot{\nu}_i &= -C_i(\nu_i)\nu_i - D_i(\nu_i)\nu_i + \tau_i + w_i, \end{cases} \quad i \in \mathscr{V}, \tag{9.3}$$

where $\eta_i = [x_i \; y_i \; \psi_i]^T$ is the position vector in the earth-fixed frame; $\nu_i = [u_i \; v_i \; r_i]^T$ is the velocity vector in the body-fixed frame; $R_i = R(\psi_i)$ is the transformation matrix; $\tau_i = [\tau_{iu} \; \tau_{iv} \; \tau_{ir}]^T$ is the control input vector with τ_{iu}, τ_{iv}, and τ_{ir} being surge and sway forces, and yaw moment, respectively; w_i is the unknown and time-varying disturbance vector describing forces and moments induced by wind, waves and ocean currents with $\|\dot{w}_i\| \le d_i^*$ (d_i^* is a positive scalar); M_i is the symmetric positive definite inertia matrix; $C_i(\nu_i)$ is the Coriolis and centripetal matrix; $D_i(\nu_i)$ is the damping matrix. In general, for $i \in \mathscr{V}$, M_i, $C_i(\nu_i)$, and $D_i(\nu_i)$ are, respectively,

$$M_i = \begin{bmatrix} m_{i11} & 0 & 0 \\ 0 & m_{i22} & m_{i23} \\ 0 & m_{i32} & m_{i33} \end{bmatrix},$$

$$C_i(\nu_i) = \begin{bmatrix} 0 & 0 & c_{i13}(v_i, r_i) \\ 0 & 0 & c_{i23}(u_i) \\ -c_{i13}(v_i, r_i) & -c_{i23}(u_i) & 0 \end{bmatrix},$$

$$D_i(\nu_i) = \begin{bmatrix} d_{i11}(u_i) & 0 & 0 \\ 0 & d_{i22}(v_i, r_i) & d_{i23}(v_i, r_i) \\ 0 & d_{i32}(v_i, r_i) & d_{i33}(v_i, r_i) \end{bmatrix},$$

where

$$\begin{aligned} m_{i11} &= m_i - X_{i\dot{u}_i}, \; m_{i22} = m_i - Y_{i\dot{v}_i}, \\ m_{i23} &= m_i x_{ig} - Y_{i\dot{r}_i}, \; m_{i32} = m_i x_{ig} - N_{i\dot{v}_i}, \\ m_{i33} &= I_{z_i} - N_{i\dot{r}_i}, \; c_{i13}(v_i, r_i) = -m_{i22} v_i - m_{i23} r_i, \\ c_{i23}(u_i) &= m_{i11} u_i, \; d_{i11}(u_i) = -X_{iu_i} - X_{i|u_i|u_i} |u_i|, \\ d_{i22}(v_i, r_i) &= -Y_{iv_i} - Y_{i|v_i|v_i} |v_i| - Y_{i|r_i|v_i} |r_i|, \\ d_{i23}(v_i, r_i) &= -Y_{ir_i} - Y_{i|v_i|r_i} |v_i| - Y_{i|r_i|r_i} |r_i|, \\ d_{i32}(v_i, r_i) &= -N_{iv_i} - N_{i|v_i|v_i} |v_i| - N_{i|r_i|v_i} |r_i|, \\ d_{i33}(v_i, r_i) &= -N_{ir_i} - N_{i|v_i|r_i} |v_i| - N_{i|r_i|r_i} |r_i|, \end{aligned}$$

with m_i being the mass of the UMV i, I_{z_i} being the inertia moment of the UMV i, and other symbols being hydrodynamic derivatives. More details can be found in [2].

The definition of cooperative target tracking is given as follows.

Definition 9.1 The N UMVs with dynamics (9.3) are said to cooperatively track the target with kinematics (9.1) if

$$p_i = \eta_0 - \eta_i - \eta_{i0}, \; i \in \mathscr{V} \tag{9.4}$$

can be made arbitrarily small. In (9.4), η_{i0} is a smooth vector function representing the ideal relative position vector between the UMV i and the target.

It is reasonable to assume that for any $i \in \mathscr{V}$, η_{i0} is bounded and is of the first and second time derivatives $\dot{\eta}_{i0}$ and $\ddot{\eta}_{i0}$.

This chapter aims to design a distributed dynamic controller for (9.3) such that the N UMVs can form a time-varying formation and cooperatively track the target with kinematics (9.1), i.e., (9.4) is arbitrarily small.

Define a distributed target tracking error as

$$e_i = a_{i0}^{\sigma_t}(\eta_0 - \eta_i - \eta_{i0}) + \sum_{j \in \mathscr{N}_i} a_{ij}^{\sigma_t}(\eta_j - \eta_i - \eta_{ij0}), \ i \in \mathscr{V}, \tag{9.5}$$

where $\eta_{ij0} = \eta_{i0} - \eta_{j0}$ represents the ideal relative position vector between the UMVs i and j.

Remark 9.1 Different from tracking errors in [3–5], the target tracking error (9.5) is in a distributed way. This kind of tracking error can cover the existing leader-follower formation pattern if the communication weights a_{i0} and a_{ij} $(\forall i, j \in \mathscr{V})$ are appropriately chosen. Thus, the distributed target tracking error (9.5) in this chapter is more general.

Denote

$$p = [p_1^T, \ p_2^T, \ldots, p_N^T]^T,$$
$$e = [e_1^T, \ e_2^T, \ldots, e_N^T]^T.$$

If p can be made arbitrarily small, then e can be made arbitrarily small, and vice versa. This is due to the fact that $e = (L_{\sigma_t} + D_{\sigma_t})p$ and $L_{\sigma_t} + D_{\sigma_t}$ is a symmetric positive definite matrix. It also means that the objective of this chapter can be converted into guaranteeing the distributed target tracking error e_i $(\forall i \in \mathscr{V})$ to be arbitrarily small.

The definition of average dwell time is introduced for later analysis.

Definition 9.2 ([6, 7]) For any constants $t_2 > t_1 \geq 0$, let $N_\sigma(t_1, \ t_2)$ denote the switching times of σ over $(t_1, \ t_2)$. T_d is called the average dwell time, if

$$N_\sigma(t_1, \ t_2) \leq N_0 + \frac{t_2 - t_1}{T_d} \tag{9.6}$$

holds for all $t_2 > t_1 \geq 0$ and a scalar $N_0 \geq 1$.

9.2 Cooperative Target Tracking Controller Design

In this section, a DESO is designed firstly to estimate the integrated unknown dynamics of the target and neighboring UMVs, and a novel kinematic controller design method is proposed accordingly. Then a disturbance observer is constructed to estimate the unknown time-varying disturbance vector w_i. A distributed dynamic

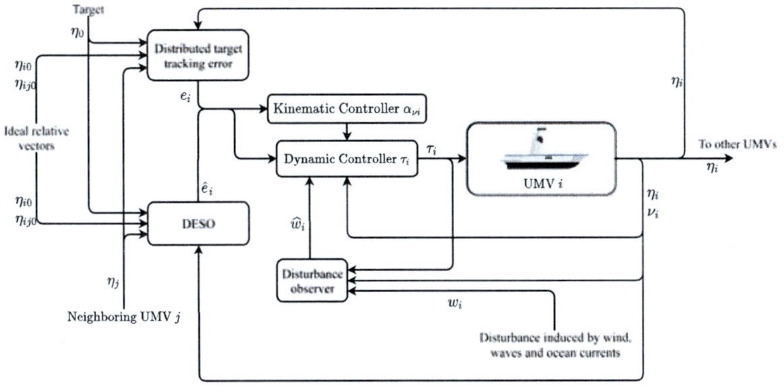

Fig. 9.1 Diagram of the design process

controller is derived to regulate the N UMVs to cooperatively track the target with kinematical equation (9.1). The design process is depicted by Fig. 9.1.

9.2.1 Estimation of Unknown Dynamics

Differentiating e_i along (9.1) and (9.3) yields

$$\dot{e}_i = \gamma_i - a_i^{\sigma_t} R_i \nu_i - a_{i0}^{\sigma_t} \dot{\eta}_{i0} - \sum_{j \in \mathcal{N}_i} a_{ij}^{\sigma_t} \dot{\eta}_{ij0}, \tag{9.7}$$

where

$$a_i^{\sigma_t} = a_{i0}^{\sigma_t} + \sum_{j \in \mathcal{N}_i} a_{ij}^{\sigma_t}, \quad \gamma_i = a_{i0}^{\sigma_t} R_0 \nu_0 + \sum_{j \in \mathcal{N}_i} a_{ij}^{\sigma_t} R_j \nu_j.$$

Remark 9.2 If (9.5) is treated as the coordinate transformation from η_i to e_i, the kinematics of the UMV i is transformed into the distributed pattern (9.7). Similar transformation strategies can be found in [3, 8]. This facilitates the subsequent kinematic controller design in this chapter.

Remark 9.3 For the UMV i, the unknown dynamics γ_i ($\forall i \in \mathcal{V}$) is relevant to its neighbors' velocity dynamics and yaw angles, which are unknown. This means that the unknown dynamics is derived from the cooperative target tracking system. It is intrinsically different from the unknown inputs and impulsive outliers in [9–11] since they are externalities. The ways of dealing with these two kinds of unknowns are different.

In practical applications, the changing rates of yaw angles, linear and angular velocities of marine targets and UMVs are bounded. This means that $\dot{\gamma}_i$ is always bounded. Thus, it is reasonable to assume that there exists a positive constant γ_i^*, such that $\|\dot{\gamma}_i\| \leq \gamma_i^*$, $\forall i \in \mathcal{V}$ (see also [3] for similar assumption).

For any $i \in \mathcal{V}$, γ_i is unknown because the UMV i can not obtain linear and angular velocities from the target and other UMVs. The UMV i can only receive position vector information from its neighbors. Considering that γ_i contains neighbors' information of the UMV i, we design a DESO to estimate γ_i as follows

$$\begin{cases} \dot{\widehat{e}}_i = -k_{i1}(\widehat{e}_i - e_i) + \widehat{\gamma}_i - a_{i0}^{\sigma_t}\dot{\eta}_{i0} - \sum\limits_{j \in \mathcal{N}_i} a_{ij}^{\sigma_t}\dot{\eta}_{ij0} - a_i^{\sigma_t} R_i \nu_i, \\ \dot{\widehat{\gamma}}_i = -k_{i2}\Big(a_{i0}^{\sigma_t}(\widehat{e}_i - e_i) + \sum\limits_{j \in \mathcal{N}_i} a_{ij}^{\sigma_t}\big((\widehat{e}_i - e_i) - (\widehat{e}_j - e_j)\big)\Big), \end{cases} \tag{9.8}$$

where $\widehat{e}_i$ and $\widehat{\gamma}_i$ are estimations of e_i and γ_i, respectively; k_{i1} and k_{i2} ($\forall i \in \mathcal{V}$) are positive parameters. It should be mentioned that k_{i1} and k_{i2} are determined by solving some inequalities, which will be given later.

Define estimation error vectors as

$$\widetilde{e}_i = \widehat{e}_i - e_i, \; \widetilde{\gamma}_i = \widehat{\gamma}_i - \gamma_i.$$

Define

$$\begin{aligned} e &= [e_1^T, e_2^T, \ldots, e_N^T]^T, \; \gamma = [\gamma_1^T, \gamma_2^T, \ldots, \gamma_N^T]^T, \\ \widetilde{e} &= [\widetilde{e}_1^T, \widetilde{e}_2^T, \ldots, \widetilde{e}_N^T]^T, \; \widetilde{\gamma} = [\widetilde{\gamma}_1^T, \widetilde{\gamma}_2^T, \ldots, \widetilde{\gamma}_N^T]^T, \\ K_1 &= diag\{k_{11}, k_{21}, \ldots, k_{N1}\}, \; K_2 = diag\{k_{12}, k_{22}, \ldots, k_{N2}\}. \end{aligned}$$

From (9.5), (9.7), and (9.8), one has that

$$\begin{cases} \dot{\widetilde{e}} &= -(K_1 \otimes I_3)\widetilde{e} + \widetilde{\gamma}, \\ \dot{\widetilde{\gamma}} &= -\big((K_2(L_{\sigma_t} + D_{\sigma_t})) \otimes I_3\big)\widetilde{e} - \dot{\gamma}. \end{cases} \tag{9.9}$$

Then define

$$\xi = [\widetilde{e}^T, \widetilde{\gamma}^T]^T, \; \dot{\vartheta} = [0_{3N}^T, \dot{\gamma}^T]^T.$$

Since $\|\dot{\gamma}_i\| \leq \gamma_i^*$, $\forall i \in \mathcal{V}$, there exists a positive constant ϑ^* such that $\|\dot{\vartheta}\| \leq \vartheta^*$. The estimation error system (9.9) can be rewritten as

$$\dot{\xi} = A_{\sigma_t}\xi - \dot{\vartheta}, \tag{9.10}$$

where

$$A_{\sigma_t} = \begin{bmatrix} -K_1 \otimes I_3 & I \\ -(K_2(L_{\sigma_t} + D_{\sigma_t})) \otimes I_3 & 0 \end{bmatrix}.$$

Then, the following lemma is given to show that the matrix A_{σ_t} is a Hurwitz matrix for any $\sigma_t \in \Omega$. Accordingly, the estimation error ξ is proved to be uniformly ultimately bounded under an average dwell time condition.

Lemma 9.1 *For any $\sigma_t \in \Omega$, the matrix A_{σ_t} is a Hurwitz matrix.*

Proof Let

$$P_{\sigma_t} = \begin{bmatrix} P_1 \otimes I_3 & c_1 P_1 \otimes I_3 \\ * & (c_1 K_1 P_1) \otimes I_3 + (K_2(L_{\sigma_t} + D_{\sigma_t})P_1) \otimes I_3 \end{bmatrix},$$

where P_1 is a symmetric positive definite matrix; c_1 is a positive and small enough constant; $*$ denotes the entries of a matrix implied by symmetry.

One can see that P_{σ_t} is a symmetric positive definite matrix by choosing a small enough c_1.

Further, by simple calculation, one can also prove that

$$A_{\sigma_t} P_{\sigma_t} + P_{\sigma_t} A_{\sigma_t}^T < 0.$$

Thus, A_{σ_t} is a Hurwitz matrix. The proof is completed. □

Then, for a positive scalar δ, there exists a unique symmetric positive definite matrix Q_{σ_t}, such that

$$A_{\sigma_t}^T Q_{\sigma_t} + Q_{\sigma_t} A_{\sigma_t} = -\delta I. \tag{9.11}$$

For any $\iota \in \Omega$, we consider the Lyapunov-Krasovskii functional candidate

$$V_{\xi_\iota} = \xi^T Q_\iota \xi. \tag{9.12}$$

From (9.10) and (9.11), one can obtain that

$$\begin{aligned} \dot{V}_{\xi_\iota} &= \xi^T (A_\iota^T Q_\iota + Q_\iota A_\iota)\xi - 2\xi^T Q_\iota \dot{\vartheta} \\ &= -\delta \xi^T \xi - 2\xi^T Q_\iota \dot{\vartheta}. \end{aligned} \tag{9.13}$$

In terms of the complete square inequality, one has that

$$-2\xi^T Q_\iota \dot{\vartheta} \le k_\xi \xi^T Q_\iota \xi + \frac{1}{k_\xi} \dot{\vartheta}^T Q_\iota \dot{\vartheta},$$

where k_ξ is a positive constant.

Then one can get that

$$\dot{V}_{\xi_\iota} \le -\rho_\xi V_{\xi_\iota} + c_\xi, \tag{9.14}$$

where

$$\rho_\xi = \delta \min_{\iota \in \Omega}\{\lambda_{\min}(Q_\iota^{-1})\} - k_\xi, \tag{9.15}$$

$$c_\xi = \frac{1}{k_\xi}\vartheta^{*2} \max_{\iota \in \Omega}\{\lambda_{\max}(Q_\iota)\}. \tag{9.16}$$

Thus, the following theorem is obtained readily.

Theorem 9.1 *Given positive scalars $\mu_\xi \geq 1$ and $N_0 \geq 1$, the estimation error ξ can exponentially converge to a ball centered at the origin with the radius being*

$$\mathscr{R}_\xi = \sqrt{\frac{c_\xi e^{\rho_\xi T_d N_0}}{\rho_\xi(1 - e^{-(\rho_\xi T_d - \ln \mu_\xi)}) \min_{\iota \in \Omega}\{\lambda_{\min}(Q_\iota)\}}}$$

for any switching signal σ_t with average dwell time T_d satisfying

$$T_d > \frac{\ln \mu_\xi}{\rho_\xi},$$

if

$$\rho_\xi > 0;\ Q_\iota \leq \mu_\xi Q_\kappa, \forall \iota, \kappa \in \Omega. \tag{9.17}$$

The ball can be made arbitrarily small by appropriately choosing k_{i1}, k_{i2} ($\forall i \in \mathscr{V}$), δ, k_ξ, μ_ξ, and N_0.

Proof Define switching instants as $t_1, t_2, t_3, \ldots, t_n, \ldots$. Consider the following Lyapunov-Krasovskii functional candidate

$$V_\xi = V_{\xi_{\sigma_t}}.$$

For $t_n \leq t < t_{n+1}$, from (9.6), (9.14), and (9.17), we have

$$\begin{aligned}
V_\xi(t) \leq & \left(V_{\xi_{\sigma_{t_n}}}(t_n^+) - \frac{c_\xi}{\rho_\xi}\right) e^{-\rho_\xi(t-t_n)} + \frac{c_\xi}{\rho_\xi} \\
\leq & \mu_\xi V_{\xi_{\sigma_{t_{n-1}}}}(t_n^-) e^{-\rho_\xi(t-t_n)} - \frac{c_\xi}{\rho_\xi} e^{-\rho_\xi(t-t_n)} + \frac{c_\xi}{\rho_\xi} \\
\leq & \mu_\xi V_{\xi_{\sigma_{t_{n-1}}}}(t_{n-1}^+) e^{-\rho_\xi(t-t_{n-1})} - \mu_\xi \frac{c_\xi}{\rho_\xi} e^{-\rho_\xi(t-t_{n-1})} \\
& + \mu_\xi \frac{c_\xi}{\rho_\xi} e^{-\rho_\xi(t-t_n)} - \frac{c_\xi}{\rho_\xi} e^{-\rho_\xi(t-t_n)} + \frac{c_\xi}{\rho_\xi} \\
\leq & \cdots \\
\leq & \mu_\xi^{N_\sigma(0,\,t)} V_{\xi_{\sigma_0}}(0) e^{-\rho_\xi t} + \frac{c_\xi}{\rho_\xi}\Big((1 - e^{-\rho_\xi(t-t_n)}) + \mu_\xi e^{-\rho_\xi(t-t_n)} \\
& \times (1 - e^{-\rho_\xi(t_n - t_{n-1})}) + \cdots + \mu_\xi^{N_\sigma(0,\,t)+1} e^{-\rho_\xi(t-t_1)}(1 - e^{-\rho_\xi t_1})\Big)
\end{aligned}$$

$$
\begin{aligned}
&\leq \mu_\xi^{N_0} V_{\xi_{\sigma_0}}(0) e^{-(\rho_\xi - \frac{\mu_\xi}{T_d})t} + \frac{c_\xi}{\rho_\xi}\left(1 + \mu_\xi e^{-\rho_\xi(t-t_n)} + \cdots + \mu_\xi^{N_\sigma(0,\, t)+1} e^{-\rho_\xi(t-t_1)}\right) \\
&\leq \mu_\xi^{N_0} V_{\xi_{\sigma_0}}(0) e^{-(\rho_\xi - \frac{\mu_\xi}{T_d})t} + \frac{c_\xi}{\rho_\xi} e^{\rho_\xi T_d N_0} \\
&\quad \times \left(1 + e^{-(\rho_\xi T_d - \ln \mu_\xi)} + e^{-2(\rho_\xi T_d - \ln \mu_\xi)} + \cdots + e^{-n(\rho_\xi T_d - \ln \mu_\xi)}\right) \\
&\leq \mu_\xi^{N_0} V_{\xi_{\sigma_0}}(0) e^{-(\rho_\xi - \frac{\mu_\xi}{T_d})t} + \frac{c_\xi e^{\rho_\xi T_d N_0}}{\rho_\xi \left(1 - e^{-(\rho_\xi T_d - \ln \mu_\xi)}\right)},
\end{aligned}
$$

which means that V_ξ is uniformly ultimately bounded. It is easy to see that V_ξ exponentially converges to a ball centered at the origin. The radius of the ball is $\frac{c_\xi e^{\rho_\xi T_d N_0}}{\rho_\xi \left(1 - e^{-(\rho_\xi T_d - \ln \mu_\xi)}\right)}$. Thus, the estimation error ξ exponentially converges to a ball centered at the origin with the radius being $\mathscr{R}_\xi$. The radius can be made arbitrarily small by appropriately choosing k_{i1}, k_{i2} ($\forall i \in \mathscr{V}$), δ, k_ξ, μ_ξ, and N_0 satisfying the condition (9.17). This completes the proof. □

9.2.2 Kinematic Controller Design

This subsection focuses on designing a novel kinematic controller.

Consider the following Lyapunov-Krasovskii functional candidate

$$
V_1 = \frac{1}{2}\sum_{i=1}^{N} e_i^T e_i + \frac{1}{2}\sum_{i=1}^{N} \widehat{e}_i^T \widehat{e}_i. \tag{9.18}
$$

The time derivative of V_1 is

$$
\begin{aligned}
\dot{V}_1 &= \sum_{i=1}^{N} e_i^T \Big(\gamma_i - a_i^{\sigma_t} R_i v_i - a_{i0}^{\sigma_t} \dot{\eta}_{i0} - \sum_{j\in\mathscr{N}_i} a_{ij}^{\sigma_t} \dot{\eta}_{ij0}\Big) \\
&\quad + \sum_{i=1}^{N} \widehat{e}_i^T \Big(- k_{i1}(\widehat{e}_i - e_i) + \widehat{\gamma}_i - a_i^{\sigma_t} R_i v_i - a_{i0}^{\sigma_t} \dot{\eta}_{i0} - \sum_{j\in\mathscr{N}_i} a_{ij}^{\sigma_t} \dot{\eta}_{ij0}\Big).
\end{aligned}
$$

The kinematic controller can be designed as

$$
\alpha_{v_i} = \frac{1}{a_i^{\sigma_t}} R_i^T \Big(\widehat{\gamma}_i + l_{i1} e_i + (k_{i1} - l_{i1})\widehat{e}_i - a_{i0}^{\sigma_t} \dot{\eta}_{i0} - \sum_{j\in\mathscr{N}_i} a_{ij}^{\sigma_t} \dot{\eta}_{ij0}\Big). \tag{9.19}
$$

where $l_{i1} > 0$ ($\forall i \in \mathscr{V}$) is the control gain.

Remark 9.4 Different from [3], this chapter introduces not only the estimated error $\widehat{e}_i$ but also the actual error e_i in the kinematic controller design, see in (9.19). The motivation is twofolds. First, for the UMV i, the distributed target tracking error e_i is known. The designed kinematic controller (9.19) can take full advantage of known information. Second, it can avoid the approximation of some virtual control vectors, which will be embodied later.

Define an error vector

$$z_{\nu_i} = \nu_i - \alpha_{\nu_i}.$$

Then

$$\dot{V}_1 = -\sum_{i=1}^{N} l_{i1} e_i^T e_i - \sum_{i=1}^{N} (2k_{i1} - l_{i1}) \widehat{e}_i^T \widehat{e}_i - \sum_{i=1}^{N} a_i (e_i^T + \widehat{e}_i^T) R_i z_{\nu_i} - \sum_{i=1}^{N} e_i^T \widetilde{\gamma}_i. \tag{9.20}$$

Coupling items $-\sum_{i=1}^{N} a_i (e_i^T + \widehat{e}_i^T) R_i z_{\nu_i}$ and $-\sum_{i=1}^{N} e_i^T \widetilde{\gamma}_i$ will be dealt with in the following.

9.2.3 *Disturbance Observer Design*

In this subsection, we consider the following disturbance observer for w_i.

$$\begin{cases} \dot{\beta}_i & = \widehat{w}_i - C_i(\nu_i)\nu_i - D_i(\nu_i)\nu_i + \tau_i, \\ \widehat{w}_i & = -K_{w_i}\beta_i + K_{w_i} M_i \nu_i, \end{cases} \tag{9.21}$$

where $\widehat{w}_i$ denotes an estimation of w_i; β_i is an auxiliary vector; $K_{w_i} > 0$ is an observer gain matrix.

Define the disturbance observer error vector as

$$\widetilde{w}_i = w_i - \widehat{w}_i. \tag{9.22}$$

From (9.3), (9.21), and (9.22), one has

$$\dot{\widetilde{w}}_i = \dot{w}_i - K_{w_i}\widetilde{w}_i. \tag{9.23}$$

Consider the following Lyapunov-Krasovskii functional candidate

$$V_{w_i} = \frac{1}{2}\widetilde{w}_i^T \widetilde{w}_i. \tag{9.24}$$

Its first time derivative is

$$\dot{V}_{w_i} = -\widetilde{w}_i^T K_{w_i}\widetilde{w}_i + \widetilde{w}_i^T \dot{w}_i. \tag{9.25}$$

For any positive scalar k_{w_i}, one has that

$$\widetilde{w}_i^T \dot{w}_i \leq \frac{1}{2} k_{w_i} \widetilde{w}_i^T \widetilde{w}_i + \frac{1}{2k_{w_i}} \dot{w}_i^T \dot{w}_i.$$

Then one can get that

$$\begin{aligned} \dot{V}_{w_i} &\leq -2(2\lambda_{\min}(K_{w_i}) - k_{w_i})V_{w_i} + \frac{1}{2k_{w_i}} d_i^{*2} \\ &= -\rho_{w_i} V_{w_i} + c_{w_i}, \end{aligned} \tag{9.26}$$

where

$$\rho_{w_i} = 2(2\lambda_{\min}(K_{w_i}) - k_{w_i}), \tag{9.27}$$

$$c_{w_i} = \frac{1}{2k_{w_i}} d_i^{*2}. \tag{9.28}$$

Similar to the analysis of Theorem 9.1, the disturbance observer (9.21) can ensure that the estimation error $\widetilde{w}_i$ is uniformly ultimately bounded by choosing appropriate K_{w_i} and k_{w_i} such that $\rho_{w_i} > 0$.

9.2.4 Distributed Dynamic Controller Design

In this subsection, a distributed dynamic controller is designed according to the DESO (9.8), the kinematic controller (9.19), and the disturbance observer (9.21).

Note that R_i is given in (9.2), and $\dot{\psi}_i = r_i$. It is easy to obtain that the transformation matrix $R_i = R(\psi_i)$ satisfies

$$\dot{R}_i = \dot{R}(\psi_i) = \begin{bmatrix} -r_i \sin(\psi_i) & -r_i \cos(\psi_i) & 0 \\ r_i \cos(\psi_i) & -r_i \sin(\psi_i) & 0 \\ 0 & 0 & 0 \end{bmatrix} = R_i S_i, \tag{9.29}$$

where

$$S_i = S(r_i) = \begin{bmatrix} 0 & -r_i & 0 \\ r_i & 0 & 0 \\ 0 & 0 & 0 \end{bmatrix}.$$

From (9.3), (9.5), (9.8), and (9.19), one has

$$\begin{aligned}
\dot{z}_{v_i} &= \dot{v}_i - \dot{\alpha}_{v_i} \\
&= M_i^{-1}\big(\tau_i + w_i - C_i(v_i)v_i - D_i(v_i)v_i\big) \\
&\quad -(R_iS_i)^T\Big(\widehat{\gamma}_i + l_{i1}e_i + (k_{i1}-l_{i1})\widehat{e}_i - a_{i0}^{\sigma_t}\dot{\eta}_{i0} - \sum_{j\in\mathcal{N}_i} a_{ij}^{\sigma_t}\dot{\eta}_{ij0}\Big) \\
&\quad -R_i^T\Big(-k_{i2}\big(a_{i0}^{\sigma_t}(\widehat{e}_i - e_i) + \sum_{j\in\mathcal{N}_i} a_{ij}^{\sigma_t}((\widehat{e}_i - e_i) - (\widehat{e}_j - e_j))\big)\Big) \\
&\quad + l_{i1}(\gamma_i - a_i^{\sigma_t}R_iv_i - a_i^{\sigma_t}\dot{\eta}_{i0} - \sum_{j\in\mathcal{N}_i} a_{ij}^{\sigma_t}\dot{\eta}_{ij0}) \\
&\quad + (k_{i1}-l_{i1})\big(-k_{i1}(\widehat{e}_i - e_i) + \widehat{\gamma}_i - a_{i0}^{\sigma_t}\dot{\eta}_{i0} - \sum_{j\in\mathcal{N}_i} a_{ij}^{\sigma_t}\dot{\eta}_{ij0} - a_i^{\sigma_t}R_iv_i\big) - a_{i0}^{\sigma_t}\ddot{\eta}_{i0} - \sum_{j\in\mathcal{N}_i} a_{ij}^{\sigma_t}\ddot{\eta}_{ij0}\Big).
\end{aligned} \tag{9.30}$$

Consider the following Lyapunov-Krasovskii functional candidate

$$V = V_\xi + V_1 + \sum_{i=1}^{N} V_{w_i} + \frac{1}{2}\sum_{i=1}^{N} z_{vi}^T z_{vi}. \tag{9.31}$$

From (9.13), (9.20), and (9.25), one can obtain that

$$\begin{aligned}
\dot{V} =& -\delta\xi^T\xi - 2\xi^T Q_{\sigma_t}\dot{\vartheta} - \sum_{i=1}^{N} l_{i1}e_i^Te_i - \sum_{i=1}^{N}(2k_{i1} - l_{i1})\widehat{e}_i^T\widehat{e}_i - \sum_{i=1}^{N} e_i^T\widetilde{\gamma}_i \\
& - \sum_{i=1}^{N}\widetilde{w}_i^T K_{w_i}\widetilde{w}_i + \sum_{i=1}^{N}\widetilde{w}_i^T\dot{w}_i + \sum_{i=1}^{N} z_{v_i}^T\left(\dot{z}_{v_i} - a_i^{\sigma_t}R_i^T(e_i + \widehat{e}_i)\right).
\end{aligned} \tag{9.32}$$

Then the distributed dynamic controller is designed as

$$\begin{aligned}
\tau_i =& -k_{z_i}M_iz_{v_i} + a_i^{\sigma_t}M_iR_i^T(e_i + \widehat{e}_i) - \widehat{w}_i + C_i(v_i)v_i + D_i(v_i)v_i \\
& + M_i(R_iS_i)^T\Big(\widehat{\gamma}_i + l_{i1}e_i + (k_{i1}-l_{i1})\widehat{e}_i - a_{i0}^{\sigma_t}\dot{\eta}_{i0} - \sum_{j\in\mathcal{N}_i} a_{ij}^{\sigma_t}\dot{\eta}_{ij0}\Big) \\
& + M_iR_i^T\Big(-k_{i2}\big(a_{i0}^{\sigma_t}(\widehat{e}_i - e_i) + \sum_{j\in\mathcal{N}_i} a_{ij}^{\sigma_t}((\widehat{e}_i - e_i) - (\widehat{e}_j - e_j))\big)\Big) \\
& + l_{i1}(\widehat{\gamma}_i - a_i^{\sigma_t}R_iv_i - a_i^{\sigma_t}\dot{\eta}_{i0} - \sum_{j\in\mathcal{N}_i} a_{ij}^{\sigma_t}\dot{\eta}_{ij0}) + (k_{i1}-l_{i1})\big(\widehat{\gamma}_i \\
& - k_{i1}(\widehat{e}_i - e_i) - a_i^{\sigma_t}R_iv_i - a_{i0}^{\sigma_t}\dot{\eta}_{i0} - \sum_{j\in\mathcal{N}_j} a_{ij}^{\sigma_t}\dot{\eta}_{ij0}\big) - a_{i0}^{\sigma_t}\ddot{\eta}_{i0} - \sum_{j\in\mathcal{N}_j} a_{ij}^{\sigma_t}\ddot{\eta}_{ij0}\Big),
\end{aligned} \tag{9.33}$$

where $k_{z_i} > 0$ is the control gain.

Then from (9.30) and (9.33), one has that

$$\begin{aligned}\dot{V} = &-\delta\xi^T\xi - 2\xi^T Q_{\sigma_t}\dot{\vartheta} - \sum_{i=1}^{N} l_{i1}e_i^T e_i - \sum_{i=1}^{N}(2k_{i1}-l_{i1})\widehat{e}_i^T\widehat{e}_i - \sum_{i=1}^{N} e_i^T\widetilde{\gamma}_i \\ &- \sum_{i=1}^{N} k_{z_i} z_{v_i}^T z_{v_i} + \sum_{i=1}^{N} z_{v_i}^T M_i^{-1}\widetilde{w}_i + \sum_{i=1}^{N} l_{i1} z_{v_i}^T R_i^T\widetilde{\gamma}_i - \sum_{i=1}^{N}\widetilde{w}_i^T K_{w_i}\widetilde{w}_i + \sum_{i=1}^{N}\widetilde{w}_i^T\dot{w}_i.\end{aligned} \tag{9.34}$$

From the complete square inequality, one can get that

$$e_i^T\widetilde{\gamma}_i \le \frac{1}{2}e_i^T e_i + \frac{1}{2}\widetilde{\gamma}_i^T\widetilde{\gamma}_i, \tag{9.35}$$

$$z_{v_i}^T M_i^{-1}\widetilde{w}_i \le \frac{1}{2}\lambda_{\max}(M_i^{-1}) z_{v_i}^T z_{v_i} + \frac{1}{2}\lambda_{\max}(M_i^{-1})\widetilde{w}_i^T\widetilde{w}_i, \tag{9.36}$$

$$z_{v_i}^T R_i^T\widetilde{\gamma}_i \le \frac{1}{2}z_{v_i}^T R_i^T R_i z_{v_i} + \frac{1}{2}\widetilde{\gamma}_i^T\widetilde{\gamma}_i = \frac{1}{2}z_{v_i}^T z_{v_i} + \frac{1}{2}\widetilde{\gamma}_i^T\widetilde{\gamma}_i. \tag{9.37}$$

Combining (9.14), (9.26), (9.35), (9.36), and (9.37) yields that

$$\begin{aligned}\dot{V} \le &-\sum_{i=1}^{N}(l_{i1} - \frac{1}{2})e_i^T e_i - \sum_{i=1}^{N}(2k_{i1} - l_{i1})\widehat{e}_i^T\widehat{e}_i \\ &- \Big(\delta\lambda_{\min}(Q_{\sigma_t}^{-1}) - k_\xi - \frac{1}{2}\lambda_{\max}(Q_{\sigma_t}^{-1})(1 + \max_i\{l_{i1}\})\Big)\xi^T Q_{\sigma_t}\xi \\ &- \sum_{i=1}^{N}(k_{z_i} - \frac{1}{2}l_{i1} - \frac{1}{2}\lambda_{\max}(M_i^{-1}))z_{v_i}^T z_{v_i} \\ &- \sum_{i=1}^{N}(\lambda_{\min}(K_{w_i}) - \frac{1}{2}k_{w_i} - \frac{1}{2}\lambda_{\max}(M_i^{-1}))\widetilde{w}_i^T\widetilde{w}_i + \frac{1}{k_\xi}\dot{\vartheta}^T Q_{\sigma_t}\dot{\vartheta} + \sum_{i=1}^{N}\frac{1}{2k_{w_i}}\dot{w}_i^T\dot{w}_i \\ \le &-\min\{h_1, h_2, h_3, h_4, h_5\}V + \frac{1}{k_\xi}\max_{\iota\in\Omega}\{\lambda_{\max}(Q_\iota)\}\vartheta^{*2} + \sum_{i=1}^{N}\frac{1}{2k_{w_i}}d_i^{*2} \\ = &-\rho V + c,\end{aligned} \tag{9.38}$$

where

$$\begin{aligned}&\rho = \min\{h_1, h_2, h_3, h_4, h_5\}, \\ &c = \frac{1}{k_\xi}\max_{\iota\in\Omega}\{\lambda_{\max}(Q_\iota)\}\vartheta^{*2} + \sum_{i=1}^{N}\frac{1}{2k_{w_i}}d_i^{*2}, \\ &h_1 = 2l_{i1} - 1,\ h_2 = 2(2k_{i1} - l_{i1}), \\ &h_3 = \delta\min_{\iota\in\Omega}\{\lambda_{\min}(Q_\iota^{-1})\} - k_\xi - \frac{1}{2}\max_{\iota\in\Omega}\{\lambda_{\max}(Q_\iota^{-1})\}(1 + \max_i\{l_{i1}\}), \\ &h_4 = 2k_{z_i} - l_{i1} - \lambda_{\max}(M_i^{-1}),\ h_5 = 2\lambda_{\min}(K_{w_i}) - k_{w_i} - \lambda_{\max}(M_i^{-1}).\end{aligned}$$

Thus, one has the following theorem.

Theorem 9.2 *Given positive scalars $\mu \geq 1$ and $N_0 \geq 1$, for any switching signal σ_t with average dwell time T_d satisfying*

$$T_d > \frac{\ln \mu}{\rho},$$

the N UMVs with the distributed dynamic controller being designed as (9.33) (together with (9.8) and (9.21)) can cooperatively track the target with desired accuracy by appropriately choosing parameters k_{i1}, k_{i2}, l_{i1}, K_{w_i}, k_{w_i}, k_{z_i} ($\forall i \in \mathscr{V}$), δ, k_ξ, μ, N_0, if

$$\rho > 0;\ Q_\iota \leq \mu Q_\kappa, \forall \iota, \kappa \in \Omega.$$

All the errors of the closed-loop target tracking system are uniformly ultimately bounded.

Proof Based on the analysis in Theorem 9.1, the proof is straightforward. □

9.3 Performance Analysis and Discussion

9.3.1 The Effectiveness of the Tracking Controller Design

In this subsection, we consider a maritime target tracking system with five UMVs. The model parameters of UMVs are taken from [12]. Without loss of generality, initial conditions of the five UMVs are given as $\eta_1(0) = [-60 \quad -100 \quad -1]^T$, $\eta_2(0) = [-30 \quad -140 \quad 1.3]^T$, $\eta_3(0) = [-40 \quad -210 \quad 1]^T$, $\eta_4(0) = [-20 \quad -260 \quad 0.5]^T$, $\eta_5(0) = [-20 \quad -320 \quad 0.7]^T$, $\nu_1(0) = \nu_2(0) = \nu_3(0) = \nu_4(0) = \nu_5(0) = [0 \quad 0 \quad 0]^T$. Disturbances of UMVs are given as $w_1 = [0.2\sin(0.02t) \quad 0.15\sin(0.1t) \quad 0.07\cos(0.3t)]^T$, $w_2 = [0.17\sin(0.02t)\ 0.2\sin(0.01t)\ 0.1\cos(1.3t)]^T$, $w_3 = [0.1\cos(0.2t)\ 0.2\sin(0.01t)\ -0.05\cos(1.3t)]^T$, $w_4 = [0.04\sin(0.1t) \quad -0.1\cos(0.2t) \quad 0.1\cos(0.3t)]^T$, while $w_5 = [-0.12\sin(0.11t)\ 0.03\sin(0.1t)\ 0.04\cos(0.21t)]^T$.

The target to be tracked moves with linear and angular velocities being $\nu_0 = [2 \ \ 0 \ \ 0.01]^T$, which is unknown to the five UMVs. The initial position vector of the target is $\eta_0(0) = [200\sin(0.2) \ \ -200\cos(0.2) \ \ 0.2]^T$. The switching interaction topologies are shown in Fig. 9.2.

The parameters of the DESO are chosen as $k_{11} = k_{21} = k_{31} = k_{12} = k_{22} = k_{32} = 1$. Then the control gains of the kinematic controller can be set as $l_{11} = l_{21} = l_{31} = \frac{3}{4}$. The disturbance observer gain matrices can be given as

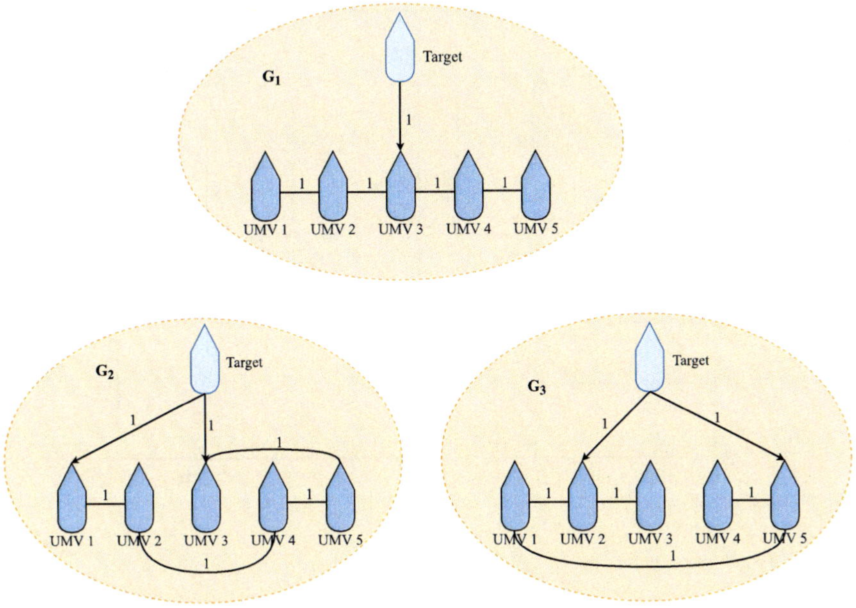

Fig. 9.2 Switching interaction topologies

$$K_{w_1} = K_{w_2} = K_{w_3} = \begin{bmatrix} 6 & 1 & 1 \\ 1 & 8 & 1 \\ 1 & 1 & 7 \end{bmatrix}.$$

The control gain parameters of the distributed dynamic controller are $k_{z_1} = k_{z_2} = k_{z_3} = k_{z_4} = k_{z_5} = 2$. Other parameters are taken as $\delta = 100$, $k_\xi = 0.01$, $k_{w_1} = k_{w_2} = k_{w_3} = k_{w_4} = k_{w_5} = 4$, $\mu = 3.4$. One can obtain matrices Q_1, Q_2, Q_3 through the Eq. (9.11). It is easy to obtain $\rho = 0.12$. The average dwell time satisfies $T_d > T_d* = 10.20s$. The switching pattern of the network topology is illustrated in Fig. 9.3.

For convenience, define $e_i = [e_{i1}\ e_{i2}\ e_{i3}]^T$, $\widetilde{\gamma}_i = [\widetilde{\gamma}_{i1}\ \widetilde{\gamma}_{i2}\ \widetilde{\gamma}_{i3}]^T$, $\widetilde{e}_i = [\widetilde{e}_{i1}\ \widetilde{e}_{i2}\ \widetilde{e}_{i3}]^T$, $\forall i \in \mathscr{V}$.

In the following, we first consider the case that the ideal relative vector function η_{i0} ($\forall i \in \mathscr{V}$) is time-varying.

The five time-varying ideal relative position vectors are described by

$$\eta_{10} = \begin{bmatrix} 200\sin(0.01(t+20)) - 260\sin(0.01t) \\ -200\cos(0.01(t+20)) + 260\cos(0.01t) \\ 0.2 \end{bmatrix}, \tag{9.39a}$$

$$\eta_{20} = \begin{bmatrix} 200\sin(0.01(t+20)) - 230\sin(0.01t) \\ -200\cos(0.01(t+20)) + 230\cos(0.01t) \\ 0.2 \end{bmatrix}, \tag{9.39b}$$

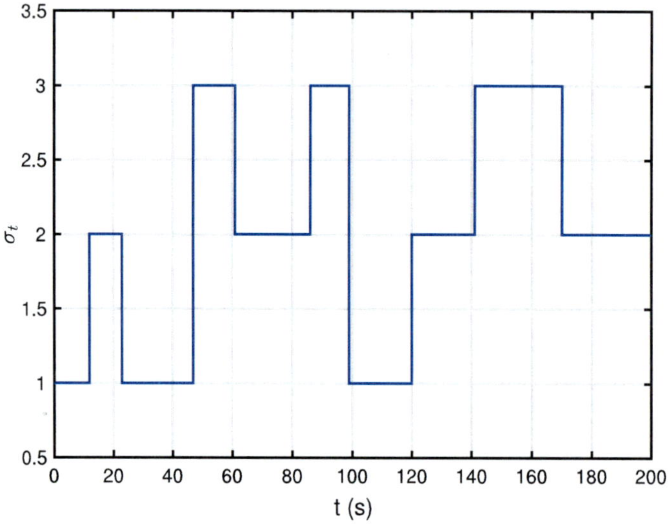

Fig. 9.3 Switching signal

$$\eta_{30} = \begin{bmatrix} 200\sin(0.01(t+20)) - 200\sin(0.01t) \\ -200\cos(0.01(t+20)) + 200\cos(0.01t) \\ 0.2 \end{bmatrix}, \tag{9.39c}$$

$$\eta_{40} = \begin{bmatrix} 200\sin(0.01(t+20)) - 170\sin(0.01t) \\ -200\cos(0.01(t+20)) + 170\cos(0.01t) \\ 0.2 \end{bmatrix}, \tag{9.39d}$$

$$\eta_{50} = \begin{bmatrix} 200\sin(0.01(t+20)) - 140\sin(0.01t) \\ -200\cos(0.01(t+20)) + 140\cos(0.01t) \\ 0.2 \end{bmatrix}. \tag{9.39e}$$

Based on the above-mentioned parameters and initial conditions, trajectories of the target and the five UMVs, and distributed errors of the DESO are illustrated in Figs. 9.4 and 9.5, respectively. From Figs. 9.4 and 9.5, one can see that the five UMVs can cooperatively track the target with desired accuracy under the influence of the switching topologies.

Then, we consider the case that the ideal relative vector function η_{i0} ($\forall i \in \mathscr{V}$) is time-invariant.

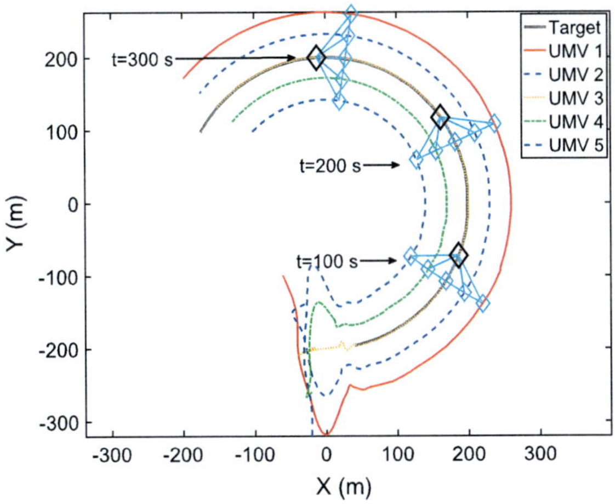

Fig. 9.4 Trajectories of the target and the five UMVs, and their snapshots at two time instants

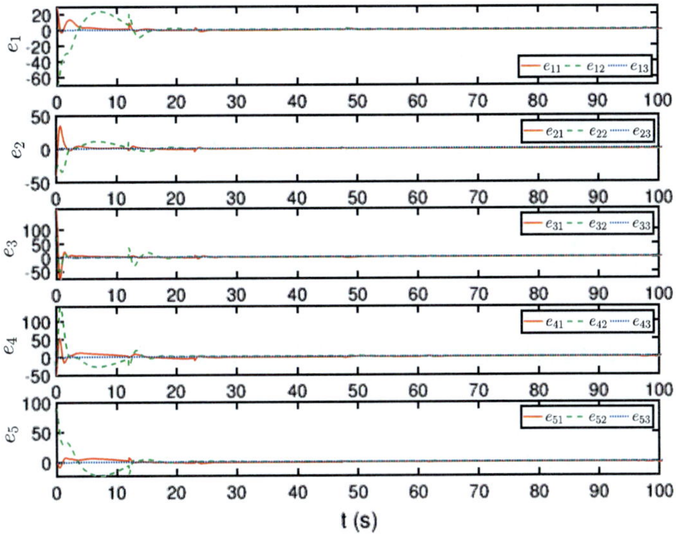

Fig. 9.5 Distributed tracking errors

The five time-invariant ideal relative position vectors are described by

$$
\begin{aligned}
\eta_{10} &= \begin{bmatrix} -60 \\ 30 \\ 0 \end{bmatrix}, \ \eta_{20} = \begin{bmatrix} -30 \\ 30 \\ 0 \end{bmatrix}, \ \eta_{30} = \begin{bmatrix} 0 \\ 30 \\ 0 \end{bmatrix}, \\
\eta_{40} &= \begin{bmatrix} 30 \\ 30 \\ 0 \end{bmatrix}, \ \eta_{50} = \begin{bmatrix} 60 \\ 30 \\ 0 \end{bmatrix}.
\end{aligned}
\tag{9.40}
$$

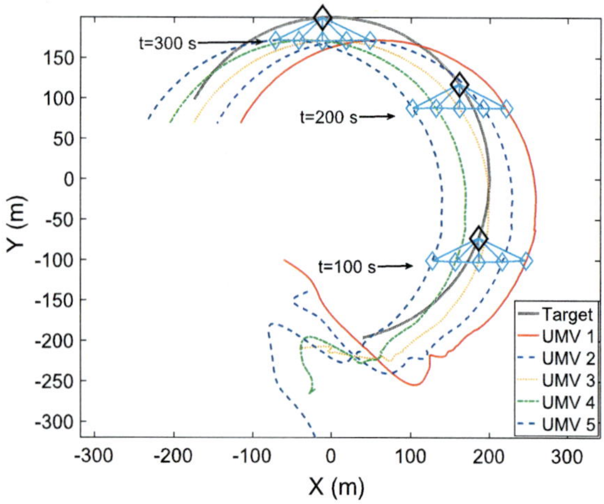

Fig. 9.6 Trajectories of the target and the five UMVs, and their snapshots at three time instants

Based on the above-mentioned parameters and initial conditions, the corresponding results are shown in Fig. 9.6. From Fig. 9.6, one can conclude that under the influence of switching topologies, the cooperative target tracking is achieved.

Next, comparing Figs. 9.4 and 9.6, one can see that the two line formations formed by the five UMVs is in different reference frames. The formation in Fig. 9.4 is in the body-fixed frame of the target, while the formation in Fig. 9.6 is in the earth-fixed frame. This means that we can achieve cooperative target tracking in different reference frames by adjusting the ideal relative vector η_{i0} ($\forall i \in \mathscr{V}$).

9.3.2 Performance Comparison for Different Design Schemes

In this subsection, we compare the cooperative target tracking performance between the DESO-based tracking controller designed in this chapter and the tracking controller designed in [13]. Consider a maritime target tracking system with three UMVs. For comparison, the fixed interaction topology in Fig. 9.7 is taken into account.

Model parameters of each UMV are taken from [12]. Initial conditions of the three UMVs are given as $\eta_1(0) = [-17 \quad -200 \quad -1]^T$, $\eta_2(0) = [-33 \quad -130 \quad 1.3]^T$, $\eta_3(0) = [-40 \quad -100 \quad 1]^T$, $\nu_1(0) = \nu_2(0) = \nu_3(0) = 0$. The disturbance vectors of the three UMVs are given as $\omega_1(t) = [0.2\sin(0.02t) \quad 0.15\sin(0.1t) \quad 0.07\cos(0.3t)]^T$, $\omega_2(t) = [0.17\sin(0.02t) \;\; 0.2\sin(0.01t) \;\; 0.1\cos(1.3t)]^T$, $\omega_3(t) = [0.1\cos(0.2t) \quad 0.2\sin(0.01t) \quad -0.05\cos(1.3t)]^T$.

The trajectory of the target is the same as the one in Sect. 9.3.1. The three time-varying ideal relative position vectors are described by

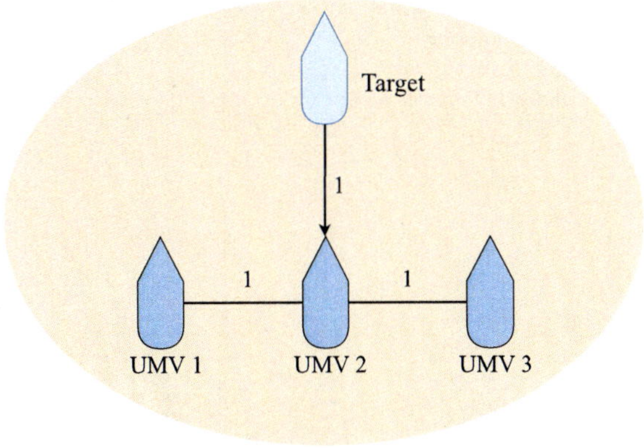

Fig. 9.7 Interaction topology of one target and three UMVs

$$\eta_{10} = \begin{bmatrix} 200\sin(0.01(t+20)) - 230\sin(0.01t) \\ -200\cos(0.01(t+20)) + 230\cos(0.01t) \\ 0.2 \end{bmatrix}, \tag{9.41a}$$

$$\eta_{20} = \begin{bmatrix} 200\sin(0.01(t+20)) - 200\sin(0.01t) \\ -200\cos(0.01(t+20)) + 200\cos(0.01t) \\ 0.2 \end{bmatrix}, \tag{9.41b}$$

$$\eta_{30} = \begin{bmatrix} 200\sin(0.01(t+20)) - 170\sin(0.01t) \\ -200\cos(0.01(t+20)) + 170\cos(0.01t) \\ 0.2 \end{bmatrix}. \tag{9.41c}$$

For the target to be tracked, its velocity is unavailable to all the UMVs, while its position information can only be measured/received by some of the UMVs. In this case, this chapter uses coordinate transformation from η_i to e_i and estimates the unknown dynamics γ_i through a DESO. However, in [13], each UMV needs to estimate the target's position and velocity in the earth-fixed frame by a distributed observer. The difference makes the tracking performance different in this chapter and [13]. The trajectories of the three UMVs under the tracking controllers in this chapter and [13] are presented in Fig. 9.8. From Fig. 9.8, one can see that the control scheme in this chapter provides a smaller target tracking error.

9.4 Conclusions

The cooperative target tracking has been addressed under switching interaction topologies. A DESO has been designed to integrally estimate unknown target dynamics and neighboring UMVs' dynamics. A novel kinematic controller, which can make full use of known information and avoid the approximation of some virtual control vectors, has been designed. A disturbance observer has been presented to estimate unknown time-varying environmental disturbances. A distributed dynamic controller

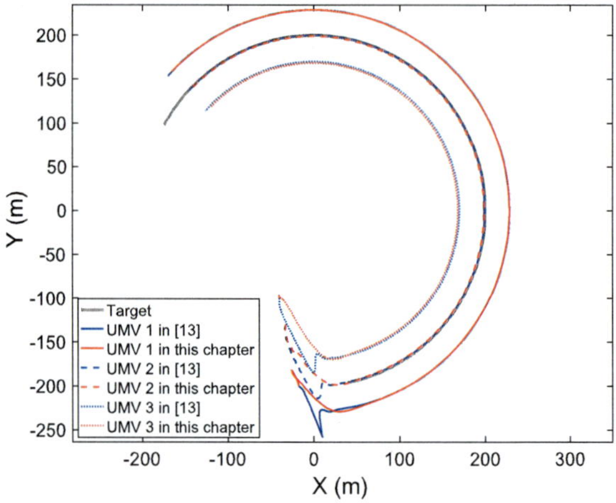

Fig. 9.8 Trajectories of the three UMVs under the tracking controllers in this chapter and [13]

has been designed to regulate the N UMVs to cooperatively track the target. Under switching interaction topologies, the effectiveness of the derived results has been demonstrated through cooperative target tracking performance analysis for a maritime target tracking system composed of five interacting UMVs.

9.5 Notes

Maritime target tracking has received much attention in the literature [14–17]. Note that some maritime target tracking missions require the cooperation of a fleet of UMVs. When dealing with the problem of cooperative target tracking, it is usually difficult to obtain the accurate velocity of the target [18, 19]. Thus, how to track a target by using only its position information needs further investigation. The interaction between the target and the UMVs can be described by a communication network, and the connectivity of the corresponding interaction topology may be switching. Motivated by these facts, the cooperative target tracking under switching interaction topologies is investigated in this chapter. The results in this chapter are based mainly on [20]. Future research includes the cooperative target tracking of multiple UMVs under actuator faults [21].

References

1. Q. Song, F. Liu, J. Cao, W. Yu, M-matrix strategies for pinning-controlled leader-following consensus in multiagent systems with non-linear dynamics. IEEE Trans. Cybern. **43**(6), 1688–1697 (2013)
2. T.I. Fossen, *Handbook of Marine Craft Hydrodynamics and Motion Control* (Wiley, Chichester, U.K., 2011)

3. L. Liu, D. Wang, Z. Peng, C.L.P. Chen, T. Li, Bounded neural network control for target tracking of underactuated autonomous surface vehicles in the presence of uncertain target dynamics. IEEE Trans. Neural Netw. Learn. Syst. **30**(4), 1241–1249 (2019)
4. Y. Yang, J. Du, H. Liu, C. Guo, A. Abraham, A trajectory tracking robust controller of surface vessels with disturbance uncertainties. IEEE Trans. Control Syst. Technol. **22**(4), 1511–1518 (2014)
5. S. Dai, S. He, M. Wang, C. Yuan, Adaptive neural control of underactuated surface vessels with prescribed performance guarantees. IEEE Trans. Neural Netw. Learn. Syst. **30**(12), 3686–3698 (2019)
6. L. Ma, Y.-L. Wang, Q.-L. Han, Event-triggered dynamic positioning for mass-switched unmanned marine vehicles in network environments. IEEE Trans. Cybern. **52**(5), 3159–3171 (2022)
7. J.P. Hespanha, A.S. Morse, Stability of switched systems with average dwell-time, in *Proceedings of the 38th IEEE Conference on Decision and Control*, Phoenix, USA, Dec. 1999, pp. 2655–2660
8. R. Cui, S.S. Ge, B.V.E. How, Y.S. Choo, Leader-follower formation control of underactuated autonomous underwater vehicles. Ocean Eng. **37**(17–18), 1491–1502 (2010)
9. L. Zou, Z. Wang, J. Hu, D. Zhou, Moving horizon estimation with unknown inputs under dynamic quantization effects. IEEE Trans. Autom. Control **65**(12), 5368–5375 (2020)
10. L. Zou, Z. Wang, J. Hu, H. Dong, Ultimately bounded filtering subject to impulsive measurement outliers. IEEE Trans. Autom. Control **67**(1), 304–319 (2022)
11. L. Zou, Z. Wang, Q.-L. Han, D. Yue, Tracking control under round-robin scheduling: handling impulsive transmission outliers. IEEE Trans. Cybern. **53**(4), 2288–2300 (2023)
12. R. Skjetne, T.I. Fossen, P.V. Kokotović, Adaptive maneuvering, with experiments, for a model ship in a marine control laboratory. Automatica **41**(2), 289–298 (2005)
13. S. Gao, Z. Peng, L. Liu, H. Wang, D. Wang, Coordinated target tracking by multiple unmanned surface vehicles with communication delays based on a distributed event-triggered extended state observer. Ocean Eng. **227**, Article no. 108283 (2021)
14. Z. Peng, J. Wang, D. Wang, Q.-L. Han, An overview of recent advances in coordinated control of multiple autonomous surface vehicles. IEEE Trans. Ind. Inf. **17**(2), 732–745 (2021)
15. O. Elhaki, K. Shojaei, Neural network-based target tracking control of underactuated autonomous underwater vehicles with a prescribed performance. Ocean Eng. **167**, 239–256 (2018)
16. O. Namaki-Shoushtari, A.P. Aguiar, A. Khaki-Sedigh, Target tracking of autonomous robotic vehicles using range-only measurements: a switched logic-based control strategy. Int. J. Robust Nonlinear Control **22**(17), 1983–1998 (2011)
17. K. Shojaei, Three-dimensional neural network tracking control of a moving target by underactuated autonomous underwater vehicles. Neural Comput. Appl. **31**, 509–521 (2019)
18. K. Choi, S.J. Yoo, J.B. Park, Y.H. Choi, Adaptive formation control in absence of leader's velocity information. IET Control Theory Appl. **4**(4), 521–528 (2010)
19. Z. Peng, D. Wang, Robust adaptive formation control of autonomous surface vehicles with uncertain dynamics. IET Control Theory Appl. **5**(12), 1378–1387 (2011)
20. L. Ma, Y.-L. Wang, Q.-L. Han, Cooperative target tracking of multiple autonomous surface vehicles under switching interaction topologies. IEEE/CAA J. Automatica Sinica. **10**(3), 673–684 (2023)
21. C. Gao, Z. Wang, X. He, D. Yue, Sampled-data-based fault-tolerant consensus control for multi-agent systems: a data privacy preserving scheme. Automatica **133**, Article no. 109847 (2021)

Index

Y.-L. Wang et al., *Network-Based Control of Unmanned Marine Vehicles*,
https://doi.org/10.1007/978-3-031-28605-6